Theory and Methodology of Electromagnetic Ultrasonic Guided Wave Imaging

Songling Huang · Yu Zhang · Zheng Wei ·
Shen Wang · Hongyu Sun

Theory and Methodology of Electromagnetic Ultrasonic Guided Wave Imaging

清華大学出版社
TSINGHUA UNIVERSITY PRESS

Springer

Songling Huang
Tsinghua University
Beijing, China

Yu Zhang
Tsinghua University
Beijing, China

Zheng Wei
Tsinghua University
Beijing, China

Shen Wang
Tsinghua University
Beijing, China

Hongyu Sun
Tsinghua University
Beijing, China

ISBN 978-981-13-8604-6 ISBN 978-981-13-8602-2 (eBook)
https://doi.org/10.1007/978-981-13-8602-2

Jointly published with Tsinghua University Press
The print edition is not for sale in China. Customers from China please order the print book from: Tsinghua University Press.

This Springer imprint is published by the registered company Springer Nature Singapore Pte Ltd.
The registered company address is: 152 Beach Road, #21-01/04 Gateway East, Singapore 189721, Singapore

Preface

The electromagnetic ultrasonic guided wave has the characteristics of small attenuation, long propagation distance, 100% coverage of the ultrasonic field, easy adjustment of the guided wave mode, and no coupling medium. Single-ended excitation, long-distance detection, and continuous monitoring of metal materials in complex structures and service environments can be achieved. The application fields of the electromagnetic ultrasonic guided waves are expanded to petroleum, chemical, automotive, aerospace, railway, and shipbuilding, which cover almost all fields. Especially for the applications where the service environment cannot be contacted directly, electromagnetic ultrasonic guided wave detection can play an important role because of its unique advantages.

As the requirements for the safety of metal structures increase, the detection engineering is no longer satisfied with the conventional judgment of defect existence and acquisition of defect equivalent dimension level. The quantitative description of the defects must be developed in the direction of defect contour shape, high-precision imaging of defects, and visualization of the detection results.

This book introduces the identification and quantification of electromagnetic ultrasonic guided wave detection defects, and the imaging theory of the implementation technology. It mainly includes directional and reliability electromagnetic ultrasonic transducers, electromagnetic ultrasonic guided wave detection signal time-of-flight extraction method, tomography method, and scattering imaging method.

The content of this book is a summary of the author's continuous research and practical application in the field of electromagnetic ultrasonic guided wave detection in the past 10 years. In the practical applications of related technology, significant support was provided by colleagues and engineers from related companies and institutes of Sinopec, PetroChina, and CNOOC. We express our sincere gratitude to them for helping to improve the technology in real practice.

In preparing this book, Prof. Songling Huang contributed Chap. 1; Dr. Shen Wang contributed Chap. 2.1; Dr. Hongyu Sun contributed Chap. 2.2; Dr. Yu Zhang contributed Chaps. 3 and 5; and Dr. Zheng Wei contributed Chap. 4.

With the increasing requirement for nondestructive testing technology, the research work of electromagnetic ultrasonic guided wave detection technology attracted more and more attention, and industrial applications are more and more extensive.

We hope that this book will be used as a reference in electromagnetic guided wave imaging by individuals at any level and by graduate students. It is also hoped that this book will expand and promote the use of electromagnetic guided waves. If there are any errors in this book, please contact us without hesitation.

Beijing, China
December 2018

Songling Huang
Yu Zhang
Zheng Wei
Shen Wang
Hongyu Sun

Contents

Chapter 1
Introduction

1.1 Overview

Compared with the ultrasonic body wave, the ultrasonic guided wave has the characteristics of slow attenuation and wave field distribution in the whole plate thickness [1–4]. Therefore, the ultrasonic guided wave detection technology has the advantages of inheriting the advantages of the traditional body wave detection technology, such as good directivity, high efficiency, and no harm to the human body [5]. It also has the advantages of long detecting distance and the ability to detect the plate surface and the internal depth [6, 7]. Also, the ultrasonic guided wave has multimodal characteristics [8, 9], and the ultrasonic guided waves of different modes are sensitive to the defects of different types and sizes. When using different modal guided waves, the detected waveforms can provide abundant defect information and help to analyze and master the detailed feature information of various types and sizes of defects more accurately. Because the ultrasonic guided wave has the above advantages, the ultrasonic guided wave detection technology is used to detect the defect of the structural parts. It can not only realize the rapid and accurate detection of the defects of different parts, different types and different sizes in the structural parts but also can detect the defects in the untouchable parts of the structure [10].

Ultrasonic guided wave detection technology based on the piezoelectric ultrasonic transducer is commonly used to detect the defects and damage of metal plate online. This technology needs to use a piezoelectric ultrasonic transducer to stimulate ultrasonic vibration [11] and propagate ultrasonic vibration to the internal [12] of metal plate relying on coupling agent. Thus, the ultrasonic guided wave detection is realized. However, the ultrasonic guided wave detection technology based on the piezoelectric ultrasonic transducer is limited by the principle of the piezoelectric transducer. The ultrasonic vibration should be coupled to the metal plate by couplant, so it is difficult to apply to the detection under special conditions such as non-contact and high temperature. The micro-machined air-coupled transducer has the advantages of non-contact and independent of couplant. However, the generated ultrasonic wave will suffer more serious energy attenuation through the air and the interface

S. Huang et al., *Theory and Methodology of Electromagnetic Ultrasonic Guided Wave Imaging*, https://doi.org/10.1007/978-981-13-8602-2_1

between the air and the measured solid material. The electromagnetic ultrasonic transducer relies on the electromagnetic coupling to complete the energy conversion of the alternating current in the transducer coil to the mechanical vibration of the measured material [13, 14]. It can produce ultrasonic guided waves directly in the metal plate without couplant and easily adjust the ultrasonically guided wave mode and can be used in special conditions such as non-contact and high-temperature testing. Therefore, it is necessary to use a magnetic acoustic transducer as the excitation source and receiver for ultrasonic guided wave detection of steel plate defects, so as to achieve non-liquid coupling detection. The electromagnetic ultrasonic transducer EMAT is used to complete the excitation and reception of ultrasonic guided waves, which is called electromagnetic guided waves.

Most of the metal plate structural parts can only determine the position of the defects, but it is more important to obtain the quantized information, such as the size of the defects and the shape of the contour. The quantitative information is an important basis for evaluating the health status of the metal plate structure and guiding the maintenance and maintenance work [15, 16]. The defect quantization usually uses a direct quantization method based on the defect equivalent size. The size of the standard defect, which is the same as the real defect waveform, is the approximate size of the real defect, and the standard defect database of the guided wave detection can be established according to the requirement. This direct quantification method is limited by the amount of information available and can only get approximate solution results, sometimes the gap between the real size and the defect is different. With the increase in the strict requirements for the safety of metal plate components, the detection engineering has not been satisfied with the conventional judgment defect with the defect equivalent dimension level. The quantitative defect description must be described to the shape of the defect contour, the defect high-precision imaging, and the visualization of the defect detection results [17, 18].

Low-frequency guided waves propagating in industrial sheet metal structures are generally used for defect inspection of large-area and long-distance structural parts [19, 20]. For a long time, it is generally necessary to use other methods if we want to meticulously detect or image defects detected by ultrasonic guided waves. In recent years, the ultrasonic guided wave imaging detection technology provides a solution for the high-precision detection of the defect, which makes the ultrasonic guided wave scanning for the large-area and long-distance defect of the industrial metal plate structure, and it is expected to realize the high-precision imaging detection of the defect. However, due to the low efficiency of the transducer, the signal-to-noise ratio of the defect detection signal is low, which is not benefit for the defect imaging. Therefore, it is necessary to establish a guided wave imaging detection method based on the magnetic, acoustic array to detect the guided wave in the surrounding area of the array from multiple angles, to provide more abundant and accurate defect information for the high-precision imaging of the defect.

When ultrasonic guided waves encounter defects, it can be divided into two parts of energy guided wave according to the difference of its direction of propagation: A part of the energy guided wave propagates along the original direction of propagation or bypassing the defect, and the other part of the energy guide wave changes the direction of propagation and propagates in other directions. To facilitate a unified description, the first kind of energy guided wave is called a guided wave, and second kind of energy guided waves is called scattered guided waves. This book will study and discuss the imaging methods, respectively. Also, the defect imaging detection object of this book is an industrial aluminum plate and steel plate. This book aims to focus on the theory and method of electromagnetic ultrasonic guided wave imaging. As for the detection of other material or shape structure parts, the electromagnetic ultrasonic exchange can be designed to meet the requirements. The method of EMAT and its array is used to detect defects in different materials and shapes.

Ultrasonic Lamb wave is a common ultrasonic guided wave. Conventional Lamb wave detection technology usually uses a single or a pair of Lamb wave transducers to carry out pulse reflection detection, determines the defects in the aluminum plate, and determines the position and size of the defect coarsely. However, in order to master the severity and development trend of defects in aluminum plate more accurately, it is not enough to use only conventional Lamb wave detection results. The exact shape, contour size, and distribution of the defect need to be mastered [21]. Therefore, the Lamb wave imaging technology has become a research hotspot in the field of Lamb wave detection. The commonly used Lamb wave imaging methods include phased array imaging method [22, 23], tomography, migration [24–26], and time reversal method [27–30]. Among them, the Lamb wave tomography method originates from the computed tomography (CT) method of X-ray, and the guided wave is used. Using Lamb wave tomography, the image of the defect can be obtained without destroying the structure and physical characteristics of the measured aluminum plate, and the detailed feature information of the defect can be obtained accurately. The implementation process of the method is as follows: First, the transducer array is used to excite and receive Lamb waves along different directions (i.e., Lamb waves are projected along different directions) to obtain a series of Lamb wave detection waveforms with defect information; then, the projected data required for imaging are extracted from the detection waveforms; and finally, the specific image is obtained. The algorithm reconstructed the defect image in the imaging area surrounded by the transducer array in the aluminum plate. In recent years, Lamb wave tomography has attracted wide attention from researchers. This is mainly because the method has the following advantages [31]:

(1) Compared with other Lamb wave imaging methods, the Lamb wave tomography method has higher reconstruction accuracy and can obtain the detailed feature information of complex defects accurately.
(2) The transducer array combination and projection methods are flexible and can be used for defect imaging of aluminum plates with various specifications and shapes.
(3) The excitation mode is simple, and there is no need for complex timing control and excitation waveform calculation for the excitation transducer array.

When guided waves encounter defects, a strong scattering occurs, and the influence and effect of the scattering are dominant. The scattering will cause more pseudoscopic image [32] in the reconstructed defect image of the traditional method and produce the detection blind area, which seriously affects the defect location and imaging precision of the metal material structure. Although signal processing methods are utilized to weaken the impact of scattering [33], the effect is not obvious, and it does not solve the problem fundamentally. On the other hand, the shape and scattering characteristics of the actual defects are varied. The traditional fixed and regular sensor array structures cannot guarantee the best matching degree and sensitivity to the scattering characteristics of the actual defects, resulting in a very limited imaging precision for the actual defects. Therefore, it is necessary to study the magneto-acoustic array imaging method based on guided wave scattering. Due to the complex mechanism of the guided wave and the defect, and the diversity of the shape and scattering characteristics of the defect, the research of the field of guided wave scattering imaging of metal sheet defects lacks complete model and theoretical guidance. In particular, there are still a large number of basic theoretical problems that need to be studied and solved in the aspects of guided wave scattering defect imaging models and algorithms, the relationship between array structure and defect shape, and the time of flight extraction method for guided wave detection signal. The in-depth study and resolution of these core theoretical problems are the basis for guided wave scattering imaging detection of metal plate structures.

1.2 Research Status of Ultrasonic Guided Wave Imaging

In this section, the research status of electromagnetic acoustic transducer EMAT, ultrasonic guided wave tomography, and the guided wave scattering imaging are discussed in the following three aspects.

1.2.1 Research Status of Electromagnetic Acoustic Transducer

The Lamb wave has multimodal characteristics. Under the condition of a certain thickness of the plate and the frequency of the ultrasonic wave, there may be a variety of modes of Lamb wave with different modes of vibration and different wave velocities in the plate [34]. According to the different modes of vibration, the modes of Lamb waves can be divided into two main categories: symmetrical mode (S mode) and antisymmetric mode (A mode). For each mode of Lamb wave, the phase velocity is the propagation velocity of the Lamb wave at a single frequency, and the group velocity is the propagation velocity of the synthetic wave (i.e., wave packet), which is superimposed by a series of wavelengths of the mode Lamb wave with similar frequencies. Also, the Lamb wave also has the dispersion characteristics: When a certain mode Lamb wave propagates in the aluminum plate, the phase velocity and the group velocity vary with the frequency, the thickness of the plate, and the material characteristics of the aluminum plate.

Due to the existence of multimodal characteristics, on the one hand, different modal Lamb waves are sensitive to the defects of different types and sizes in aluminum plates. Therefore, in practical applications, it is necessary to receive appropriate modal Lamb waves selectively by modal control means for specific detection conditions and requirements in order to make full use of the advantages of each mode of Lamb wave. On the other hand, the detection waveform of Lamb wave is very complex and contains a large number of detection signals of different modal waves that overlap each other, so that the useful information in the waveform is difficult to be identified and extracted. Therefore, the mode control should be used to suppress the generation of interference modes to improve the useful information in the Lamb wave detection waveform. Therefore, realizing modal control is one of the most important problems in the research and design of Lamb wave EMAT.

Also, for large aluminum plate, electromagnetic ultrasonic Lamb wave tomography is usually carried out in the way of scanning. The EMAT array is shifted relative to the large area of the aluminum plate and the Lamb wave tomography is carried out at a fixed space interval during the translation process, and the results of each image are spliced to get a complete weight and build images. In order to meet the requirement of scanning speed, each imaging needs to be completed in a limited time, so it is not possible to gradually complete the excitation and reception of the Lamb waves along each direction by rotation of the various EMAT in the array. However, the omnidirectional EMAT is developed with high accuracy and efficiency, which can receive and receive Lamb waves along the 360-degree direction (i.e., the omnidirectional) [35].

In conclusion, it is necessary to use omnidirectional mode controllable Lamb wave EMAT for electromagnetic tomography Lamb wave tomography of aluminum plates. The research of this kind of EMAT mainly involves two aspects: (1) the mode control method of the omnidirectional Lamb wave; (2) the energy exchange mechanism and the design method of the omnidirectional mode-controlled Lamb wave EMAT. Among them, the former constitutes part of the research base of the latter. The following is a review of the above two aspects.

In actual detection, whether the Lamb wave can be excited and received to the correct mode depends on many factors, such as the wave structure, dispersion, and attenuation characteristics of each mode Lamb wave [36, 37]. Based on the theory of elasticity, the Rose of Penn State University, USA, analyzed and expounded the theoretical basis of Lamb wave mode selection. On this basis, they deeply study the relationship between the structure and size of the vibrosource in the plate and the excited Lamb wave mode and propose that the structure and size of the transducer used to stimulate the Lamb wave have an important influence on the type of the generated Lamb waves [38, 39]. The traditional mode controllable Lamb wave transducers mainly include oblique incidence [40], interdigitated [41], and comb transducers [42]. By using these transducers, the excitation and reception of different modal Lamb waves can be realized by changing the incidence angle of the ultrasonic wave or the space structure and size of the transducer. However, these traditional mode controllable transducers are directional Lamb wave transducers and cannot achieve the omnidirectional transceiver of Lamb modes.

The dual excitation method [43] is a relatively simple omnidirectional mode-controlled Lamb wave excitation method. By arranging a disk piezoelectric transducer on the upper and lower surfaces of the flat plate, it produces symmetric or antisurface displacements on the upper and lower surfaces of the plate and selectively excites the S mode or the A mode omnidirectional Lamb wave. The modal control method is also applicable to omnidirectional Lamb wave EMAT. However, the method requires the two-sided arrangement of the transducer on the flat plate, resulting in certain restrictions on its application.

Since the 1990s, with the rapid development of structural health monitoring (SHM) technology, the omnidirectional Lamb wave transducer has attracted more and more attention because it is suitable for the implementation of high efficiency SHM for large-area plates. However, the research on the theoretical model of the omnidirectional Lamb wave transmission and transmission process is relatively rare, and in practical application, the selection of all the parameters of the omnidirectional Lamb wave lacks the support of the mathematical basis. In 2004, Raghavan and Cesnik established and analyzed the three-dimensional analytical model of the omnidirectional Lamb wave sound field produced by the circular piezoelectric source based on the basic equations of three-dimensional elastic mechanics and verified the validity of the model by numerical calculation and experimental means. Then, they set up a three-dimensional analytical model of the omnidirectional Lamb wave sound field produced by a small rectangular piezoelectric source. The validity of the model is simulated and tested, and the size of the omnidirectional Lamb wave piezoelectric transducer is optimized by using the model [44, 45]. However, the above study does not give a modal control method for omnidirectional Lamb waves.

By in-depth study and analysis of the analytical model of Lamb wave sound field, the omnidirectional mode-controlled Lamb wave excitation is realized by Giurgiutiu et al. using the method of frequency tuning of the piezoelectric transducer. The validity of the mode control method is verified by the test, and the omnidirectional Lamb wave piezoelectric transducer based on this method is applied to Lamb wave phased array defect detection and time reversal defect detection [46, 47]. The results show that this method can control the mode of the excited Lamb wave to a certain extent and reduce the difficulty of extracting a useful signal in the detection waveform. However, this method can only enhance or weaken the amplitude of a certain mode Lamb wave to a certain extent, and when the Lamb wave with more modes is contained in the plate, the method cannot simultaneously suppress the appearance of all other modal Lamb waves outside the useful mode, that is, the limitation of its mode control ability.

In recent years, some researchers have tried to use array piezoelectric transducers to realize omnidirectional Lamb wave mode control. Based on theoretical analysis, an omnidirectional Lamb wave mode control method is proposed by Glushkov et al. By applying different amplitude excitation signals to multiple independent annular energy transfer units in an array piezoelectric transducer, the omnidirectional Lamb wave [48] of a specific mode is excited. However, a large number of transducers are required for tomography, and there are several independent energy exchange units in each transducer. Therefore, it is difficult to control such a large number of independent energy exchange units accurately in practical applications. Koduru and others will introduce an array omnidirectional Lamb wave piezoelectric transducer with the cross-finger electrode structure and comb electrode structure in the directivity piezoelectric transducer so that the mode controlled [49] can be realized only by applying the same amplitude excitation signal to each energy exchange unit in the array transducer. However, the above electrode structure and Lamb wave mode control method are only applicable to piezoelectric omnidirectional Lamb wave transducers but cannot be used in omnidirectional mode controllable Lamb wave EMAT.

In conclusion, the researchers have carried out much research on the Lamb wave mode control method and the mode-controlled Lamb wave transducer, and the research on the omnidirectional Lamb wave mode control method has also made some progress, which lays a good foundation for the study of the omnidirectional mode-controlled Lamb wave EMAT. However, the existing omnidirectional Lamb wave mode control methods still have some limitations, such as the need to arrange the transducer on both sides of the plate or not to completely suppress the appearance of all interference modal waves. Also, most of the existing methods for piezoelectric transducers cannot be used directly to guide the design of omnidirectional mode controllable Lamb wave EMAT. Therefore, it is still necessary to further study the omnidirectional Lamb wave mode control method suitable for EMAT.

EMAT excitation and reception of Lamb waves in aluminum plates are mainly based on the Lorenz force mechanism [50, 51]. Much research has been done on the energy transfer mechanism and design method of Lorenz power EMAT.

The research on the mechanism of Lorenz power EMAT energy exchange began in the 1970s. In 1973, Thompson pioneered the Lamb wave EMAT structure with important application value and studied the energy exchange mechanism and characteristic [52] of Lorenz force EMAT through theoretical analysis and experiment. However, his EMAT is a directional Lamb wave EMAT and can only transmit Lamb waves in a specific direction. In the early 1990s, Ludwig and Dai et al. derived the system control equation of Lorentz force EMAT, established the two-dimensional finite element model of Lorenz force EMAT, and introduced the solution of the model in detail [53–57]. The model they set up consists of five parts: (1) the distribution of static magnetic field provided by electromagnet; (2) the distribution of eddy current on the surface of aluminum plate; (3) distribution of Lorenz force; (4) Lorenz force produces ultrasonic wave; and (5) the induction electromotive force is produced in the receiving EMAT by the ultrasonic signal. However, the model they built is only for the ultrasonic surface wave and does not involve the Lamb wave's sound field model. In the process of studying Lorenz force EMAT by the analytic method and finite element method, Shapoorabadi focuses on the skin effect and proximity effect between each turn conductor of the EMAT coil and improves the accuracy of theoretical analysis and simulation calculation [58–61]. However, the study only calculated the electromagnetic coupling relationship in EMAT and did not involve ultrasonic field. The Dhayalan application finite element analysis software COMSOL Multiphysics and ABAQUS are combined to realize the simulation analysis of the electromagnetic field of Lorenz force EMAT and the simulation analysis [62, 63] of the Lamb wave propagating in the tested sample. However, the EMAT studied by him is EMAT with directivity rather than omnidirectional Lamb wave EMAT.

Hao and others at Tsinghua University have carried out the three-dimensional analytical modeling [64] for directional controllable Lamb wave EMAT. In order to improve the signal intensity and mode control ability of EMAT detection, Harbin Institute of Technology's Kang and others have optimized the parameter optimization design of the mode-controlled Lamb wave EMAT with directivity [65].

According to the researchers above, the study of Lorenz power EMAT mechanism involves only surface wave EMAT and directivity Lamb wave EMAT and does not involve omnidirectional Lamb wave EMAT. In order to use Lorenz force EMAT to transmit and receive omnidirectional Lamb wave in aluminum plate, Wilcox proposed a design method of Lorenz force omnidirectional Lamb wave EMAT and analyzed the principle of Lamb wave mode control for EMAT [66]. The structure of the EMAT (referred to "traditional EMAT" as follows) designed by them, shown in Fig. 1.1a, is mainly composed of a cylindrical permanent magnet, a dense spiral coil [67] with a pulse excitation current I and three parts of the measured aluminum plate. The EMAT based on the A0 mode Lamb wave is dominated by the off-plane displacement [68], and the S0 mode Lamb wave is dominated by the in-plane displacement. By making the direction of the static magnetic field produced by the permanent magnet perpendicular to the surface of the aluminum plate, the vibration displacement based on the Lorenz force is only the internal component of the plane, so that the excited omnidirectional Lamb wave is dominated by the S0 wave. However, on the one hand, the EMAT can only receive Lamb waves based on S0 waves; on the other hand, A0 waves also contain a small amount of in-plane displacement components. Nagy proposed another design method of Lorenz force omnidirectional mode-controlled Lamb wave EMAT: Based on the EMAT structure shown in Fig. 1.1, by adjusting the diameter of the cylindrical permanent magnet and control the angle of the static magnetic field and the Lorenz force, the displacements in the aluminum plate are enhanced, and the in-plane displacement is suppressed, so that the Lamb wave is excited, and the lamb wave is dominated by A0 waves [69]. The principle and limitations of the EMAT are similar to those of traditional EMAT.

It is concluded that most of the research on the Lorenz force EMAT energy transfer mechanism involves only surface wave EMAT and Lamb wave EMAT with directivity, rarely including omnidirectional Lamb wave EMAT, and more rarely involves omnidirectional mode-controlled Lamb wave EMAT. Moreover, for the design method of omnidirectional mode-controlled Lamb wave EMAT based on Lorenz force, the design of EMAT's modal control capability has some limitations. Therefore, it is difficult to meet the requirement of accurately extracting useful Lamb wave detection information in tomography. Therefore, it is necessary to carry out corresponding research on the energy transfer mechanism and effective design method of omnidirectional mode controllable Lamb wave EMAT based on Lorenz force.

Based on the principle of magnetostrictive, EMAT is generally used to excite horizontal shear wave (shear-horizontal guided waves, SH guided wave) [70] in the steel plate. Due to the restriction of the structure of piezoelectric transducer and

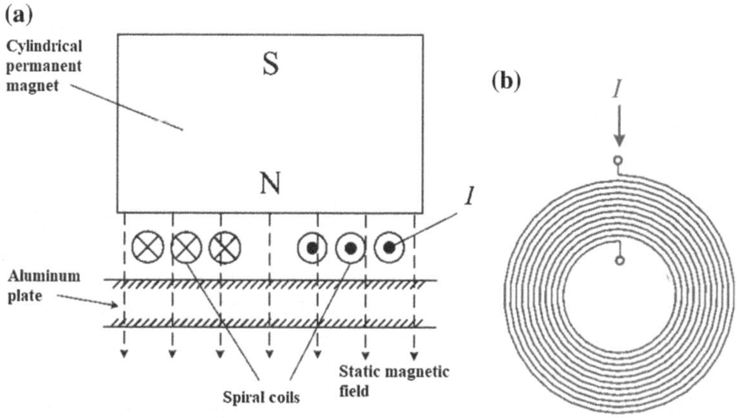

Fig. 1.1 Schematic diagram of the structure of traditional EMAT: **a** EMAT structure principle (side view cross section); **b** coil (overlook chart)

the principle of energy exchange, the SH guide wave is not easy to be produced by the piezoelectric transducer, which is more suitable for the production of EMAT. In practice, EMAT is often used as a transducer for exciting and receiving SH guided waves and for defect detection. The SH0 guided wave in the metal plate is one of the most commonly used guided modes. Compared with the Lamb wave in the metal plate, the strength attenuation of the guided wave signal is less [71] when the scattering process is occurring the defect. The mode conversion [72] does not occur during the propagation process, and the wave velocity does not change with the frequency of the work. There is no dispersion phenomenon [73]. The single-mode control can be realized in the frequency domain. It can simplify the analysis process of the electromagnetic ultrasonic guided wave detection signal to a certain extent [74]. It can provide a single and pure SH0 mode guided wave in the metal plate. It is beneficial to improve the analysis efficiency of the guided wave detection signal and the detection efficiency of the guided wave of the metal plate. Also, the SH wave does not generate particle displacement outside the plane. Therefore, particle motion and wave propagation are not affected by the steel bearing medium.

The directional EMAT of SH guide wave of the steel plate is designed to explore the directional problem of the EMAT excited SH guided wave, including the direction control of the excited SH guided wave, the direction characteristic of the SH guided wave in the process of the plate propagation, and the receiving characteristic of the EMAT on the guided wave of the steel plate. The directional problem of SH guided wave EMAT of steel plate is an important part of the defect contour guided wave scattering imaging model and method in this book. It can provide a priori conditions for solving the imaging model of plate defect contour guided wave scattering imaging and improve the precision of the defect contour imaging of steel plate.

The research on EMAT or other types of ultrasonic guided wave transducers is mostly focused on how to improve the performance of transducer [75, 76], which also includes some research on the direction of the ultrasonic guided waves in the propagation process. Simulation and experimental method have been used by Xie to study the direction of the Rayleigh beam excited by the bending coil [77, 78]. By changing the length of the bending coil in different parts, the focusing of the Rayleigh wave propagation process in the aluminum plate is enhanced, and the signal intensity of the sidelobe is reduced [79–81]. Hill adopts the EMAT based on the periodic permanent magnet, which makes full use of the periodic characteristics of PPM EMAT and changes the angle of guided wave emission by changing the frequency of the input signal of the coil in the EMAT [82, 83]. Seung uses a pair of permanent ring magnets to provide bias magnetic field based on the magnetostrictive principle of ferromagnetic materials; an omnidirectional SH guided wave EMAT is developed. Wilcox aimed at the requirement of fast scan detection for large-area metal plate structures, and an omnidirectional EMAT array based on S0 mode guided waves was designed [84]. By changing the interdigital transducer (IDT) of a given piezoelectric element (IDT), an IDT type Lamb wave transducer with certain controllability of working frequency and directivity is designed, and it is used to detect the length and angle of the artificial crack defect of the aluminum plate [85]. The piezoelectric transducer has certain controllability in working frequency and excitation Lamb wave direction. Using numerical simulation, a circular transducer array is designed by Masuyama to stimulate the narrow beam of the given radiation direction and realize the controllability of the radiation direction of the narrow beam [86]. Dixon uses a plasma burner to excite the ultrasonic wave and uses a stable Michelson interferometer with a wide bandwidth to measure the displacement response of the ultrasonic vibration and uses EMAT to measure and study the directional mode [87] of the generated longitudinal waves. Bernstein uses a laser line light source to produce an ultrasonic transverse wave in the aluminum plate and uses EMAT to receive ultrasonic transverse wave propagating in the aluminum plate. The amplitude directional problem of the transverse wave produced by the laser line source and the point source [88] is compared. Ogi designed the line focusing EMAT that produces vertical shear elastic waves in the steel plate, trying to realize the linear propagation of the vertical shear elastic wave by changing the distance between the various turns of the bending coil, and the relationship between the detection sensitivity and the liftoff value [89–91] is also studied. Wu and others of the Institute of the Acoustics of the Academy of Sciences used EMAT to produce ultrasonic transverse waves on the surface of the aluminum block. By analyzing the distribution of the nonuniform horizontal shear force on the surface of the aluminum block, the directional mode of the surface ultrasonic transverse wave was studied [92].

In conclusion, the research on the direction of EMAT to excite or receive ultrasonic guided waves is mostly the production of linear ultrasonic EMAT with omnidirectional EMAT or single direction. It is lack of research on the accuracy and controllability of the direction of ultrasonic guided waves by EMAT, and whether the direction of the EMAT excited ultrasonic guided wave can be accurately controlled is directly affected by the lack of the direction of the ultrasonic guided wave. The sensitivity of the detection and the accuracy and reliability of the defect guided wave detection signal. Therefore, it is necessary to develop an accurate and controllable EMAT to stimulate or receive the direction of ultrasonic guided wave, so as to provide accurate and abundant guided wave detection signal for the defect contour guided wave scattering imaging, which is of great significance for improving the precision of the imaging of the defect contour guided wave scattering imaging.

1.2.2 Research Status of Ultrasonic Guided Wave Tomography

The word "tomography," originally derived from the medical field, specifically refers to X-ray computed tomography (CT), is an image reconstruction technique using the information obtained from X-ray scanning to calculate the distribution of media in a scanned object. After that, tomography technology is mainly applied in the field of medicine and geophysics. The latter uses acoustic waves to scan objects, which is different from the former.

In 1990, Jansen and Hutchins first introduced tomography technology into the field of Lamb wave detection [93]. They use the filtered back projection (FBP) algorithm commonly used in the field of medical CT to reconstruct the image of the pass defects on the aluminum plate. In each reconstruction process, the projection data are obtained by parallel projection, that is, through the rotation converter array, the projection from dozens of different directions is carried out. Each projection contains dozens of parallel Lamb wave rays, as shown in Fig. 1.2. The Lamb wave tomography method based on FBP and parallel projection is called Lamb wave parallel FBP tomography. In the study, they selected the energy attenuation of the Lamb wave signal and the time-of-flight as the projection data. The research results show that the latter parameter can be used as projection data to achieve higher accuracy of defect reconstruction. Then, Hutchins used the offset of the central frequency of the Lamb wave signal (referred to "frequency shift") as the projection data to reconstruct the pass defect image. The results show that the defect reconstruction accuracy of the two-projection data is higher than that of the pre-used projection data [94]. Wright et al. took attenuation and frequency shift as projection data and studied the imaging effect of Lamb wave parallel FBP tomography for slot defects [95]. Zhao et al. introduced the interpolation method to Lamb wave parallel FBP tomography to reduce the amount of projection data needed for imaging [96, 97].

The limitation of the Lamb wave parallels FBP tomography method mentioned above is that it needs to rotate the transducer array continuously to obtain enough projection data. Therefore, the imaging efficiency is low, and the workload is huge. In this regard, in 1999, McKeon and Hinders, etc. applied the cross-hole projection method commonly used in the field of geophysical tomography to Lamb wave tomography to form a Lamb wave cross-hole tomography method [98]. The principle of the cross-hole projection method is shown in Fig. 1.3. The transceiver–transducer is arranged on both sides of the imaging region in a straight line, and any pair of transceiver–transducer combinations are used for Lamb wave projection. Compared

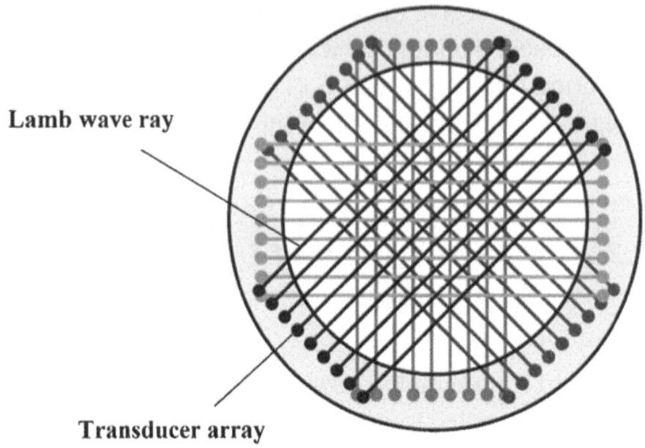

Fig. 1.2 Schematic diagram of the principle of parallel projection

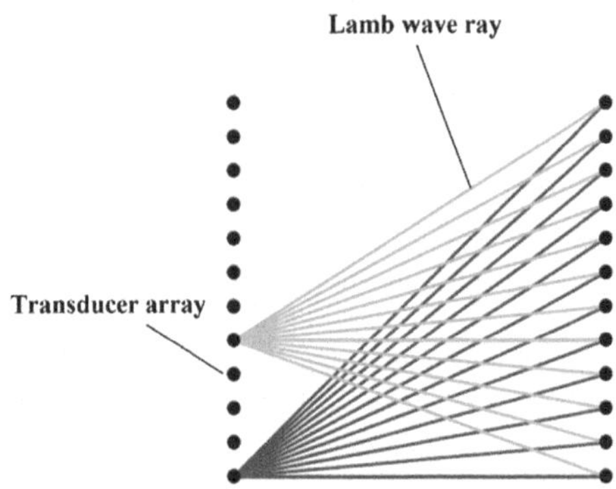

Fig. 1.3 Schematic diagram of cross-hole projection principle

to the Lamb wave parallel FBP tomography, the Lamb wave cross-hole tomography method does not need a rotating transducer array and the number of projection is less, and it can work under low or uneven projection conditions. Lamb wave cross-hole tomography is unable to use the traditional FBP algorithm because of fewer projections. It needs to use iterative methods to reconstruct the image. The commonly used iterative methods mainly include algebraic reconstruction (ART) and Simultaneous Iterative Reconstruction Technique (SIRT) [99]. Also, this method usually takes the TOF of Lamb wave as projection data. Prasad has improved and optimized the distance and number of transducers in Lamb transducer for cross-hole tomography [100].

In recent years, Reconstruction Algorithm for Probabilistic Inspection of Damage (RAPID) has attracted much attention in the field of Lamb wave tomography [101]. The Lamb wave tomography method based on the algorithm uses the signal difference coefficient (SDC) of Lamb wave to detect the signal as projection data. SDC can be obtained by calculating the cross-covariance between the detected signal and defect-free detection signal [102]. The Lamb wave RAPID tomography method uses a circular, square, or irregular shape surround transducer array for projection. The advantage is that only a small amount of transducer can be used to complete the image reconstruction. Rose et al. studied the Lamb wave RAPID tomography method and applied it to the detection, location, and development monitoring of aircraft wing defects [103–105]. Wang et al. introduced the virtual sensitive path and digital damage fingerprinting technology to Lamb wave RAPID tomography to improve the quality of image reconstruction [106–108].

Compared to Lamb wave parallel FBP and RAPID tomography, Lamb wave cross-hole tomography has a high accuracy of defect reconstruction, and it can complete imaging only in a narrow linear region on both sides of the imaging region. It is very suitable for the defect imaging of the aluminum plate under the condition of the large surface covering or covering. Therefore, it has important research and application value. This book focuses on the theoretical and experimental research of Lamb wave cross-hole tomography.

The basic theory of the Lamb wave cross-hole tomography method is to divide the imaging region into several pixel grids. Then, according to the TOF of each Lamb wave ray extracted from the detected waveform and the propagation path of each Lamb wave ray obtained from the calculation, the residual plate thickness in each pixel grid is iteratively solved. Thus, the defect image reconstruction is realized. Therefore, the accuracy of Lamb wave ray TOF extraction and propagation path determination directly determines the success or failure of the Lamb wave cross-hole tomography. Also, due to the sensitivity of different modal Lamb waves to different types and sizes, how to make full and effective use of the advantages of different modal Lamb waves is also a key issue in the Lamb wave cross-hole tomography. Therefore, the study of Lamb wave cross-hole tomography is mainly involved in three aspects: (1) How to accurately extract the travel time of Lamb wave; (2) how to improve the imaging performance by implementing multimodal Lamb wave tomography; and (3) how to solve the propagation path of Lamb wave accurately. Focusing on the above three aspects, researchers have done a lot of theoretical and experimental research.

The TOF information of the electromagnetic ultrasonic guided wave refers to the time from the excitation transducer to the ultrasonic guided wave to receive the ultrasonic guided wave, including the propagation process of the ultrasonic guided wave and the process of interaction with the defect [109–111]. The TOF of guided waves is one of the most important features of electromagnetic ultrasonic guided wave detection signals [112, 113]. It can provide the most direct information for defect location based on ultrasonic guided wave detection [114–116]. At the same time, it is also the direct input of various ultrasonic guided wave imaging methods developed in recent years [117–119]. Its accuracy directly affects the accuracy of defect location and the defect imaging accuracy of guided wave imaging models and algorithms [120, 121]. Therefore, it is urgent to study the travel time extraction method of the electromagnetic ultrasonic guided wave detection signal, to improve the accuracy of the time delay extraction of the electromagnetic ultrasonic guided wave detection signal.

The efficiency of EMAT is lower than that of piezoelectric transducers and laser transducers [122–125]. In order to improve the signal intensity of the ultrasonic guided wave and the signal-to-noise ratio of the guided wave detection signal, the excitation voltage of the EMAT coil is input to a multiperiod tone-burst excitation signal when the EMAT excitation of the narrowband electromagnetic ultrasonic guided wave is used [126, 127]. The guided wave detection signal received by the receiving transducer has a wider time scale in the time domain [128]. Most of them exist in the form of wave packets, so it is difficult to determine the specific arrival time of the guided wave detection signal. On the other hand, in the commonly used electromagnetic ultrasonic guided wave detection system, in order to improve the signal-to-noise ratio of the guided wave detection signal, the excitation signal exerted on the excitation transducer is more powerful [129, 130]. When the excitation signal is applied instantaneously, the receiving transducer receives the space induced pulse wave from the excitation transducer and its excitation signal [131]. The waveform of the guided wave detected by the receiving transducer shows the initial pulse wave, and the time scale range in the time domain is also wider. Since the initial pulse wave is formed by the space induction of the excitation transducer, which contains the time information of the ultrasonic guided wave excited by the excitation transducer in the medium, the timing information of the ultrasonic guided wave detection signal is contained in the initial pulse wave with a certain time-domain width. Therefore, the extraction of the TOF data is hard. The above problem is an important reason for the difficulty and accuracy of ultrasonic guided wave TOF data detection. It is a difficult problem in the field of electromagnetic ultrasonic guided wave detection.

There are some related researches on the TOF extraction of ultrasonic guided wave detection signals, mainly focused on the estimation of the time of ultrasonic guided wave signal detection. Legendre uses a wavelet transform algorithm to extract time information from the main peak of the guided wave detection signal [132]. In order to improve the imaging quality and accuracy of Lamb wave tomography algorithm, Leonard proposed a dynamic wavelet transform fingerprint method (DWFP) to extract the TOF of the Lamb wave detection signal [133]. Some scholars also use wavelet neural network to measure the TOF difference between ultrasonic pulse and target echo [134]. Based on the concept of information entropy, Li chose the optimal wavelet generating a function to determine the specific location of the guided wave packet [135]. For the band-pass filter signal obtained by the optical fiber sensor, the signal envelope is obtained by the Hilbert transform of the signal, and then, the signal feature extraction and defect detection are carried out [136]. In order to improve the spatial resolution of defect imaging, some scholars have proposed a warped frequency transform (WFT) method, which mainly aims at correcting the frequency shift of the Lamb wave detection signal generated by the piezoelectric transducer array, reducing the distortion of the wave packet of the guided wave detection signal [137, 138]. Moll based on the matching pursuit decomposition algorithm, a time-varying reversal filter is proposed to process the guided wave detection signal, which improves the time resolution of the first wave detection signal and obtains the required guided wave arrival time [139]. Scholars at Xi'an Jiao Tong University have used the chirp wavelet transform and instantaneous frequency estimation to calculate the arrival time of ultrasonic guided waves in aluminum plates.

As for the study of TOF extraction of ultrasonic guided wave detection signals, all of them have their specific application conditions. At present, there is a lack of a universal and accurate TOF extraction method for electromagnetic ultrasonic guided wave detection signals. Therefore, it is necessary to study a universal method of TOF extraction for electromagnetic ultrasonic guided wave detection signal, to accurately extract the TOF of guided wave detection signal, and provide accurate and reliable reference or input for defect location, defect imaging algorithm and model. It improves the detection accuracy and defect imaging accuracy of defect electromagnetic ultrasonic guided wave detection.

Different modes of Lamb wave are sensitive to different types and size defects, and the use of multiple modes of Lamb wave tomography cannot only help to realize effective imaging of different types of defects. It is also helpful to achieve high-quality image reconstruction and accurate feature extraction for defects with complex features such as variable-depth defects. Therefore, how to make full and effective use of the advantages of modal Lamb waves is one of the important research contents in the field of Lamb wave tomography.

Hutchins et al. applied A0, S0, A1, and S1 mode Lamb waves to parallel FBP tomography, respectively. Ho uses S0, A1, and S1 waves to perform parallel FBP tomography tests on circular hole defects. The results show that the use of A1 wave can obtain higher accuracy of image reconstruction accuracy [140]. Koduru has compared the sensitivity of A1 and S1 waves to water loading on aluminum plates in RAPID tomography [141]. The principle of Lamb wave parallel FBP and RAPID tomography is different from the principle of Lamb wave cross-hole tomography, and the above results are relatively small for the study of multimodal Lamb wave cross-hole tomography. Leonard uses the wide-band excitation signal to excite the Lamb wave along the direction and separate the A0, S0, and A1 modal Lamb waves from each detection wave, and extract the travel time of the three waves. Then, the cross-hole tomography experiments were carried out separately by these three waves, and the results were compared [142]. By the above research, a walking time classification algorithm is proposed to reduce the adverse effects caused by the interference of each mode wave, thus improving the accuracy of the travel time extraction of each modal wave. Subsequently, the above three kinds of waves were used to conduct cross-hole tomographic imaging of different sizes of circular hole defects, and the effect of defect reconstruction was compared.

In conclusion, the existing research on multimodal Lamb wave cross-hole tomography method is limited to comparing the imaging results of different modal Lamb waves through experimental means, in which only a certain mode of Lamb wave is used in each imaging, and the principle and square of the coordination of different modal waves in multimodal Lamb wave cross-hole tomography are not given. Therefore, multimodal Lamb wave cross-hole tomography is not realized. Moreover, there are still some problems need to be further studied for multimodal Lamb wave cross-hole tomography. On the one hand, it is necessary to study the regularity of the characteristic parameters associated with the common modal Lamb wave and the cross-hole tomography, so that the Lamb waves of these modes can be used more fully and effectively in the cross-hole tomography. On the other hand, it is necessary to study the use of multiple modal Lamb waves for cross-hole tomography, and we should focus on the principles and methods of mutual coordination between Lamb waves.

In the previous studies, the Lamb wave cross-hole tomography is based on the straight-ray model, that is, the Lamb wave always propagate along the straight line. In the actual situation, when the Lamb wave ray in the aluminum plate meets the defect, the direction of the propagation will be deflected, that is, the Lamb wave will be propagated along the curved path. Therefore, in the defective aluminum plate, the Lamb wave ray propagation path based on the direct ray model has a certain error compared with the actual situation, which will affect the performance [143] of the cross-hole tomography of the direct ray Lamb wave. An effective solution to the above problem is to introduce ray tracing (RT) technology in cross-hole tomography: First, the approximate distribution of the defects is obtained by direct ray cross-hole tomography, and then, based on the distribution of the defects, the RT technique is used to search for the curved Lamb ray which is closer to the true condition. Finally, we use the modified path to obtain cross-hole tomography.

The conventional ultrasound RT algorithm is mainly divided into two categories: One is the bending RT algorithm for the boundary value problem, and the other is the RT algorithm for the initial value problem [144–146]. The bending RT algorithm with good application effect mainly includes finite difference method, the linear interpolation method, and shortest path method. Also, the simulated annealing algorithm is also introduced into the bending RT algorithm [147–150] to overcome the problem that the traditional bending RT algorithm is too dependent on the accuracy of the initial model [151]. The principle of the RT algorithm is to calculate a ray path according to the given angle of incidence and to find the ray path of the connecting transceiver by constant correction of the incident angle. Through numerical method and angular displacement calculation, Andersen et al. give a numerical algorithm to test RT. Used the Ferma principle to test RT by ultrasonic body wave in the anisotropic medium and realized the cross-hole tomography of ultrasonic body wave based on the test of RT algorithm by simulation and experiment.

The above studies focus on the RT algorithm of acoustic or ultrasonic body waves, rather than the Lamb wave RT algorithm. The Lamb wave has the characteristics of dispersion and multimodal, and the scattering characteristics of different modes Lamb waves are different from that of the defects. Therefore, the propagation characteristics and the RT principle in the defective aluminum plates are also more complex. Zhang et al. improved the simulated annealing method [152] and a linear interpolation method in the bending RT algorithm, respectively, and formed two Lamb wave bending RT algorithms, and applied them to the Lamb wave cross-hole tomography. Compared with the Lamb wave bending RT algorithm, the Lamb wave test RT algorithm has greater advantages in global search and adaptive complex scattering models. Malyarenko et al. proposed a Lamb wave test RT algorithm [153, 154]. The algorithm uses the extrapolation method to calculate the direction of the Lamb wave step by step according to the gradient direction of the Lamb wave velocity distribution obtained by the vertical ray cross-hole tomography. Because of the discrete Lamb wave velocity distribution, they use the bilinear interpolation (Bilinear Interpolation, BI) [155] to solve the phase velocity gradient because of the direct ray cross-hole tomography. However, in each step, the accuracy of the phase velocity gradient obtained by the BI method is low because of the low resolution of the discrete Lamb wave velocity distribution obtained by the straight-ray cross-hole tomography, so the accuracy of the Lamb wave ray path obtained by this gradient is also low. This leads to the failure of searching for the Lamb wave path connecting the transmitter and receiver during the Lamb wave test RT, failing in the RT process. Therefore, the reliability of the RT algorithm based on the Lamb wave test based on BI is low, which will lead to the low reliability of the curved ray Lamb wave cross-hole tomography method using the algorithm.

In view of the above problems, it is necessary to study the mechanism of the influence of defect distribution on the propagation path of Lamb wave and study the higher accuracy of phase velocity gradient method, and then study the higher accuracy and reliability of the Lamb wave test RT algorithm and the corresponding curved ray Lamb wave cross-hole tomography method.

1.2.3 Research Status of Ultrasonic Guided Wave Scattering Imaging

There is less research on defect imaging using guided wave scattering. Most of the defect detection based on guided wave scattering is limited to the defect location and most of the defects are artificial defects of the standard contour shape, and the transmission guided wave is used to carry out the defect imaging research. However, the analysis and processing of the guided wave detection signal are encountered by the guided wave. The scattering effect after the defect is greatly influenced, which leads to a large error in the characteristics of the guided wave signal from the travel time and amplitude extracted from the guided wave detection signal, which has a negative impact on the accuracy of the defect imaging. Therefore, the direct use of guided wave scattering for defect detection and imaging has become a potential research direction, which is of great significance for improving the imaging accuracy of ultrasonic guided wave detection and improving the theory and methods of imaging defects in guided wave detection. The model of guided wave scattering defect imaging is the foundation of defect imaging using scattered guided waves. The transducer array structure is complementary to the defect imaging model, and its structure depends to some extent on the requirements of the guided wave scattering imaging model. The structure of the transducer will also affect the accuracy of the defect imaging. Next, the research status and the problems to be solved in two aspects of the structure of the magnetic, acoustic array based on the guided wave scattering and the model and method of the guided wave scattering defect reconstruction are sorted out.

In order to obtain sufficient information for defect imaging, sensor array must be used to detect and scan the defects in different directions. Through a certain algorithm, the image of the defect of the metal sheet is reconstructed, and the location and size of the defect are obtained. In 1993, Hutchins and others first tried to use the electromagnetic ultrasonic transducer for non-contact ultrasonic guided wave tomography [156, 157], which was limited to the energy efficiency of the magneto-acoustic transducer with low imaging precision. The research group of Tsinghua University has made a continuous and in-depth study on the analytical calculation model of the electromagnetic ultrasonic transducer EMAT, the Lorenz force mechanism, and the magnetostriction mechanism [158, 159]. The Tsinghua University has made many important research achievements in the tomography method and optimization design based on the EMAT magnetic, acoustic array [160–163]. At present, the defects imaging of metal sheet based on the ultrasonic guided wave is still based on the piezoelectric transducer and laser transducer, and the study of ultrasonic guided wave imaging with the magnetic acoustic transducer is very rare.

The topology of the transducer array in traditional ultrasonic guided wave imaging detection is relatively fixed and has a regular shape, which mainly includes parallel distribution, moment distribution, circular distribution, and so on, as shown in Fig. 1.4 [164–166]. However, there is a serious mismatch between the geometric structure of a specific rule shape sensor array and the multiple scattering characteristics of the actual defects. The geometric structure of a specific sensor array has the best matching degree and sensitivity to the defects of the specific scattering characteristics. The shape of the actual defect and its scattering characteristics are varied. The array detection method of traditional specific geometric structure cannot keep high sensitivity to all kinds of scattering characteristics, so the accuracy of imaging detection for actual defects is very limited. Chen has studied the sensitivity of several kinds of piezoelectric transducer arrays to the scatters in specific directions and positions [167]. Michaels attempts to adjust the geometry and location of piezoelectric transducer arrays to improve the accuracy of defect image reconstruction [168]. The three-dimensional finite element method (Fromme) is used to simulate and study the influence of the standard crack defects on the detection results relative to the sensor array element.

At present, the elements in the ultrasonic guided wave detection array are mainly piezoelectric transducer and laser transducer, and the array structure is relatively fixed and regular. The study of the relationship between the array structure of the transducer and the accuracy of the defect imaging is insufficient, and the dynamic

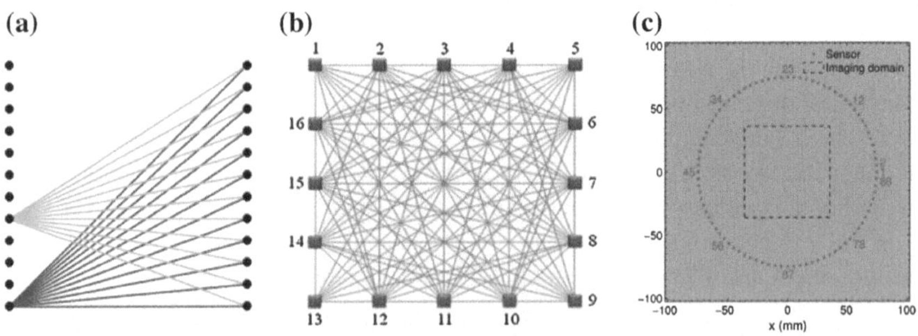

Fig. 1.4 Schematic diagram of traditional transducer array detected by ultrasonic guided wave imaging: **a** parallel distribution; **b** moment form distribution; and **c** circumferential distribution

adjustment and performance optimization of the geometric topology of the transducer array has not yet been studied. Therefore, the optimization of the geometry structure of the magnetic, acoustic array based on the guided wave scattering is very important for improving the defect detection ability and imaging accuracy.

The actual defects in sheet metal structures are very irregular, so the discontinuity of defects is more complicated [169]. The scattering of guided waves is dominant. At the beginning of this century, Professor Rose and other researchers simulated the interaction between guided waves and defects by boundary element method and made a preliminary study on the classification of defects [170]. Immediately after that, the Cawley Professor project team of Imperial College London in England studied the scattering phenomenon of guided waves in the pipeline through experiments [171–173]. Based on the simplified normal mode expansion method, Professor Cho studied the modal transformation and scattering of ultrasonic Lamb waves in the defect [174, 175]. The energy distribution of each conversion mode was studied from the angle of conservation of energy. The scattering of ultrasonic Lamb waves in welding and bonding was studied by this method. McKeon solved the scattering from the regularly shaped hole through the analytical solution of the plane ultrasonic Lamb wave in the point source [176]. Santos and other guided waves are used to detect defects in the lap joints of structural parts [177]. In recent years, some project groups have used a scanning laser Doppler vibrometers (SLDV) to study the propagation of the guided wave and the wave field images when the guided wave meets the defect scattering [178–180]. The project group of Professor Trevelyan of the Durham University, Britain, uses the extended equal geometric boundary element method and the extended double boundary element method to analyze the scattering process of guided wave and the stress analysis in the defect [181–183].

The existing ultrasonic guided wave imaging methods for metal material defects mainly include phased array imaging, tomography, migration imaging, and delayed superposition imaging. The phased array imaging usually uses a piezoelectric transducer to make up a dense array. Because the piezoelectric transducer needs a liquid coupling agent or the surface of the material to be measured, it cannot be applied to the detection of high-temperature materials [22, 184, 185]. The tomography method uses multiple projection data to reconstruct the defect image [35], without considering the scattering of the guided wave and the defect. The use of the process has a negative impact on the accuracy of the inversion of the defect imaging [186]. The migration imaging method and the delayed superposition imaging method, to some extent, use the scattering signal of the guided wave and the defect, but the array structure is relatively fixed, and the relationship between the scattering intensity of the guided wave and the shape of the defect is not considered, which leads to the inability of the array structure to adapt to the structure. The defect detection accuracy is limited due to the change of defect shape.

Some scholars began to study the defect reconstruction method based on irregular transducer array. Levine and so on, based on the Lamb wave, set up a general linear scattering model for each defect position and use the block sparse reconstruction algorithm to divide the Lamb wave detection signal into the position-based component and reconstruct the defect image [187]. Based on the minimum variance imaging method, Hall constructed an irregular guided wave array [188] for defect feature extraction. The Michaels project team designed the array of transducer arrays based on the characteristics of artificial defects in the laboratory, trying to use the sparse reconstruction method to perform imaging for artificial defects [189], which aims to reduce the number of the transducer by the quality of the imaging [190]. Professor Harley also tries to reduce the number of elements needed in the guided wave imaging dictionary matrix, thus reducing the computational burden and effectively extracting the characteristic [191, 192], and other characteristics of the guided wave dispersion. The above research mainly uses sparse reconstruction method to detect defects. The research focuses on improving the algorithm. In addition, the existing guided wave scattering defect reconstruction model needs to compare the guided wave detection signal with the defect when the plate is not defective, and a guided wave detection should be carried out when the plate is not defective, which cannot be realized in many practical testing projects because the measuring plate is treated. When testing, the defect has already existed in its specific position, which leads to the failure to obtain the guided wave detection signal when the plate is without defect, which makes the model of the wave scattering defect reconstruction difficult to be applied in the engineering practice, and the detection efficiency of the guided wave is reduced to a certain extent by using the signal contrast of the two guided wave detection signals. A variety of unknown interference factors, including the consistency of the transducer position, the transducer excitation, and the consistency of the detection parameters, are not conducive to the provision of accurate guided wave detection data. Therefore, it is necessary to study the defect contour reconstruction model and imaging method based on the guided wave scattering and study the model and algorithm of the defect contour reconstruction directly using the guided wave scattering signal. The imaging algorithm directly determines the imaging precision of the metal sheet defect and uses the guided wave scattering to carry out the defect contour. The establishment of the model method and the solution of the related problems are of great significance for improving the imaging precision of ultrasonic guided wave detection and improving the theory and method of ultrasonic guided wave detection for metal sheet defects.

References

1. S.L. Huang, Y. Tong, W. Zhao, A denoising algorithm for an electromagnetic acoustic transducer (EMAT) signal by envelope regulation. Meas. Sci. Technol. **21**(8), 1–6 (2010)
2. S. Wang, S.L. Huang, W. Zhao et al., Approach to lamb wave lateral crack quantification in elastic plate based on reflection and transmission coefficients surfaces. Res. Nondestr. Eval. **21**(4), 213–223 (2010)
3. S.L. Huang, S. Wang, W.B. Li et al., in *Electromagnetic Ultrasonic Guided Waves* (Springer, Berlin, 2016)
4. S.L. Huang, S. Wang, in *New Technologies in Electromagnetic Non-Destructive Testing* (Springer, Berlin, 2016)
5. A. Benammar, R. Drai, A. Guessoum, Detection of delamination defects in CFRP materials using ultrasonic signal processing. Ultrasonics **48**(8), 731–738 (2008)

6. J.L. Rose, A baseline and vision of ultrasonic guided wave inspection potential. J. Press. Vess.Technol.-T. ASME **124**, 273–282 (2002)
7. Z.Q. Su, L. Ye, Y. Lu, Guided lamb waves for identification of damage in composite structures: A review. J. Sound Vib. **295**, 753–780 (2006)
8. R. Sicard, J. Goyette, D. Zellouf, A SAFT algorithm for lamb wave imaging of isotropic plate-like structures. Ultrasonics **39**, 487–494 (2001)
9. S. Wang, S.L. Huang, W. Zhao, Plate and pipe circumferential lamb wave dispersion and wave structure characteristics. J. Tsinghua Univ. **49**, 925–928 (2009)
10. C. Pei, P. Xiao, S. Zhao et al., Development of a flexible film electromagnetic acoustic transducer for nondestructive testing. Sensors Actuat. A Phys. **258**(WAC), 68–73 (2017)
11. R.Z. Hou, D. Hutson, K.J Kirk, Investigation of thin-film ultrasonic transducers for high-temperature applications. Insight **54**(2), 68–71 (2012)
12. J. Szelazek, Sets of piezoelectric probeheads for stress evaluation with subsurface waves. J. Nondestr. Eval. **32**(2), 188–199 (2013)
13. H.M. Seung, C. Il Park, Y.Y. Kim, An omnidirectional shear-horizontal guided wave EMAT for a metallic plate. Ultrasonics **69**, 58–66 (2016)
14. F. Passarelli, G. Alers, R. Alers, Electromagnetic transduction of ultrasonic waves [M], in *Review of Progress in Quantitative Nondestructive Evaluation*, vols. 31a and 31b, vol. 1430, ed, by D.O. Thompson, D.E. Chimenti (American Institute of Physics, Melville, 2012), pp. 28–37
15. O. Mesnil, H. Yan, M. Ruzzene et al., Fast wavenumber measurement for accurate and automatic location and quantification of defect in composite. Struct. Health Monit. **15**(2), 223–234 (2016)
16. S. Ranjit, M. Choi, W. Kim, Quantification of defects depth in glass fiber reinforced plastic plate by infrared lock-in thermography. J. Mech. Sci. Technol. **30**(3), 1111–1118 (2016)
17. F.C. Tenoudji, J.M. Citerne, H. Dutilleul et al., Non-contact ultrasonic defect imaging in composites [M], in *42nd Annual Review of Progress in Quantitative Nondestructive Evaluation: Incorporating the 6th European-American Workshop on Reliability of NDE*, vol. 1706, ed. by D.E. Chimenti, L.J. Bond (American Institute of Physics, Melville, 2016)
18. E. Yahaghi, A. Movafeghi, N. Mohmmadzadeh, Enhanced radiographic imaging of defects in aircraft structure materials with the dehazing method. Nondestruct. Test. Eval. **30**(2), 138–146 (2015)
19. J.L. Rose, J.J. Ditri, A. Pilarski et al., A guided-wave inspection technique for nuclear steam-generator tubing. NDT E Int. **27**(6), 307–310 (1994)
20. D.N. Alleyne, B. Pavlakovic, M.J.S. Lowe et al., Rapid long-range inspection of chemical plant pipework using guided waves. Insight **43**(2), 93 (2001)
21. H.H. Xu, C.G. Xu, Y.T. Zhang, Ultrasonic transducer array tomography for detecting defect of metal plate based on Hilbert transformation. J. Beijing Inst. Technol. **20**(4), 514–519 (2011)
22. P. Fromme, P.D. Wilcox, M.J.S. Lowe et al., On the development and testing of guided ultrasonic wave array for structural integrity monitoring. IEEE Trans. Ultrason. Ferroelectr. Freq. Control **53**(4), 777–785 (2006)
23. L. Yu, V. Giurgiutiu, In situ 2-D piezoelectric wafer active sensor arrays for guided wave damage detection. Ultrasonics **48**(2), 117–134 (2008)
24. X. Lin, F.G. Yuan, Damage detection of a plate using migration technique. J. Intell. Mater. Syst. Struct. **12**(7), 469–482 (2001)
25. X. Lin, F.G. Yuan, Experimental study applying a migration technique in structural health monitoring. Structural health monitoring **4**(4), 341–353 (2005)
26. L. Wang, F.G. Yuan, Damage identification in a composite plate using prestack reverse-time migration technique. Struct. Health Monit. **4**(3), 195–211 (2005)
27. M. Fink, Time reversal mirror. J. Phys. D Appl. Phys. **26**, 1333–1350 (1993)
28. N. Charkroun, M. Fink, F. Wu, Time reversal processing in ultrasonic nondestructive testing. IEEE Trans. Ultrason. Ferroelectr. Freq. Control **42**(6), 1087–1098 (1995)

29. R.K. Ing, M. Fink, Time-reversed lamb waves. IEEE Trans. Ultrason. Ferroelectr. Freq. Control **45**(4), 1032–1043 (1998)
30. G. Derveaux, G. Papanicolaou, C. Tsogka, Time reversal imaging for sensor networks with optimal compensation in time. J. Acoust. Soc. Am. **121**(4), 2071–2085 (2007)
31. P. Huthwaitea, F. Simonetti, High-resolution guided wave tomography. Wave Motion **50**, 979–993 (2013)
32. T. Clarke, P. Cawley, Enhancing the defect localization capability of a guided wave SHM system applied to a complex structure. Struct. Health Monit. **10**(3), 247–259 (2011)
33. J.S. Hall, P. McKeon, L. Satyanarayan et al., Minimum variance guided wave imaging in a quasi-isotropic composite plate. Smart Mater. Struct. **20**(2), 8 (2011)
34. S. Wang, S.L. Huang, A. Velichko et al., A multi-objective structural optimization of an omnidirectional electromagnetic acoustic transducer [J]. Ultrasonics **81**, 23–31 (2017)
35. Z. Wei, S.L. Huang, W. Zhao et al., A large ratio data compression method for pipeline inspection based on EMAT-generated guided wave. Int. J. Appl. Electromagnet. Mech. **45**, 511–517 (2014)
36. D.N. Alleyne, P. Cawley, Optimization of lamb wave inspection techniques. NDT E Int. **25**, 11–22 (1992)
37. P.D. Wilcox, M.J.S. Lowe, P. Cawley, Mode and transducer selection for long range lamb wave inspection. J. Intell. Mater. Syst. Struct. **12**, 553–565 (2001)
38. J.J. Ditri, J.L. Rose, Excitation of guided waves in generally anisotropic layers using finite sources. J. Appl. Mech. **61**, 330–338 (1994)
39. S.P. Pelts, J.L. Rose, Source influence parameters on elastic guided waves in an orthotropic plate. J. f Acoust. Soc. Am. **99**, 2124–2129 (1996)
40. J.L. Rose, *Ultrasonic Waves in Solid Media* (Cambridge University Press, New York, 1999)
41. R.S.C. Monkhouse, P.W. Wilcox, M.J.S. Lowe et al., The rapid monitoring of structures using interdigital lamb wave transducers. Smart Mater. Struct. **9**, 304–309 (2000)
42. J.L. Rose, S.P. Pelts, M.J. Quarry, A comb transducer model for guided wave NDE. Ultrasonics **36**, 163–169 (1998)
43. X. Lin, F.G. Yuan, in *Diagnostic Lamb Waves in an Integrated Piezoelectric Sensor/Actuator Plate: Analytical and Experimental Studies*, vol. 10 (Institute of Physics Publishing, 2001) pp. 907–913
44. A. Raghavan, C.E.S. Cesnik, Piezoelectric-actuator excited-wave field solutions for guided-wave structural health monitoring. Smart Struct. Mater. **5765**, 313–321 (2005)
45. A. Raghavan, C.E.S. Cesnik, Finite-dimensional piezoelectric transducer modeling for guided wave based structural health monitoring. Smart Struct. Mater. **14**, 1448–1461 (2005)
46. V. Giurgiutiu, Tuned lamb wave excitation and detection with piezoelectric wafer active sensors for structural health monitoring. J. Intell. Mater. Syst. Struct. **16**, 291–305 (2005)
47. B. Xu, V. Giurgiutiu, Single mode tuning effects on lamb wave time reversal with piezoelectric wafer active sensors for structural health monitoring. J. Nondestruct. Eval. **26**, 123–134 (2007)
48. E.V. Glushkov, N.V. Glushkova, O.V. Kvasha et al., Selective lamb mode excitation by piezoelectric coaxial ring actuators. Smart Mater. Struct. **19**, 035018:1–035018:7 (2010)
49. J.P. Koduru, J.L. Rose, Transducer arrays for omnidirectional guided wave mode control in plate like structures. Smart Mater. Struct. **19**, 015010:1–015010:10 (2013)
50. J. Tu, T. Chen, Z. Xiong et al., Calculation of Lorentz force in planar EMAT for thickness measurement of steel plate [J]. Compel Int. J. Comput. Math. Electr. Electron. Eng. **36**(4), 1257–1269 (2017)
51. S. Choi, H. Cho, M.S. Lindsey et al., Electromagnetic acoustic transducers for robotic nondestructive inspection in harsh environments [J]. Sensors **18**(1), 193 (2018)
52. R.B. Thompson, A model for the electromagnetic generation and detection of rayleigh and lamb waves. IEEE Trans. Sonics Ultrason. **20**(4), 340–346 (1973)
53. X.W. Dai, R. Ludwig, R. Palanisamy, Numerical simulation of pulsed eddy-current nondestructive testing phenomena. IEEE Trans. Magn. **26**(6), 3089–3096 (1990)
54. R. Ludwig, X.W. Dai, Numerical and analytical modeling of pulsed eddy currents in a conducting half-space. IEEE Trans. Magn. **26**(1), 299–307 (1990)

55. R. Ludwig, Theoretical basis for a unified conservation law description of the electromagnetic acoustic transduction process. IEEE Trans. Ultrason. Ferroelectr. Freq. Control **39**(4), 476–480 (1992)
56. R. Ludwig, Z. You, R. Palanisamy, Numerical simulations of an electromagnetic acoustic transducer-receiver system for NDT applications. IEEE Trans. Magn. **29**(3), 2081–2089 (1993)
57. X.W. Dai, Electromagnetic acoustic transducer: Physical principles and finite element modeling [D] (Worcester Polytechnic Institute, 1990)
58. R. Shapoorabadi, A. Konrad, A. Sinclair, Improved FEM for EMAT analysis and design. IEEE Trans. Magn. **37**(4), 2821–2823 (2001)
59. R. Shapoorabadi, Electromagnetic acoustic transducer analysis by the finite element method [D] (University of Toronto, 2002)
60. R. Shapoorabadi, A. Konrad, A. Sinclair, Comparison of three formulations for eddy current and skin effect problems. IEEE Trans. Magn. **38**(2), 617–620 (2002). R. Shapoorabadi, A. Konrad, A. Sinclair, Computation of current densities in the receiving mode of EMATs. J. Appl. Phys. **97**(10Q106), 1–3 (2005)
61. R. Shapoorabadi, A. Konrad, A. Sinclair, The governing electrodynamic equations of EMATs. J. Appl. Phys. **97**(10E102), 1–3 (2005)
62. R. Dhayalan, K. Balasubramaniam, C.V. Krishnamurthy, *Numerical Simulation of Pulsed Meander Coil EMAT* (American Institute of Physics, USA, 2010)
63. R. Dhayalan, K. Balasubramaniam, A hybrid finite element model for simulation of electromagnetic acoustic transducer (EMAT) based plate waves. NDT E Int. **43**(6), 519–526 (2010)
64. Y. Zhang, S. L. Huang, S. Wang, et al., Time-Frequency Energy Density Precipitation Method for Time-Of-Flight Extraction of Narrowband lamb Wave Detection Signals. Review of Scientific Instruments, 2016, 87(5): 054702, 1–8
65. L. Kang, S.J. Wang, T. Jiang et al., in Optimal design of lamb wave EMATs. *Proceedings of Symposium on Ultrasonic Electronics*, vol. 31 (2010), pp. 313–314
66. P.D. Wilcox, J.S.M. Lowe, P. Cawley, The excitation and detection of lamb waves with planar coil electromagnetic acoustic transducers. IEEE Trans. Ultrason. Ferroelectr. Freq. Control **52**(12), 2370–2383 (2005)
67. W.Z. Shi et al., Comparison of three formulations for eddy-current problems in a spiral coil electromagnetic acoustic transducer [J]. J. Cent. South Univ. **23**(4), 817–824 (2016)
68. Z. Liu, Y. Hu, M. Xie et al., Development of omnidirectional A0 mode EMAT employing a concentric permanent magnet pairs with opposite polarity for plate inspection [J]. NDT E Int. (2017)
69. P.B. Nagy, F. Simonetti, G. Instanes, Corrosion and erosion monitoring in plates and pipes using constant group velocity lamb wave inspection. Ultrasonics **54**, 1832–1841 (2014)
70. M. Clough, M. Fleming, S. Dixon, Circumferential guided wave EMAT system for pipeline screening using shear horizontal ultrasound [J]. NDT E Int. (2016)
71. P.A. Petcher, S.E. Burrows, S. Dixon, Shear horizontal (SH) ultrasound wave propagation around smooth corners. Ultrasonics **54**(4), 997–1004 (2014)
72. C. Rouge, A. Lhemery, D. Segur et al., Modal solutions for SH guided waves radiated by an EMAT in a ferromagnetic plate [M], in *10th Anglo-French Physical Acoustics Conference*, Iop Publishing Ltd, Bristol, vol. 353 (2012)
73. P.A. Petcher, S. Dixon, Weld defect detection using PPM EMAT generated shear horizontal ultrasound. NDT E Int. **74**, 58–65 (2015)
74. Z. Wei, S.L. Huang, S. Wang et al., Magnetostriction-based omni-directional guided wave transducer for high-accuracy tomography of steel plate defects. IEEE Sens. J. **15**(11), 6549–6558 (2015)
75. R. Ribichini, F. Cegla, P.B. Nagy et al., Experimental and numerical evaluation of electromagnetic acoustic transducer performance on steel materials. NDT E Int. **45**(1), 32–38 (2012)
76. S. Dixon, M.P. Fletcher, G. Rowlands, The accuracy of acoustic birefringence shear wave measurements in sheet metal. J. Appl. Phys. **104**(11), 5 (2008)

77. Y.D. Xie, Z.H. Liu, L.Y. Yin et al., Directivity analysis of meander-line-coil EMATs with a wholly analytical method. Ultrasonics **73**, 262–270 (2017)
78. L. Kang, C. Zhang, S. Dixon et al., Enhancement of ultrasonic signal using a new design of rayleigh-wave electromagnetic acoustic transducer [J]. NDT E Int. (2016)
79. Y.D. Xie, L.Y. Yin, Z.H. Liu et al., A novel variable-length meander-line-coil EMAT for side lobe suppression. IEEE Sens. J. **16**(16), 6279–6287 (2016)
80. Y.D. Xie, L.Y. Yin, Z.H. Liu et al., Simulation of ultrasonic and EMAT arrays using FEM and FDTD. Ultrasonics **66**, 154–165 (2016)
81. Y.D. Xie, S. Rodriguez, W.B. Zhang et al., Simulation of an electromagnetic acoustic transducer array by using analytical method and FDTD. J. Sens. **2016**, 10
82. S. Hill, S. Dixon, Localisation of defects with time and frequency measurements using pulsed arrays. NDT E Int. **67**, 24–30 (2014)
83. S. Hill, S. Dixon, Frequency dependent directivity of periodic permanent magnet electromagnetic acoustic transducers. NDT E Int. **62**, 137–143 (2014)
84. P.D. Wilcox, M. Lowe, P. Cawley, Omnidirectional guided wave inspection of large metallic plate structures using an EMAT array. IEEE Trans. Ultrason. Ferroelectr. Freq. Control **52**(4), 653–665 (2005)
85. Y. Roh, J. Kim, Detection of cracks on a plate by piezoelectric interdigital transducers [M], in *Sensors and Smart Structures Technologies for Civil, Mechanical, and Aerospace Systems 2011*, vol. 7981, eds. by M. Tomizuka, C.B. Yun, V. Giurgiutiu, J.P. Lynch (Spie-Int Soc Optical Engineering, Bellingham, 2011)
86. H. Masuyama, K. Mizutani, K. Nagai, Sound source with direction-variable beam using annular transducer array [M], in *2002 IEEE Ultrasonics Symposium Proceedings*, vols. 1 and 2, eds. by D.E. Yuhas, S.C. Schneider (IEEE, New York, 2002), pp. 1097–1100
87. S. Dixon, C. Edwards, S.B. Palmer et al., Ultrasonic generation using a plasma igniter. J. Phys. D-Appl. Phys. **34**(7), 1075–1082 (2001)
88. J.R. Bernstein, J.B. Spicer, Line source representation for laser-generated ultrasound in aluminum. J. Acoust. Soc. Am. **107**(3), 1352–1357 (2000)
89. H. Ogi, M. Hirao, T. Ohtani, Line-focusing electromagnetic acoustic transducers for the detection of slit defects. IEEE Trans. Ultrason. Ferroelectr. Freq. Control **46**(2), 341–346 (1999)
90. H. Ogi, M. Hirao, T. Ohtani, Line-focusing of ultrasonic SV wave by electromagnetic acoustic transducer. J. Acoust. Soc. Am. **103**(5), 2411–2415 (1998)
91. X. Song, G. Qiu, Optimization of a focusable and rotatable shear-wave periodic permanent magnet electromagnetic acoustic transducers for plates inspection [J]. Sensors **17**(12), 2722 (2017)
92. D. Wu, M.X. Li, X.M. Wang, Shear wave field radiated by an electromagnetic acoustic transducer. Chin. Phys. Lett. **23**(12), 3294–3296 (2006)
93. D.P. Jansen, D.A. Hutchins, Lamb wave tomography, in *IEEE 1990 Ultrasonics Symposium Proceedings*, pp. 1017–1020 (1990)
94. D.A. Hutchins, D.P. Jansen, C. Edwards, Non-contact lamb wave tomography, in *IEEE 1992 Ultrasonics Symposium Proceedings*, pp. 883–886 (1992)
95. W. Wright, D.A. Hutchins, D.P. Jansen et al., Air-coupled lamb wave tomography. IEEE Trans. Ultrason. Ferroelectr. Freq. Control **44**(1), 53–59 (1997)
96. X. Zhao, J.L. Rose, Ultrasonic tomography for density gradient determination and defect analysis. J. Acoust. Soc. Am. **117**, 2546 (2005)
97. X. Zhao, J.L. Rose, H. Gao, Determination of density distribution in ferrous powder compacts using ultrasonic tomography. IEEE Trans. Ultrason. Ferroelectr. Freq. Control **53**, 360–369 (2006)
98. J.C.P. McKeon, M.K. Hinders, Parallel projection and crosshole lamb wave contact scanning tomography. J. Acoust. Soc. Am. **106**(5), 2568–2577 (1999)
99. E.V. Malyarenko, M.K. Hinders, Fan beam and double crosshole lamb wave tomography for mapping flaws in aging aircraft structures. J. Acoust. Soc. Am. **108**(4), 1631–1639 (2000)

100. S.M. Prasad, K. Balasubramaniam, C.V. Krishnamurthy, Structural health monitoring of composite structures using lamb wave tomography. Smart Mater. Struct. **13**, 73–79 (2004)
101. Y. Cho, B. Sheen, A study on quantitative lamb wave tomogram via modified rapid algorithm with shape factor optimization. Int. J. Precis. Eng. Manuf. **13**, 671–677 (2012)
102. Y. Cho, Model-based guided wave NDE: The evolution of guided wave NDE from "magic" to "physically based engineering tool". J. Nondestr. Eval. **31**, 324–338 (2012)
103. X. Zhao, H. Gao, G. Zhang et al., Active health monitoring of an aircraft wing with embedded piezoelectric sensor/actuator network: I. Defect detection, localization and growth monitoring. Smart Mater. Struct. **16**, 1208–1217 (2007)
104. T.R. Hay, R.L. Royer, H. Gao et al., A comparison of embedded sensor lamb wave ultrasonic tomography approaches for material loss detection. Smart Mater. Struct. **15**, 946–951 (2006)
105. X. Zhao, R.L. Royer, S.E. Owens et al., Ultrasonic lamb wave tomography in structural health monitoring. Smart Mater. Struct. **20**, 105002:1–105002:10 (2011)
106. D. Wang, L. Ye, Y. Lu, A probabilistic diagnostic algorithm for identification of multiple notches using digital damage fingerprints (DDFs). J. Intell. Mater. Syst. Struct. **20**, 1439–1450 (2009)
107. D. Wang, L. Ye, Y. Lu et al., A damage diagnostic imaging algorithm based on the quantitative comparison of lamb wave signals. Smart Mater. Struct. **19**, 065008:1–065008:12 (2010)
108. D. Wang, L. Ye, Su Z et al., Probabilistic damage identification based on correlation analysis using guided wave signals in aluminum plates. Struct. Health Monit. **9**, 133–144 (2010)
109. S. Periyannan, P. Rajagopal, K. Balasubramaniam, Ultrasonic bent waveguides approach for distributed temperature measurement. Ultrasonics **74**, 211–220 (2017)
110. B. Alem, A. Abedian, K. Nasrollahi-Nasab, Reference-free damage identification in plate-like structures using lamb-wave propagation with embedded piezoelectric sensors. J. Aerospace Eng. **29**(6), 13 (2016)
111. M. Zhao, L. Zeng, J. Lin et al., Mode identification and extraction of broadband ultrasonic guided waves. Meas. Sci. Technol. **25**(11), 15 (2014)
112. A.V. Harish, P. Ray, P. Rajagopal et al., Detection of fundamental shear horizontal mode in plates using fibre Bragg gratings. J. Intell. Mater. Syst. Struct. **27**(16), 2229–2236 (2016)
113. S. Zeroug, S. Bose, B. Sinha et al., Sonic and ultrasonic measurement applications for cased oil wells. Insight **58**(8), 423–430 (2016)
114. H.Z. Hosseinabadi, B. Nazari, R. Amirfattahi et al., Wavelet network approach for structural damage identification using guided ultrasonic waves. IEEE Trans. Instrum. Meas. **63**(7), 1680–1692 (2014)
115. B.S. Ben, B.A. Ben, K.A. Vikram et al., Damage identification in composite materials using ultrasonic based lamb wave method. Measurement **46**(2), 904–912 (2013)
116. N. Hu, T. Shimomukai, H. Fukunaga et al., Damage identification of metallic structures using A(0) mode of lamb waves. Struct. Health Monit. **7**(3), 271–285 (2008)
117. S.L. Huang, Z. Wei, W. Zhao et al., A new omni-directional EMAT for ultrasonic lamb wave tomography imaging of metallic plate defects. Sensors **14**(2), 3458–3476 (2014)
118. J. Deboucq, M. Duquennoy, M. Ouaftouh et al., Development of interdigital transducer sensors for non-destructive characterization of thin films using high frequency Rayleigh waves. Rev. Sci. Instrum. **82**(6), 7 (2011)
119. P. Belanger, P. Cawley, Feasibility of low frequency straight-ray guided wave tomography. NDT E Int. **42**(2), 113–119 (2009)
120. D.Y. Gao, Z.J. Wu, L. Yang et al., Guide waves-based multi-damage identification using a local probability-based diagnostic imaging method. Smart Mater. Struct. **25**(4), 9 (2016)
121. H. Li, R.H. Li, B. Hu et al., Application of guided waves and probability imaging approach for insulation damage detection of large generator stator bar. IEEE Trans. Dielectr. Electr. Insul. **22**(6), 3216–3225 (2015)
122. S. Wang, S.L. Huang, Y. Zhang et al., Modeling of an omni-directional electromagnetic acoustic transducer driven by the Lorentz force mechanism. Smart Mater. Struct. **25**(12), 12 (2016)

123. S. Wang, S.L. Huang, Y. Zhang et al., Multiphysics modeling of a Lorentz force-based meander coil electromagnetic acoustic transducer via steady-state and transient analyses. IEEE Sens. J. **16**(17), 6641–6651 (2016)
124. Y.G. Wang, X.J. Wu, P.F. Sun et al., Enhancement of the excitation efficiency of a torsional wave PPM EMAT array for pipe inspection by optimizing the element number of the array based on 3-D FEM. Sensors **15**(2), 3471–3490 (2015)
125. X.C. Song, Y.W. Wang, Effects of AC coils parameters on transduction efficiency of EMAT for steel plate inspection. Math. Probl. Eng. **2014**, 11 (2014)
126. S.L. Huang, W. Zhao, Y.S. Zhang et al., Study on the lift-off effect of EMAT. Sens. Actuators A Phys. **153**(2), 218–221 (2009)
127. M. Cong, X.J. Wu, C.Q. Qian, A longitudinal mode electromagnetic acoustic transducer (EMAT) based on a permanent magnet chain for pipe inspection. Sensors **16**(5), 18 (2016)
128. X. Ding, X.J. Wu, Y.G. Wang, Bolt axial stress measurement based on a mode-converted ultrasound method using an electromagnetic acoustic transducer. Ultrasonics **54**(3), 914–920 (2014)
129. M. Seher, P. Huthwaite, M.J.S. Lowe et al., Model-based design of low frequency lamb wave EMATs for mode selectivity. J. Nondestr. Eval. **34**(3), 16 (2015)
130. P.F. Sun, X.J. Wu, J. Li et al., An EMAT for inspecting heat exchanger tubes of stainless steel using longitudinal guided waves based on non-uniform static magnetic fields. Insight **57**(4), 221–228 (2015)
131. D. Rueter, T. Morgenstern, Ultrasound generation with high power and coil only EMAT concepts. Ultrasonics **54**(8), 2141–2150 (2014)
132. S. Legendre, D. Massicotte, J. Goyette et al., Wavelet-transform-based method of analysis for lamb-wave ultrasonic NDE signals. IEEE Trans. Instrum. Meas. **49**(3), 524–530 (2000)
133. K.R. Leonard, M.K. Hinders, Multi-mode Lamb wave tomography with arrival time sorting. J. Acoust. Soc. Am. **117**(4), 2028–2038 (2005)
134. D. Grimaldi, Time-of-flight measurement of ultrasonic pulse echoes using wavelet networks. IEEE Trans. Instrum. Meas. **55**(1), 5–13 (2006)
135. F.C. Li, G. Meng, L. Ye et al., Dispersion analysis of lamb waves and damage detection for aluminum structures using ridge in the time-scale domain. Meas. Sci. Technol. **20**(9), 10 (2009)
136. F.C. Li, H. Murayama, K. Kageyama et al., Multiple damage assessment in composite laminates using a doppler-effect-based fiber-optic sensor. Meas. Sci. Technol. **20**(11), 10 (2009)
137. L. De Marchi, M. Ruzzene, B.L. Xu et al., Warped basis pursuit for damage detection using lamb waves. IEEE Trans. Ultrason. Ferroelectr. Freq. Control **57**(12), 2734–2741 (2010)
138. C. Zhang, J.H. Qiu, H.L. Ji et al., Damage localization using warped frequency transform in active structural health monitoring. Int. J. Appl. Electromagnet. Mech. **47**(4), 897–909 (2015)
139. J. Moll, C. Heftrich, C.P. Fritzen, Time-varying inverse filtering of narrowband ultrasonic signals. Struct. Health Monit. **10**(4), 403–415 (2011)
140. K.S. Ho, D.R. Billson, D.A. Hutchins, Ultrasonic lamb wave tomography using scanned EMATs and wavelet processing. Nondestuct. Test. Eval. **22**, 19–34 (2007)
141. J.P. Koduru, J.L. Rose, Mode controlled guided wave tomography using annular array transducers for SHM of water loaded plate like structures. Smart Mater. Struct. **22**, 125021:1–125021:10 (2013)
142. J. Hou, K.R. Leonard, M.K. Hinders, Automatic multi-mode lamb wave arrival time extraction for improved tomographic reconstruction. Inverse Probl. **20**(6), 1873–1888 (2004)
143. H.Y. Zhang, X.L. Sun, X.R. Qi et al., A modified LTI ray tracing algorithm in Lamb wave tomography, in *International Conference on Audio, Language and Image Processing*, 2008, pp. 1059–1062
144. J.E. Vidale, Finite-difference calculations of travel times in three dimensions. Geophysics **55**, 521–526 (1990)
145. E. Asakawa, T. Kawanaka, Seismic ray tracing using linear travel-time interpolation. Geophys. Prospect. **41**, 99–111 (1993)

146. S.L. Huang, Y. Zhang, S. Wang et al., Multi-mode electromagnetic ultrasonic lamb wave tomography imaging for variable-depth defects in metal plates. Sensors **16**(5), 628, 1–10 (2016)
147. A.H. Andersen, A.C. Kak, Digital ray tracing in two-dimensional refractive fields. J. Acoust. Soc. Am. **72**, 1593–1606 (1982)
148. A.H. Andersen, Ray tracing for reconstructive tomography in the presence of object discontinuity boundaries: A comparative analysis of recursive schemes. J. Acoust. Soc. Am. **89**, 574–582 (1991)
149. R.J. Lytle, K.A. Dines, Iterative ray tracing between boreholes for underground image reconstruction. IEEE Trans. Geosci. Remote Sens. **18**, 234–240 (1980)
150. Y.Q. Wang, R.A. Kline, Ray tracing in isotropic and anisotropic materials: Application to tomographic image reconstruction. J. Acoust. Soc. Am. **95**, 2525–2532 (1994)
151. D.R. Velis, T.J. Ulrych, Simulated annealing ray tracing in complex three-dimensional media. Geophys. J. Int. **145**, 447–459 (2001)
152. H.Y. Zhang, X.L. Sun, X.R. Qi et al., Lamb wave tomography combining simulated annealing and simultaneous iterative reconstruction technique, in *Proceeding of China-Japan Joint Microwave Conference* 2008, pp. 741–744
153. E.V. Malyarenko, M.K. Hinders, Ultrasonic lamb wave diffraction tomography. Ultrasonics **39**, 269–281 (2001)
154. E.V. Malyarenko, Lamb wave diffraction tomography [D] (College of William and Mary, 2000)
155. A. Balvantin, A. Baltazar, P. Rodriguez, Characterization of laser generated lamb wave modes after interaction with a thickness reduction discontinuity using ray tracing theory. Exp. Mech. **54**, 743–752 (2014)
156. D.A. Hutchins, D.P. Jansen, C. Edwards, Lamb-wave tomography using non-contact transduction. Ultrasonics **31**(2), 97–103 (1993)
157. D.P. Jansen, D.A. Hutchins, Ultrasonic rayleigh and lamb wave tomography. Acustica **79**(2), 117–127 (1993)
158. K.S. Hao, S.L. Huang, W. Zhao et al., Analytical modelling and calculation of pulsed magnetic field and input impedance for EMATs with planar spiral coils. NDT E Int. **44**(3), 274–280 (2011)
159. K.S. Hao, S.L. Huang, W. Zhao et al., Circuit-field coupled finite element analysis method for an electromagnetic acoustic transducer under pulsed voltage excitation. Chin. Phys. B **20**(6), 8 (2011)
160. S. Wang, S.L. Huang, W. Zhao et al., 3D modeling of circumferential SH guided waves in pipeline for axial cracking detection in ILI tools. Ultrasonics **56**, 325–331 (2015)
161. Y. Zhang, S.L. Huang, S. Wang et al., Time-frequency energy density precipitation method for time-of-flight extraction of narrowband lamb wave detection signals. Rev. Sci. Instrum. **87**(5), 8 (2016)
162. S.L. Huang, Y. Zhang, S. Wang et al., Multi-mode electromagnetic ultrasonic Lamb wave tomography imaging for variable-depth defects in metal plates. Sensors **16**(5), 10 (2016)
163. S. Wang, S.L. Huang, W. Zhao, Simulation of lamb wave's interactions with transverse internal defects in an elastic plate. Ultrasonics **51**(4), 432–440 (2011)
164. P. Belanger, Feasibility of thickness mapping using ultrasonic guided waves [D] (Imperial College London, 2009)
165. E. Chan, L.R.F. Rose, C.H. Wang, An extended diffraction tomography method for quantifying structural damage using numerical Green's functions. Ultrasonics **59**, 1–13 (2015)
166. W.B. Li, Y. Cho, Quantification and imaging of corrosion wall thinning using shear horizontal guided waves generated by magnetostrictive sensors. Sens. Actuators A-Phys. **232**, 251–258 (2015)
167. X. Chen, J.E. Michaels, T.E. Michaels, Design of distributed sparse arrays for lamb wave SHM based upon estimated scattering matrices[M], in *40th Annual Review of Progress in Quantitative Nondestructive Evaluation: Incorporating the 10th International Conference on Barkhausen Noise and Micromagnetic Testing*, vols. 33a and 33b, vol. 1581, eds. by D.E. Chimenti, L.J. Bond, D.O. Thompson (American Institute of Physics, Melville, 2014), pp. 248–255

168. J.E. Michaels, X. Chen, T.E. Michaels, Scattering measurements and in situ imaging with sparse guided wave arrays. Struct. Health Monit. **1** and **2**, 2177–2184 (2013)
169. M.J.S. Lowe, D.N. Alleyne, P. Cawley, Defect detection in pipes using guided waves. Ultrasonics **36**(1–5), 147–154 (1998)
170. J. Rose, S. Pelts, Y. Cho, Modeling for flaw sizing potential with guided waves. J. Nondestr. Eval. **19**(2), 55–66 (2000)
171. A. Demma, P. Cawley, M. Lowe et al., The reflection of the fundamental torsional mode from cracks and notches in pipes. J. Acoust. Soc. Am. **114**(2), 611–625 (2003)
172. A. Demma, P. Cawley, M. Lowe et al., The reflection of guided waves from notches in pipes: A guide for interpreting corrosion measurements. NDT E Int. **37**(3), 167–180 (2004)
173. J.H. Park, D.K. Kim, H.J. Kim et al., Development of EMA transducer for inspection of pipelines [J]. J. Mech. Sci. Technol. **31**(11), 5209–5218 (2017)
174. Y.H. Cho, D.D. Hongerholt, J.L. Rose, Lamb wave scattering analysis for reflector characterization. IEEE Trans. Ultrason. Ferroelectr. Freq. Control **44**(1), 44–52 (1997)
175. Y. Cho, Estimation of ultrasonic guided wave mode conversion in a plate with thickness variation. IEEE Trans. Ultrason. Ferroelectr. Freq. Control **47**(3), 591–603 (2000)
176. J.C.P. McKeon, M.K. Hinders, Lamb wave scattering from a through hole. J. Sound Vib. **224**(5), 843–862 (1999)
177. M.J. Santos, J. Perdigao, P. Faia, Ultrasonic guided waves scattering effects from defects in adhesively bonded lap joints using pitch and catch and pulse-echo techniques. J. Adhes. **84**(5), 421–438 (2008)
178. T.E. Michaels, J.E. Michaels, M. Ruzzene, Frequency-wavenumber domain analysis of guided wavefields. Ultrasonics **51**(4), 452–466 (2011)
179. S. Li, C.J. Lissenden, Composite piezoelectric strip transducer development for structural health monitoring [M], in *Health Monitoring of Structural and Biological Systems 2011*, vol. 7984, ed. by T. Kundu (Spie-Int Soc Optical Engineering, Bellingham, 2011)
180. T.E. Michaels, J.E. Michaels, S.J. Lee et al., Chirp generated acoustic wavefield images [M], in *Health Monitoring of Structural and Biological Systems 2011*, vol. 7984, ed. by T. Kundu (Spie-Int Soc Optical Engineering, Bellingham, 2011)
181. M.J. Peake, J. Trevelyan, G. Coates, Extended isogeometric boundary element method (XIBEM) for three-dimensional medium-wave acoustic scattering problems. Comput. Methods Appl. Mech. Eng. **284**, 762–780 (2015)
182. I.A. Alatawi, J. Trevelyan, A direct evaluation of stress intensity factors using the extended dual boundary element method. Eng. Anal. Bound. Elem. **52**, 56–63 (2015)
183. T.M. Foster, M.S. Mohamed, J. Trevelyan et al., Interactive three-dimensional boundary element stress analysis of components in aircraft structures. Eng. Anal. Bound. Elem. **56**, 190–200 (2015)
184. J. Rajagopalan, K. Balasubramaniam, C.V. Krishnamurthy, A single transmitter multi-receiver (STMR) PZT array for guided ultrasonic wave based structural health monitoring of large isotropic plate structures. Smart Mater. Struct. **15**(5), 1190–1196 (2006)
185. L. Yu, V. Giurgiutiu, In situ 2-D piezoelectric wafer active sensors arrays for guided wave damage detection. Ultrasonics **48**(2), 117–134 (2008)
186. P. Huthwaite, Evaluation of inversion approaches for guided wave thickness mapping, in *Proceedings of the Royal Society A-Mathematical Physical and Engineering Sciences*, 2014, vol. 470, no. 2166, pp. 28
187. R.M. Levine, J.E. Michaels, Block-sparse lamb wave structural health monitoring using generic scattering models [M], in *40th Annual Review of Progress in Quantitative Nondestructive Evaluation: Incorporating the 10th International Conference on Barkhausen Noise and Micromagnetic Testing*, vols. 33a and 33b, vol. 1581, eds. by D.E. Chimenti, L.J. Bond, D.O. Thompson (American Institute of Physics, Melville, 2014), pp. 232–239
188. J.S. Hall, P. Fromme, J.E. Michaels, Guided wave damage characterization via minimum variance imaging with a distributed array of ultrasonic sensors. J. Nondestr. Eval. **33**(3), 299–308 (2014)

189. R.M. Levine, J.E. Michaels, Block-sparse reconstruction and imaging for Lamb wave structural health monitoring. IEEE Trans. Ultrason. Ferroelectr. Freq. Control **61**(6), 1006–1015 (2014)
190. J.S. Hall, J.E. Michaels, Multipath ultrasonic guided wave imaging in complex structures. Struct. Health Monit. **14**(4), 345–358 (2015)
191. C. Kexel, J.B. Harley, J. Moll et al., Attenuation and phase compensation for guided wave based inspection using a filter approach [M], in *2015 IEEE International Ultrasonics Symposium*, IEEE, New York, 2015
192. J.B. Harley, A.C. Schmidt, J.M.F. Moura et al., *Accurate Sparse Recovery of Guided Wave Characteristics for Structural Health Monitoring* (IEEE, New York, 2012)

Chapter 2
Directivity and Controllability of Electromagnetic Ultrasonic Transducer

2.1 Omnidirectional Lamb Wave EMAT for Aluminum Plate Tomography

2.1.1 Complete Model of Transmitting and Receiving Omnidirectional Lamb Wave of Single Turn Loop EMAT

The EMAT for Lamb wave tomography needs to have the ability to excite Lamb along the omnidirectional excitation and effectively receive Lamb waves from all directions. The structure should be axisymmetric. Therefore, in this section, the principle and feasibility of the omnidirectional Lamb wave in the aluminum plate will be explained by the establishment of a complete energy exchange model for the single turn loop EMAT with a relatively simple structure and axisymmetric structure.

The single turn loop EMAT consists of three parts: a cylindrical permanent magnet, a single turn loop, and a measured aluminum plate, as shown in Fig. 2.1a. The structure of the single turn loop is shown in Fig. 2.1b. In order to facilitate discussion and analysis, a cylindrical coordinate system is established above EMAT, shown in Fig. 2.1. The origin of the coordinate system is located in the middle of the aluminum plate and just below the center of the cylindrical permanent magnet. The z-direction of the cylindrical coordinate system is perpendicular to the surface of the aluminum plate, and the r and θ are the radial and circumferential directions of the coil, respectively.

The permanent magnet of the single turn loop EMAT produces a uniform static magnetic field $\boldsymbol{B}_{\mathrm{p}}$ perpendicular to the surface of the aluminum plate. When the EMAT is used to stimulate the omnidirectional Lamb wave, the pulse excitation current is $\boldsymbol{I}_{\mathrm{exc}}$ in the coil; then, the $\boldsymbol{I}_{\mathrm{exc}}$ will induce a pulsed eddy current $\boldsymbol{J}_{\mathrm{eddy}}$ in the aluminum plate through the electromagnetic coupling. Because the working frequency of Lamb wave EMAT is usually more than 10 kHz, the depth of skin effect is shallow, and it will distribute on the surface of the aluminum plate [1]. Under the

S. Huang et al., *Theory and Methodology of Electromagnetic Ultrasonic Guided Wave Imaging*, https://doi.org/10.1007/978-981-13-8602-2_2

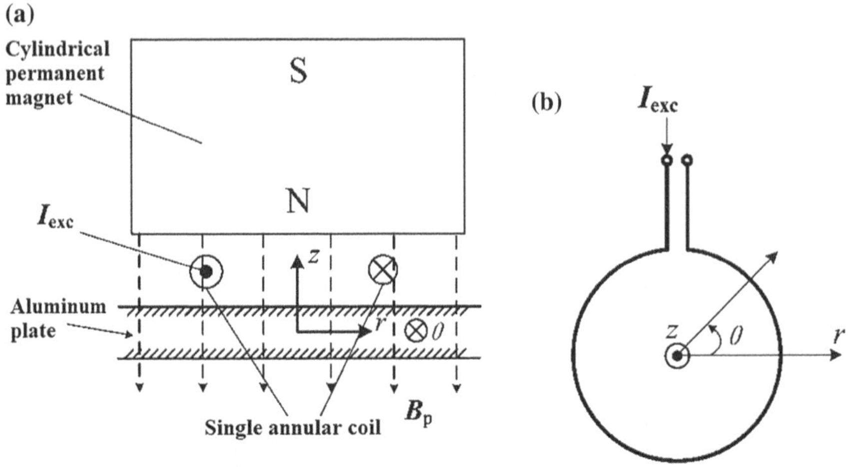

Fig. 2.1 Schematic diagram of the structure and principle of single turn loop EMAT: **a** EMAT structure principle (side view section); **b** single turn loop coil (top view)

effect of $\boldsymbol{J}_{\text{eddy}}$ and static magnetic field $\boldsymbol{B}_{\text{p}}$, the surface of the aluminum plate will be affected by Lorenz force $\boldsymbol{f}_{\text{L}}$. $\boldsymbol{f}_{\text{L}}$ will cause ultrasonic vibration on the surface of the aluminum plate, thereby stimulating the Lamb wave. The whole process of transmitting and receiving omnidirectional ultrasonic Lamb wave is theoretically modeled for the single turn loop EMAT.

In the calculation and formula derivation, the model of the single turn loop EMAT should meet the following assumptions:

(1) The medium in each region of the model is linear, homogeneous, and isotropic.
(2) In practical application, the coil is made of printed circuit board (PCB) technology and is made up of flat wire with a thin thickness. Therefore, it is assumed that the coil is a rectangular coil.
(3) Ignoring the skin effect and the proximity effect in the coil, it is considered that the current density in the coil is evenly distributed.
(4) The cross-sectional radius of the cylindrical permanent magnet is large enough. Therefore, in the EMAT working area, the magnetic flux density Bp of the static magnetic field is evenly distributed.

In calculating the Lorenz force generated by the $\boldsymbol{I}_{\text{exc}}$ in the coil, the three-dimensional geometric model used is shown in Fig. 2.2. The model is extended to the cylindrical coordinate system in Fig. 2.1 and can be divided into three solving regions along the z-axis: The region 1 is the coil, area 2 is air, and the region 3 is the aluminum plate. The thickness of the aluminum plate is d, the width of the coil is w_{c}, the thickness is h_{c}, the cross-sectional area of the coil is $S_{\text{c}} = w_{\text{c}}h_{\text{c}}$, and the density of the pulse excited current is:

$$\boldsymbol{J}_{\text{exc}} = \boldsymbol{I}_{\text{exc}}/S_{\text{c}} \tag{2.1}$$

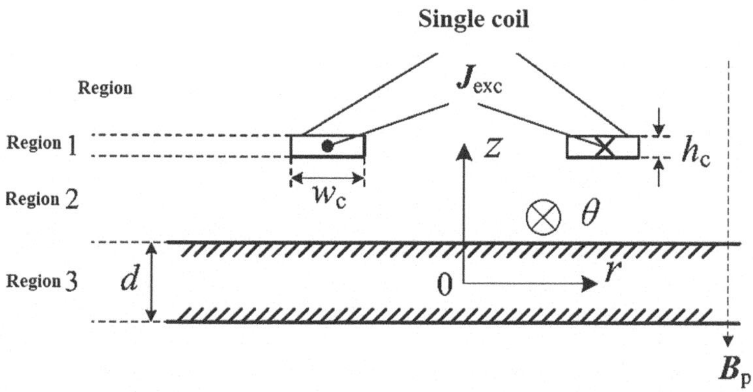

Fig. 2.2 Lorenz force three-dimensional geometric model (side cross section) generated by the pulse excitation current in the coil is obtained by the full current law and Faraday's law

According to the structure of the coil, $\boldsymbol{J}_{exc}$ has only θ-direction component: $\boldsymbol{J}_{exc} = J_{exc}\boldsymbol{\theta}$.

$$\nabla \times \boldsymbol{H} = \boldsymbol{J}_c + \frac{\partial \boldsymbol{D}}{\partial t} \tag{2.2}$$

$$\nabla \times \boldsymbol{E} = -\frac{\partial \boldsymbol{B}}{\partial t} \tag{2.3}$$

where $\boldsymbol{H}$ is magnetic field strength; $\boldsymbol{J}_c$ is conduction current density; $\boldsymbol{D}$ is electric displacement; $\boldsymbol{E}$ is electric field intensity; $\boldsymbol{B}$ is magnetic induction intensity.

In the three solution regions, the dielectric properties satisfy the following relations:

$$\boldsymbol{B} = \mu \boldsymbol{H} \tag{2.4}$$

$$\boldsymbol{J}_c = \sigma \boldsymbol{E} \tag{2.5}$$

$$\boldsymbol{D} = \varepsilon \boldsymbol{E} \tag{2.6}$$

where μ is the permeability, σ is the conductivity, and ε is the permittivity of the medium. Formula (2.4) to type (2.6) is substituted (2.2). Substituting (2.4)–(2.6) in Formula (2.2), it can be obtained:

$$\nabla \times \boldsymbol{B} = \mu\left(\sigma \boldsymbol{E} + \varepsilon \frac{\partial \boldsymbol{E}}{\partial t}\right) \tag{2.7}$$

Formulas (2.3) and (2.7) give the law of mutual coupling of electromagnetic fields in different regions.

Because the divergence of magnetic flux density $\boldsymbol{B}$ is 0, vector magnetic potential $\boldsymbol{A}$ is introduced based on vector identity $\nabla\cdot(\nabla\times\boldsymbol{A})\equiv 0$.

$$\nabla\times\boldsymbol{A}=\boldsymbol{B} \tag{2.8}$$

$$\nabla\times\left(\boldsymbol{E}+\frac{\partial \boldsymbol{A}}{\partial t}\right)=0 \tag{2.9}$$

Therefore, by introducing the scalar potential function φ_{e} according to the identity type $\nabla\times(\nabla\varphi_{\mathrm{e}})\equiv 0$, we get:

$$\boldsymbol{E}+\frac{\partial \boldsymbol{A}}{\partial t}=-\nabla\varphi_{\mathrm{e}} \tag{2.10}$$

The $\boldsymbol{A}$ and φ_{e} introduced above are a pair of auxiliary quantities for solving the distribution of electromagnetic fields, in which $\boldsymbol{A}$ is not unique, so it is necessary to further specify the divergence of the $\boldsymbol{A}$. Here, in order to simplify the calculation, the Lorenz criterion is used to stipulate the divergence of $\boldsymbol{A}$ is:

$$\nabla\cdot\boldsymbol{A}=-\mu\varepsilon\frac{\partial\varphi_{\mathrm{e}}}{\partial t} \tag{2.11}$$

According to the axisymmetric characteristics of the pulse excited current density $\boldsymbol{J}_{\mathrm{exc}}$, the two-order partial differential equation of magnetic vector potential $\boldsymbol{A}$ can be obtained.

$$\frac{1}{\mu}\nabla^2\boldsymbol{A}-\sigma\frac{\partial \boldsymbol{A}}{\partial t}-\varepsilon\frac{\partial^2\boldsymbol{A}}{\partial t^2}=-\boldsymbol{J}_{\mathrm{s}} \tag{2.12}$$

where $\boldsymbol{J}_{\mathrm{s}}$ is $\boldsymbol{J}_{\mathrm{exc}}$ in region 1, while equals 0 in regions 2 and 3. Because the $\boldsymbol{J}_{\mathrm{exc}}$ is only the component of the θ direction, the vector magnetic potential $\boldsymbol{A}$ is also only the θ-directional component; because of the axisymmetric characteristics of $\boldsymbol{J}_{\mathrm{exc}}$, the value of the $\boldsymbol{A}$ is independent of θ. Therefore, the formula is simplified to (2.12) when $\boldsymbol{A}=A_\theta(r,z)\boldsymbol{\theta}$:

$$\frac{1}{\mu}\nabla^2 A_\theta(r,z)-\sigma\frac{\partial A_\theta(r,z)}{\partial t}-\varepsilon\frac{\partial^2 A_\theta(r,z)}{\partial t^2}=-J_{\mathrm{exc}} \tag{2.13}$$

Formula (2.13) is a scalar equation, which can be expanded in the cylindrical coordinate system with ∇^2.

$$\frac{1}{\mu}\left(\frac{\partial^2}{\partial r^2}+\frac{1}{r}\frac{\partial}{\partial r}+\frac{\partial^2}{\partial z^2}-\frac{1}{r^2}\right)A_\theta(r,z)-\sigma\frac{\partial A_\theta(r,z)}{\partial t}-\varepsilon\frac{\partial^2 A_\theta(r,z)}{\partial t^2}=-J_{\mathrm{exc}} \tag{2.14}$$

In Formula (14), the conductivity σ is 0 in the region 2. If the displacement current is ignored, the expression of Formula (2.14) can be obtained in different regions.

$$\begin{cases} \frac{1}{\mu_1}\left(\frac{\partial^2}{\partial r^2}+\frac{1}{r}\frac{\partial}{\partial r}+\frac{\partial^2}{\partial z^2}-\frac{1}{r^2}\right)A_{\theta,1}(r,z)-\sigma_1\frac{\partial A_{\theta,1}(r,z)}{\partial t}=-J_{\text{exc}} \\ \frac{1}{\mu_2}\left(\frac{\partial^2}{\partial r^2}+\frac{1}{r}\frac{\partial}{\partial r}+\frac{\partial^2}{\partial z^2}-\frac{1}{r^2}\right)A_{\theta,2}(r,z)=0 \\ \frac{1}{\mu_3}\left(\frac{\partial^2}{\partial r^2}+\frac{1}{r}\frac{\partial}{\partial r}+\frac{\partial^2}{\partial z^2}-\frac{1}{r^2}\right)A_{\theta,3}(r,z)-\sigma_3\frac{\partial A_{\theta,3}(r,z)}{\partial t}=0 \end{cases} \tag{2.15}$$

The subscript 1–3 of each variable represents the area corresponding to the variable. At the outer boundary of the model, the magnetic field satisfies the magnetically insulated boundary condition.

At each interface, the following boundary conditions are met by $A_{\theta,i}(r, z)$:

$$\begin{cases} A_{\theta,i}(r,z)=A_{\theta,(i+1)}(r,z) \\ \frac{1}{\mu_i}\frac{\partial A_{\theta,i}(r,z)}{\partial z}-\frac{1}{\mu_{i+1}}\frac{\partial A_{\theta,(i+1)}(r,z)}{\partial z}=J_{\text{sf},i} \end{cases} \tag{2.16}$$

The subscript i of each variable represents the region corresponding to the variable ($i = 1, 2, 3$); $J_{\text{sf},i}$ is the surface current density at the interface between i and $i + 1$.

The equations composed of (3.15) and (3.16) describe the transformation of the electric field and magnetic field in different regions by using single variable vector magnetic potential. The magnetic vector potential generated by the current in the aluminum plate is solved in the frequency domain by using the equation set and the $\delta(r-r_0)\,\delta(z-z_0)$ in the single turn loop coil. r_0 is the radius of the coil element, and z_0 is the position of the coil element in the z-direction. Under these conditions, the equations can be expressed in the frequency domain as follows:

$$\begin{cases} \frac{1}{\mu_1}\left(\frac{\partial^2}{\partial r^2}+\frac{1}{r}\frac{\partial}{\partial r}+\frac{\partial^2}{\partial z^2}-\frac{1}{r^2}\right)A_{\theta,1}(\omega,r,z)-j\omega\sigma_1A_{\theta,1}(\omega,r,z) \\ \quad=-J_{\text{exc}}(\omega)\delta(r-r_0)\delta(z-z_0) \\ \frac{1}{\mu_2}\left(\frac{\partial^2}{\partial r^2}+\frac{1}{r}\frac{\partial}{\partial r}+\frac{\partial^2}{\partial z^2}-\frac{1}{r^2}\right)A_{\theta,2}(\omega,r,z)=0 \\ \frac{1}{\mu_3}\left(\frac{\partial^2}{\partial r^2}+\frac{1}{r}\frac{\partial}{\partial r}+\frac{\partial^2}{\partial z^2}-\frac{1}{r^2}\right)A_{\theta,3}(\omega,r,z)-j\omega\sigma_3A_{\theta,3}(\omega,r,z)=0 \\ A_{\theta,i}(\omega,r,z)=A_{\theta,(i+1)}(\omega,r,z) \\ \frac{1}{\mu_i}\frac{\partial A_{\theta,i}(\omega,r,z)}{\partial z}-\frac{1}{\mu_{i+1}}\frac{\partial A_{\theta,(i+1)}(\omega,r,z)}{\partial z}=J_{\text{sf},i}(\omega) \end{cases} \tag{2.17}$$

Using the separation variable method [2] to solve the above equations, we can get the expression of vector magnetic potential $A_{\theta,3}(\omega, r, z)$ in an aluminum plate.

$$A_{\theta,3}(\omega,r,z)=\mu_2J_{\text{exc}}(\omega)r_0\int_0^{\infty}J_1(k_pr_0)J_1(k_pr)e^{-k_pz_0}\left(P_2e^{K_3z}+P_3e^{-K_3z}\right)dk_p \tag{2.18}$$

where

$$K_3 = \sqrt{k_p^2 + j\omega\mu_3\sigma_3}$$
$$P_2 = \frac{k_p(K_3 + k_p)e^{2K_3 d}}{(K_3 + k_p)^2 e^{2K_3 d} - (K_3 - k_p)^2}$$
$$P_3 = \frac{k_p(K_3 - k_p)e^{2K_3 d}}{(K_3 + k_p)^2 e^{2K_3 d} - (K_3 - k_p)^2}$$

J_1 is a first-class Bessel function.

Using the integration of $\delta(r - r_0)\,\delta(z - z_0)$, the magnetic vector potential generated by the impulse current in the whole single turn loop can be obtained.

$$\begin{aligned} A_{\theta,3}^{c}(\omega, r, z) &= \int\limits_{S_c} A_{\theta,3}(\omega, r, z, r_0, z_0)\mathrm{d}s \\ &= \int\limits_{h_c}\int\limits_{w_c} A_{\theta,3}(\omega, r, z, r_0, z_0)\mathrm{d}r_0\mathrm{d}z_0 \end{aligned} \tag{2.19}$$

where the superscript C represents the entire coil.

The electromagnetic field generated by $\boldsymbol{I}_{\mathrm{exc}}$ in the single turn loop will induce eddy density on the surface of the aluminum plate.

$$J_{\mathrm{eddy}}(\omega, r, z) = -j\omega\sigma A_{\theta,3}^{c}(\omega, r, z) \tag{2.20}$$

According to the superposition theory, the eddy current density response in the time domain can be obtained by using Fourier inverse transform.

$$J_{\mathrm{eddy}}(t, r, z) = \frac{1}{2\pi}\int\limits_{\infty}^{-\infty} J_{\mathrm{eddy}}(\omega, r, z)\, e^{j\omega t}\mathrm{d}\omega \tag{2.21}$$

$\boldsymbol{J}_{\mathrm{eddy}}(t, r, z)$ concentrates on the skin depth of the surface of the aluminum plate under the single turn ring coil, and its value is only the function of R and Z; that is, it has the circumferential conformance. Also, the direction is consistent with the direction of the magnetic vector potential $A_{\theta,3}^{c}(\omega, r, z)\boldsymbol{\theta}$ in the aluminum plate, that is, only the direction component of the aluminum plate. It can be seen that the eddy current density vector generated by $\boldsymbol{I}_{\mathrm{exc}}$ on the surface of the single loop is:

$$\boldsymbol{J}_{\mathrm{eddy}}(t, r, z) = J_{\mathrm{eddy}}(t, r, z)\boldsymbol{\theta} \tag{2.22}$$

Under the combined action of eddy current density $\boldsymbol{J}_{\mathrm{eddy}}$ and static magnetic field $\boldsymbol{B}_{\mathrm{p}}$, the density of Lorenz force generated on the surface of the aluminum plate is:

$$\begin{aligned} \boldsymbol{f}_{\mathrm{L}} &= \boldsymbol{J}_{\mathrm{eddy}} \times \boldsymbol{B}_{\mathrm{p}} \\ &= J_{\mathrm{eddy}}(t, r, z)\boldsymbol{\theta} \times B_{\mathrm{p}}\boldsymbol{z} \\ &= J_{\mathrm{eddy}}(t, r, z) B_{\mathrm{p}}\boldsymbol{r} \end{aligned} \tag{2.23}$$

$\boldsymbol{f}_{\mathrm{L}}$ is concentrated in the skin depth of the aluminum surface under the single turn loop. According to Formula (2.23),$\boldsymbol{f}_{\mathrm{L}}$ has only r-directional component, and its value is independent of θ, which is only a function of r and z, which is the circumferential consistency.

The Lorenz force$\boldsymbol{f}_{\mathrm{L}}$ produced by the single turn loop coil EMAT has the circumferential conformance and only the R-direction component; when it excites the Lamb wave, the circular ring wave front Lamb wave propagating along the radial diffusion will be produced. In this section, the analytical model of the acoustic field of the excited end omnidirectional Lamb wave is established by deriving the displacement and stress analytic expressions of the ring wave front Lamb wave and the frequency dispersion control equation. Then, the undetermined system in the analytical model of the sound field is solved by using the circumferential uniform Lorenz force produced by the single turn loop coil EMAT as the boundary condition. The complete analytical model of the omnidirectional Lamb wave excited by EMAT generated by circumferential coherent Lorenz force is obtained.

When the single turn loop EMAT is used to excite the Lamb wave, the circular wave front Lamb wave will be generated near it. Therefore, we need to deduce the analytical expression of the physical characteristics of the Lamb wave, such as displacement, stress, and dispersion, to establish the sound field model of the Lamb wave at the excitation end. In the model, we assume that the measured aluminum plates are isotropic and they satisfy the conditions of linear elasticity and continuity. The cylindrical coordinate system is built in the infinite free aluminum plate in space. As shown in Fig. 2.3, $z = \pm\, h$ is the upper and lower free surface of the aluminum plate, and the thickness of the plate is $d = 2h$. The point source of the Lamb wave generated by the circular wave front is located near $r = 0$. The circular wave front Lamb wave propagates along the r-direction, and its vibration displacement is only z-direction and r-direction component and is uniformly distributed along the θ-direction.

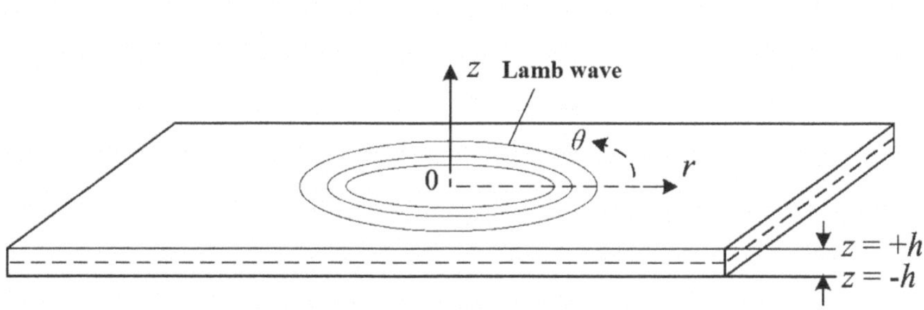

Fig. 2.3 Wave front Lamb wave in an infinite free aluminum plate

In the cylindrical coordinate system, the displacement equation $\boldsymbol{u}$ of the particle in the aluminum plate satisfies the Navier motion displacement equations [3].

$$\begin{cases} (\lambda_a + 2G_a)\frac{\partial \phi}{\partial r} - \frac{2G_a}{r}\frac{\partial \omega_z}{\partial \theta} + 2G_a\frac{\partial \omega_\theta}{\partial z} = \rho\frac{\partial^2 u_r}{\partial t^2} \\ (\lambda_a + 2G_a)\frac{1}{r}\frac{\partial \phi}{\partial \theta} - 2G_a\frac{\partial \omega_r}{\partial z} + 2G_a\frac{\partial \omega_z}{\partial r} = \rho\frac{\partial^2 u_\theta}{\partial t^2} \\ (\lambda_a + 2G_a)\frac{\partial \phi}{\partial z} - \frac{2G_a}{r}\frac{\partial (r\omega_\theta)}{\partial r} + \frac{2G_a}{r}\frac{\partial \omega_r}{\partial \theta} = \rho\frac{\partial^2 u_z}{\partial t^2} \end{cases} \tag{2.24}$$

The λ_a, G_a, and ρ are the Lame constant, the shear modulus, and the density of the medium in the aluminum plate, respectively, which are the volume invariants under the cylindrical coordinate system. The three components of the rotating vector are the ω_r, ω_θ, ω_z.

$$\begin{cases} \phi = \frac{1}{r}\frac{\partial (ru_r)}{\partial r} + \frac{1}{r}\frac{\partial u_\theta}{\partial \theta} + \frac{\partial u_z}{\partial z} \\ \omega_r = \frac{1}{2}\left(\frac{1}{r}\frac{\partial u_z}{\partial \theta} - \frac{\partial u_\theta}{\partial z}\right) \\ \omega_\theta = \frac{1}{2}\left(\frac{\partial u_r}{\partial z} - \frac{\partial u_z}{\partial r}\right) \\ \omega_z = \frac{1}{2r}\left(\frac{\partial}{\partial r}ru_\theta - \frac{\partial u_r}{\partial \theta}\right) \end{cases} \tag{2.25}$$

The Lamb wave of the circular wave front has the vibration displacement only in the z-direction and the r-direction. The displacement equation group can be simplified to:

$$\begin{cases} (\lambda_a + 2G_a)\frac{\partial \Delta}{\partial r} + G_a\nabla^2 u_r - G_a\frac{u_r}{r^2} = \rho\ddot{u}_r \\ (\lambda_a + 2G_a)\frac{\partial \Delta}{\partial z} + G_a\nabla^2 u_z = \rho\ddot{u}_z \end{cases} \tag{2.26}$$

where $\Delta = \frac{\partial u_r}{\partial r} + \frac{u_r}{r} + \frac{\partial u_z}{\partial z}$

Analogous to the solution of an elastic circular plate under static perturbation, it is assumed that the solution of the equation Δ and the above equations are as follows:

$$\Delta = A_u\frac{G_a}{\lambda_a + G_a}J_0(kr)e^{j\omega t}\begin{cases} \cosh(az) \\ \sinh(az) \end{cases} \tag{2.27}$$

$$u_r = f_1(z)J_1(kr)e^{j\omega t} \tag{2.28}$$

$$u_z = f_2(z)J_0(kr)e^{j\omega t} \tag{2.29}$$

k and ω are wavenumbers and angular frequencies of the Lamb wave, respectively; A_u and a are constants, and $J_1(kr)$ are zero-order and first-order Bessel functions.

The expression of $f_1(z)$ and $f_2(z)$ can be obtained by solution:

$$f_1(z) = \left[-j\frac{q}{k}B_s\cos(qz) + \frac{k}{q^2 - p^2}A_s\cos(pz)\right]$$

$$+\left[\frac{q}{k}B_{\mathrm{a}}\sin(qz)+j\frac{k}{q^{2}-p^{2}}A_{\mathrm{a}}\sin(pz)\right] \tag{2.30}$$

$$f_{2}(z)=\left[jB_{\mathrm{s}}\sin(qz)+\frac{p}{q^{2}-p^{2}}A_{\mathrm{s}}\sin(pz)\right]$$
$$+\left[B_{\mathrm{a}}\cos(qz)-j\frac{p}{q^{2}-p^{2}}A_{\mathrm{a}}\cos(pz)\right] \tag{2.31}$$

A_{s}, A_{a}, B_{s}, and B_{a} are constants, and:

$$p^{2}=\frac{\omega^{2}}{c_{\mathrm{L}}^{2}}-k^{2}$$
$$q^{2}=\frac{\omega^{2}}{c_{\mathrm{T}}^{2}}-k^{2}$$

$c_{\mathrm{L}}=[(\lambda_{\mathrm{a}}+G_{\mathrm{a}})/\rho]^{1/2}$ and $c_{\mathrm{T}}=(G_{\mathrm{a}}/\rho)^{1/2}$ are wave velocities of P-wave and S-wave in aluminum plates, respectively.

By replacing the expressions of $f_{1}(z)$ and $f_{2}(z)$ into formulas (2.28) and (2.29), the analytic expression of the Lamb wave displacement in the wave front can be obtained. According to the odd and even characteristics of the displacement analytic expression, it can be divided into two modes of symmetric (S) and antisymmetric (A). On this basis, the analytic expression of the stress can be obtained. Finally, the expression of the stress can be obtained. The dispersion equation of the circular wave front Lamb wave can be derived from the zero-stress boundary condition. The calculation results are as follows.

$$u_{\mathrm{r}}(t,r,z)=\left[-j\frac{q}{k}B_{\mathrm{s}}\cos(qz)+\frac{k}{q^{2}-p^{2}}A_{\mathrm{s}}\cos(pz)\right]J_{1}(kr)\mathrm{e}^{\mathrm{j}\omega t} \tag{2.32}$$

$$u_{z}(t,r,z)=\left[jB_{\mathrm{s}}\sin(qz)+\frac{p}{q^{2}-p^{2}}A_{\mathrm{s}}\sin(pz)\right]J_{0}(kr)\mathrm{e}^{\mathrm{j}\omega t} \tag{2.33}$$

$$\tau_{zr}(t,r,z)=G_{\mathrm{a}}\left[-j\frac{k^{2}-q^{2}}{k}B_{\mathrm{s}}\sin(qz)-\frac{2pk}{q^{2}-p^{2}}A_{\mathrm{s}}\sin(pz)\right]J_{1}(kr)\mathrm{e}^{\mathrm{j}\omega t} \tag{2.34}$$

$$\sigma_{zz}(t,r,z)=G_{\mathrm{a}}\left[j2qB_{\mathrm{s}}\cos(qz)-\frac{k^{2}-q^{2}}{q^{2}-p^{2}}A_{\mathrm{s}}\cos(pz)\right]J_{0}(kr)\mathrm{e}^{\mathrm{j}\omega t} \tag{2.35}$$

In the above calculation, the dispersion equation of the constant A_{s} and B_{s} and the Lamb wave of the circular wave front are still unknown. The boundary conditions are necessary to be further solved and derived.

In the process of Lamb wave propagation in the wave front of S mode, the zero-stress boundary condition is:

$$\tau_{zr}|_{z=\pm h}=\sigma_{zz}|_{z=\pm h}=0 \tag{2.36}$$

Formulae (2.34) and (2.35) are substituted into (2.36), and the homogeneous equations of constant A_s and B_s can be obtained.

$$\begin{vmatrix} \mathrm{j}2q\cos(qh) & \frac{q^2-k^2}{q^2-p^2}\cos(ph) \\ j\frac{q^2-k^2}{k}\sin(qh) & -\frac{2pk}{q^2-p^2}\sin(ph) \end{vmatrix} = 0 \tag{2.37}$$

By using Formula (2.37), the dispersion equation of wave front Lamb wave of S mode can be obtained.

$$\frac{\tan(qh)}{\tan(ph)} = -\frac{4k^2pq}{\left(q^2-k^2\right)^2} \tag{2.38}$$

Also, the Lorenz force generated by the single turn loop EMAT has only *r*-directional component, so Formula (2.35) is substituted (2.36).

$$\frac{B_s}{A_s} = \frac{\left(k^2-q^2\right)\cos(ph)}{j2q\left(q^2-p^2\right)\cos(qh)} \tag{2.39}$$

Formula (2.39) is replaced by the analytic expression of the displacement of the Lamb wave in the wave front of the S mode Formulas (3.32) and (3.33) to eliminate the undetermined coefficient B_s.

$$\begin{aligned} u_r(t,r,z) = {} & \frac{A_s}{2kq\left(q^2-p^2\right)\cos(qh)}\left[2k^2q\cos(qh)\cos(pz)\right. \\ & \left. -q\left(k^2-q^2\right)\cos(ph)\cos(qz)\right]J_1^{(}kr)\mathrm{e}^{\mathrm{j}\omega t} \end{aligned} \tag{2.40}$$

$$\begin{aligned} u_z(t,r,z) = {} & \frac{A_s}{2kq\left(q^2-p^2\right)\cos(qh)}[2kqp\cos(qh)\sin(pz) \\ & +k\left(k^2-q^2\right)\cos(ph)\sin(qz)]J_0(kr)\mathrm{e}^{\mathrm{j}\omega t} \end{aligned} \tag{2.41}$$

When $A'' = -A_s/\left[2kq\left(q^2-p^2\right)\cos(qh)\right]$, the displacement analytic expression is simplified to:

$$u_r(t,r,z) = A''\left[-2k^2q\cos(qh)\cos(pz) + q\left(k^2-q^2\right)\cos(ph)\cos(qz)\right]J_1(kr)\mathrm{e}^{\mathrm{j}\omega t} \tag{2.42}$$

$$u_z(t,r,z) = A''\left[-2kqp\cos(qh)\sin(pz) - k\left(k^2-q^2\right)\cos(ph)\sin(qz)\right]J_0(kr)\mathrm{e}^{\mathrm{j}\omega t} \tag{2.43}$$

Similarly, the analytical expressions for the Lamb wave displacement and stress of the A mode are derived as follows:

$$u_r(t,r,z) = \left[\frac{q}{k}B_a\sin(qz) + j\frac{k}{q^2-p^2}A_a\sin(pz)\right]J_1(kr)\mathrm{e}^{\mathrm{j}\omega t} \tag{2.44}$$

$$u_z(t,r,z) = \left[B_{\mathrm{a}} \cos(qz) - j\frac{p}{q^2 - p^2} A_{\mathrm{a}} \cos(pz) \right] J_0(kr) \mathrm{e}^{\mathrm{j}\omega t} \tag{2.45}$$

$$\tau_{zr}(t,r,z) = G_{\mathrm{a}} \left[-\frac{k^2 - q^2}{k} B_{\mathrm{a}} \cos(qz) + j\frac{2pk}{q^2 - p^2} A_{\mathrm{a}} \cos(pz) \right] J_1(kr) \mathrm{e}^{\mathrm{j}\omega t} \tag{2.46}$$

$$\sigma_{zz}(t,r,z) = G_{\mathrm{a}} \left[-2q B_{\mathrm{a}} \sin(qz) - j\frac{k^2 - q^2}{q^2 - p^2} A_{\mathrm{a}} \sin(pz) \right] J_0(kr) \mathrm{e}^{\mathrm{j}\omega t} \tag{2.47}$$

Using the boundary conditions, the dispersion equation of the Lamb wave of the A mode can be obtained.

$$\frac{\tan(qh)}{\tan(ph)} = -\frac{\left(q^2 - k^2\right)^2}{4k^2 pq} \tag{2.48}$$

The analytical expressions for the displacement of the Lamb wave of the A mode can be simplified to:

$$u_{\mathrm{r}}(t,r,z) = A''\left[-2k^2 q \sin(qh)\sin(pz) + q\left(k^2 - q^2\right)\sin(ph)\sin(qz)\right] J_1(kr) \mathrm{e}^{\mathrm{j}\omega t} \tag{2.49}$$

$$u_z(t,r,z) = A''\left[2kqp \sin(qh)\cos(pz) + k\left(k^2 - q^2\right)\sin(ph)\cos(qz)\right] J_0(kr) \mathrm{e}^{\mathrm{j}\omega t} \tag{2.50}$$

The A'' expressions in the analytical expressions for the displacement of the Lamb wave in the two modes of the modes mentioned above are determined by the initial state of the Lamb wave vibration, such as the initial shear stress. The initial shear stress in the aluminum plate is determined by the circumferential uniform Lorenz force exerted by EMAT on the surface of the aluminum plate when the Lamb wave is excited by the single turn loop coil EMAT in the aluminum plate.

In order to establish an analytical model of the circumferential uniform Lorenz force $\boldsymbol{f}_{\mathrm{L}}$ excited omnidirectional Lamb wave generated by the single turn loop coil EMAT, the relationship between the $\boldsymbol{f}_{\mathrm{L}}$ and the undetermined coefficient A'' of the Lamb wave displacement needs to be established.

For the single turn loop EMAT model shown in Fig. 2.2, the distance from the center of the coil width to the center of the coil is r_{c}. The Lorenz force $\boldsymbol{f}_{\mathrm{L}}(t, r, z)$ caused by the single turn loop at any position (r, θ, z) below the aluminum plate can be obtained by Formula (2.23). For any given angle θ_0, in the $r - z$ plane of $\theta = \theta_0$, the $\boldsymbol{f}_{\mathrm{L}}(t, r, z)$ can be integrated into its action area $\mathrm{S_f}$, and the total Lorenz force $\boldsymbol{F}_{\mathrm{L}}$ on $\mathrm{S_f}$ can be obtained.

$$\boldsymbol{F}_{\mathrm{L}} = \iint\limits_{\mathrm{S_f}} \boldsymbol{f}_{\mathrm{L}}(t,r,z)\mathrm{d}S \tag{2.51}$$

The width of the coil made by PCB is narrower than the wavelength of the Lamb wave, and the width of the coil is ignored in the model built. At the same time, considering the effect of skin effect, it is considered that $\boldsymbol{F}_{\mathrm{L}}$ is concentrated on the surface of the aluminum plate (r_c, θ_0, h). The same as $\boldsymbol{f}_{\mathrm{L}}$ (t, r, z), $\boldsymbol{F}_{\mathrm{L}}$ has only r-directional components and circumferential consistency and equals to F_{L}.

Under the effect of $\boldsymbol{F}_{\mathrm{L}}$, the shear stress and normal stress on the upper surface of the aluminum plate in the model are:

$$\begin{cases} \tau_{zr}|_{z=h} = F_{\mathrm{L}} \\ \sigma_{zz}|_{z=h} = 0 \end{cases} \tag{2.52}$$

The shear stress and the positive stress of the lower surface are as follows:

$$\begin{cases} \tau_{zr}|_{z=-h} = 0 \\ \sigma_{zz}|_{z=-h} = 0 \end{cases} \tag{2.53}$$

Formulas (2.52) and (2.53) form the boundary condition of Lamb wave in the aluminum plate under the action of circumferential uniform Lorenz force. If it is combined with symmetric and antisymmetric modes, it can be obtained as follows:

Upper surface:

$$\begin{cases} \tau_{zr}|_{z=h} = \frac{F_{\mathrm{L}}}{2} + \frac{F_{\mathrm{L}}}{2} \\ \sigma_{zz}|_{z=h} = 0 \end{cases} \tag{2.54}$$

Lower surface:

$$\begin{cases} \tau_{zr}|_{z=-h} = -\frac{F_{\mathrm{L}}}{2} + \frac{F_{\mathrm{L}}}{2} \\ \sigma_{zz}|_{z=-h} = 0 \end{cases} \tag{2.55}$$

The first half part of the upper and lower surface shear stresses forms symmetrical modes, while the latter half forms antisymmetric modes. The symmetric modal part of Formula (2.54) is replaced by the analytic expression of the Lamb wave stress in the wave front of the S mode, which is Formulas (2.34) and (2.35), and the index term can be omitted.

$$\begin{cases} G_{\mathrm{a}}\left[-j\frac{k^2-q^2}{k}B_{\mathrm{s}}\sin(qh) - \frac{2pk}{q^2-p^2}A_{\mathrm{s}}\sin(ph)\right] = \frac{F_{\mathrm{L}}}{2} \\ G_{\mathrm{a}}\left[2jqB_{\mathrm{s}}\cos(qh) - \frac{k^2-q^2}{q^2-p^2}A_{\mathrm{s}}\cos(ph)\right] = 0 \end{cases} \tag{2.56}$$

The solution of the above equations can be obtained:

$$\begin{cases} A_{\mathrm{s}} = \frac{F_{\mathrm{L}}}{2G_{\mathrm{a}}}\frac{2kq\left(q^2-p^2\right)\cos(qh)}{\left(k^2-q^2\right)^2\sin(qh)\cos(ph)+4pqk^2\cos(qh)\sin(ph)} \\ B_{\mathrm{s}} = j\frac{F_{\mathrm{L}}}{2G_{\mathrm{a}}}\frac{-k\left(k^2-q^2\right)\cos(ph)}{\left(k^2-q^2\right)^2\sin(qh)\cos(ph)+4pqk^2\cos(qh)\sin(ph)} \end{cases} \tag{2.57}$$

On this basis, the expression of A_{s} in Formula (2.57) is replaced by the definition of A''.

$$A'' = -A_s/\left[2kq\left(q^2 - p^2\right)\cos(qh)\right]$$

Therefore,

$$A'' = \frac{-F_L}{2G_a\left[\left(k^2 - q^2\right)^2 \sin(qh)\cos(ph) + 4pqk^2\cos(qh)\sin(ph)\right]} \tag{2.58}$$

A complete analytic expression of the Lamb wave displacement of two modes wave front can be obtained by replacing the (2.58) to Formulas (2.42), (2.43), (2.49), and (2.50), and combining the source position $r = r_c$.

S mode,

$$u_r(t, r, z) = \frac{F_L\left[2k^2 q\cos(qh)\cos(pz) - q\left(k^2 - q^2\right)\cos(ph)\cos(qz)\right]}{2G_a\left[\left(k^2 - q^2\right)^2 \sin(qh)\cos(ph) + 4pqk^2\cos(qh)\sin(ph)\right]} J_1(kr - kr_c)e^{j\omega t} \tag{2.59}$$

$$u_z(t, r, z) = \frac{F_L\left[2kqp\cos(qh)\sin(pz) + k\left(k^2 - q^2\right)\cos(ph)\sin(qz)\right]}{2G_a\left[(k^2 - q^2)^2 \sin(qh)\cos(ph) + 4pqk^2\cos(qh)\sin(ph)\right]} J_0(kr - kr_c)e^{j\omega t} \tag{2.60}$$

A mode,

$$u_r(t, r, z) = \frac{F_L\left[2k^2 q\sin(qh)\sin(pz) - q\left(k^2 - q^2\right)\sin(ph)\sin(qz)\right]}{2G_a\left[(k^2 - q^2)^2 \sin(qh)\cos(ph) + 4pqk^2\cos(qh)\sin(ph)\right]} J_1(kr - kr_c)e^{j\omega t} \tag{2.61}$$

$$u_z(t, r, z) = \frac{F_L\left[-2kqp\sin(qh)\cos(pz) - k\left(k^2 - q^2\right)\sin(ph)\cos(qz)\right]}{2G_a\left[(k^2 - q^2)^2 \sin(qh)\cos(ph) + 4pqk^2\cos(qh)\sin(ph)\right]} J_0(kr - kr_c)e^{j\omega t} \tag{2.62}$$

The analytic relation between the circumferential uniform Lorenz force $\boldsymbol{F}_L$ and the displacement of the wave front Lamb wave is given by Formulas (2.59)–(2.62), and the integral analytic model of the circumferential uniform Lorenz force excited omnidirectional wave is formed by the frequency dispersion equation of the wave front Lamb wave given by Formulas (2.38) and (2.48).

By using the dispersion equation in the above model, the frequency dispersion curves of the omnidirectional Lamb wave velocity CP and the group velocity CG in the aluminum plate can be drawn. As shown in Fig. 2.4, each of the curves represents a Lamb wave mode. For each curve, the wave velocity of the Lamb wave varies with the product of the frequency f and the thickness d product, that is the so-called dispersion phenomenon. The variation of Lamb wave velocity with d shows that the change of the thickness of the plate caused by the defect will lead to the change of the wave velocity of the defective Lamb wave, which will lead to the change of the time of the Lamb wave. The cross-hole tomography is the characteristic of Lamb wave. By measuring the TOF of a large number of Lamb waves, the information of

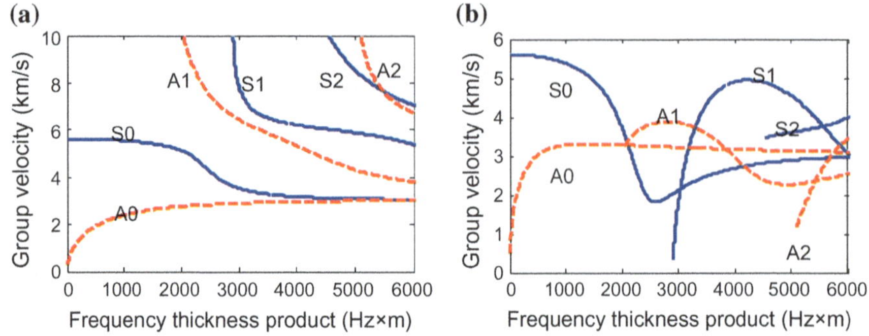

Fig. 2.4 Dispersion curves of Lamb wave in aluminum plate: **a** phase velocity; **b** group velocity

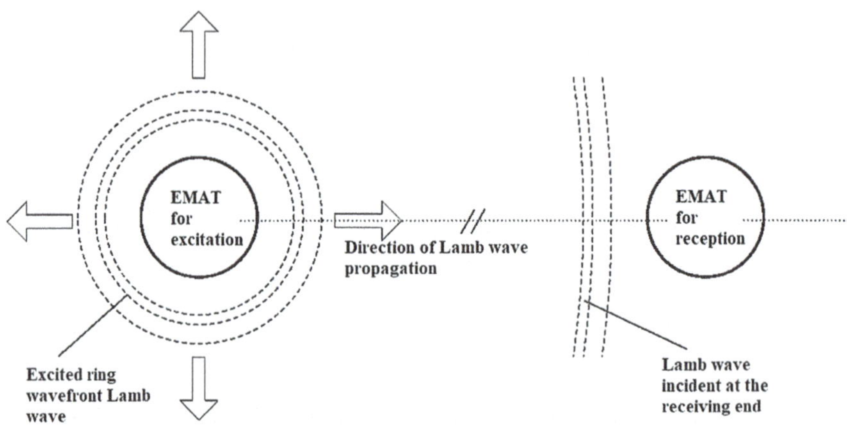

Fig. 2.5 Difference of spatial geometric models involved in transmitting and receiving Lamb waves in EMAT

the defects is obtained from various angles, and then, the defects are reconstructed by using this information.

The spatial geometric model involved in the excitation and reception of Lamb waves by the single turn loop coil EMAT is different. As shown in Fig. 2.5, when it is used to excite the Lamb wave, because its structure is axisymmetric, the circumferential wave front Lamb wave propagating along the circumferential and radial diffusion will be produced; and when it is used to receive the Lamb wave, the Lamb wave will be from the particular. The incident wave field does not have a circumferential consistency for the receiving EMAT.

In tomography, the spacing of EMAT is far greater than that of EMAT itself and Lamb wavelength λ. Therefore, it is considered that the receiving EMAT is located in the far field of the omnidirectional Lamb sound field generated by the excitation EMAT. Based on this, the analytical expression of the Lamb wave displacement of the receiver can be obtained by the far-field approximation by the analytic expression of

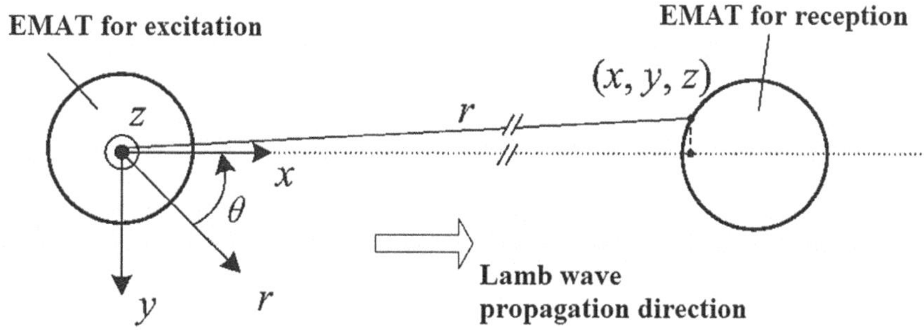

Fig. 2.6 Schematic diagram of the analytical solution for the Lamb wave displacement at the receiving terminal

the excited omnidirectional Lamb wave displacement given at the receiving terminal (2.59)–(2.62).

As shown in Fig. 2.6, the same cylindrical coordinate system as in Fig. 2.1 is established on the excitation of EMAT. On this basis, a three-dimensional Cartesian coordinate system is established: The origin coincides with the origin of the cylindrical coordinate system; the x-direction points to the receiving EMAT along the direction of the two EMAT centers; the direction of the z is perpendicular to the surface of the aluminum plate.

The Bessel function of Formulas (2.59)–(2.62) satisfies the following relations:

$$J_1(kr - kr_c) = \frac{1}{2}\left[H_1^{(1)}(kr - kr_c) + H_1^{(2)}(kr - kr_c)\right] \tag{2.63}$$

$$J_0(kr - kr_c) = \frac{1}{2}\left[H_0^{(1)}(kr - kr_c) + H_0^{(2)}(kr - kr_c)\right] \tag{2.64}$$

$H_1^{(1)}$ and $H_1^{(2)}$ are the first-order Hankel function of the first and second kind, and $H_0^{(2)}$ are zero-order Hankel function of the first and second kind. In Formula (2.63), $H_1^{(1)}$ and $H_1^{(2)}$, respectively, represent the internal wave component and the external traveling wave component of omnidirectional Lamb waves generated by EMAT coils. As the characteristics of the internal traveling wave and external traveling wave are similar, the following only external traveling waves are taken as an example for analysis. Under this condition, only $H_1^{(2)}$ is reserved on the right side of the equals of type (2.63). Similarly, on the right side of the equals of Formula (2.64), only $H_0^{(2)}$ is considered in the case of the external wave only. Under far-field conditions ($r \gg r_c, r \gg \lambda$), $H_1^{(2)}$ and $H_0^{(2)}$ should meet the following approximate relations [4]:

$$H_1^{(2)}(kr - kr_c)|_{r\gg\lambda, r\gg r_c} \approx \sqrt{\frac{2}{\pi k(r - r_c)}}\mathrm{e}^{-\mathrm{j}\left(kr - kr_c - \frac{3}{4}\pi\right)} \approx \sqrt{\frac{2}{\pi kr}}\mathrm{e}^{-\mathrm{j}\left(kr - kr_c - \frac{3}{4}\pi\right)} \tag{2.65}$$

$$H_0^{(2)}(kr - kr_{\rm c})|_{r\gg\lambda, r\gg r_{\rm c}} \approx \sqrt{\frac{2}{\pi k(r - r_{\rm c})}} {\rm e}^{-{\rm j}\left(kr - kr_{\rm c} - \frac{1}{4}\pi\right)} \approx \sqrt{\frac{2}{\pi kr}} {\rm e}^{-{\rm j}\left(kr - kr_{\rm c} - \frac{1}{4}\pi\right)} \tag{2.66}$$

As a result, in the case of only traveling waves, the analytical expressions of the excited omnidirectional Lamb wave displacement given by Formulas (2.59)–(2.62) are changed in the far-field condition.

S mode,

$$u_{\rm r}(t, r, z) = \frac{F_{\rm L}\left[2k^2 q \cos(qh)\cos(pz) - q\left(k^2 - q^2\right)\cos(ph)\cos(qz)\right]}{2G_{\rm a}\left[(k^2 - q^2)^2 \sin(qh)\cos(ph) + 4pqk^2\cos(qh)\sin(ph)\right]} \sqrt{\frac{1}{2\pi kr}} {\rm e}^{-{\rm j}(kr - \omega t) + {\rm j}\left(kr_{\rm c} + \frac{3}{4}\pi\right)} \tag{2.67}$$

$$u_z(t, r, z) = \frac{F_{\rm L}\left[2kqp \cos(qh)\sin(pz) + k\left(k^2 - q^2\right)\cos(ph)\sin(qz)\right]}{2G_{\rm a}\left[\left(k^2 - q^2\right)^2 \sin(qh)\cos(ph) + 4pqk^2\cos(qh)\sin(ph)\right]} \sqrt{\frac{1}{2\pi kr}} {\rm e}^{-{\rm j}(kr - \omega t) + {\rm j}\left(kr_{\rm c} + \frac{1}{4}\pi\right)} \tag{2.68}$$

A mode,

$$u_{\rm r}(t, r, z) = \frac{F_{\rm L}\left[2k^2 q \sin(qh)\sin(pz) - q\left(k^2 - q^2\right)\sin(ph)\sin(qz)\right]}{2G_{\rm a}\left[\left(k^2 - q^2\right)^2 \sin(qh)\cos(ph) + 4pqk^2\cos(qh)\sin(ph)\right]} \sqrt{\frac{1}{2\pi kr}} {\rm e}^{-{\rm j}(kr - \omega t) + {\rm j}\left(kr_{\rm c} + \frac{3}{4}\pi\right)} \tag{2.69}$$

$$u_z(t, r, z) = \frac{F_{\rm L}\left[-2kqp \sin(qh)\cos(pz) - k\left(k^2 - q^2\right)\sin(ph)\cos(qz)\right]}{2G_{\rm a}\left[\left(k^2 - q^2\right)^2 \sin(qh)\cos(ph) + 4pqk^2\cos(qh)\sin(ph)\right]} \sqrt{\frac{1}{2\pi kr}} {\rm e}^{-{\rm j}(kr - \omega t) + {\rm j}\left(kr_{\rm c} + \frac{1}{4}\pi\right)} \tag{2.70}$$

The omnidirectional Lamb wave excited by the single turn loop EMAT has the following characteristics:

(1) Both the S mode and the A mode Lamb wave can be decomposed into a standing wave component along the thickness of the plate and a traveling wave field along the radial direction.

(2) The traveling wave component in the displacement analytical expression contains an amplitude attenuation coefficient $1/\sqrt{r}$, which is because with the spread of the omnidirectional Lamb wave, the diffusion area is increasing, and the energy will become more and more dispersed.

As shown in Fig. 2.6, for any point near EMAT in the aluminum plate (x, y, z), the distance from the EMAT center $(0, 0, z)$ to the excitation is r, and the following conclusion can be obtained under the far-field condition $r \gg r_{\mathrm{c}}$.

(1) $r \approx x$.
(2) At (x, y, z), the x-direction component of Lamb wave displacement is $u_x \approx u_r$.
(3) At (x, y, z), the y-direction component of Lamb wave displacement is $u_y \approx 0$.

Based on Formulas (2.67)–(2.70) and the conclusion of the far-field approximation above, the analytic expression of the Lamb wave displacement of the receiver (the far-field region of the excited EMAT) can be obtained.

S mode,

$$u_x(t,x,z) = \frac{F_{\mathrm{L}}\left[2k^2 q\cos(qh)\cos(pz) - q(k^2-q^2)\cos(ph)\cos(qz)\right]}{2G_{\mathrm{a}}\left[(k^2-q^2)^2\sin(qh)\cos(ph) + 4pqk^2\cos(qh)\sin(ph)\right]} \sqrt{\frac{1}{2\pi kx}}\,\mathrm{e}^{-\mathrm{j}(kx-\omega t)+\mathrm{j}\left(kr_{\mathrm{c}}+\frac{3}{4}\pi\right)} \tag{2.71}$$

$$u_z(t,x,z) = \frac{F_{\mathrm{L}}\left[2kqp\cos(qh)\sin(pz) + k(k^2-q^2)\cos(ph)\sin(qz)\right]}{2G_{\mathrm{a}}\left[(k^2-q^2)^2\sin(qh)\cos(ph) + 4pqk^2\cos(qh)\sin(ph)\right]} \sqrt{\frac{1}{2\pi kx}}\,\mathrm{e}^{-\mathrm{j}(kx-\omega t)+\mathrm{j}\left(kr_{\mathrm{c}}+\frac{1}{4}\pi\right)} \tag{2.72}$$

A mode,

$$u_x(t,x,z) = \frac{F_{\mathrm{L}}\left[2k^2 q\sin(qh)\sin(pz) - q(k^2-q^2)\sin(ph)\sin(qz)\right]}{2G_{\mathrm{a}}\left[(k^2-q^2)^2\sin(qh)\cos(ph) + 4pqk^2\cos(qh)\sin(ph)\right]} \sqrt{\frac{1}{2\pi kx}}\,\mathrm{e}^{-\mathrm{j}(kx-\omega t)+\mathrm{j}\left(kr_{\mathrm{c}}+\frac{3}{4}\pi\right)} \tag{2.73}$$

$$u_z(t,x,z) = \frac{F_{\mathrm{L}}\left[-2kqp\sin(qh)\cos(pz) - k(k^2-q^2)\sin(ph)\cos(qz)\right]}{2G_{\mathrm{a}}\left[(k^2-q^2)^2\sin(qh)\cos(ph) + 4pqk^2\cos(qh)\sin(ph)\right]} \sqrt{\frac{1}{2\pi kx}}\,\mathrm{e}^{-\mathrm{j}(kx-\omega t)+\mathrm{j}\left(kr_{\mathrm{c}}+\frac{1}{4}\pi\right)} \tag{2.74}$$

From Formulas (2.71)–(2.74), it is known that the receiver Lamb wave sound field has the following characteristics: (1) only the displacement of x- and z-directions; (2) each displacement contains only stationary wave components along the z-direction and traveling wave components propagating along the x-direction, and the traveling wave component contains an amplitude attenuation coefficient $1/\sqrt{x}$; (3) the expression of each displacement is independent of y, and the distribution of the Lamb wave sound field along the y-direction is consistent. The above characteristics show that the

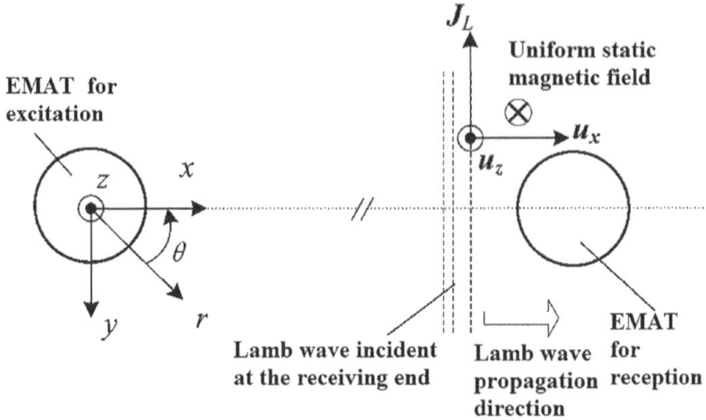

Fig. 2.7 Schematic diagram of the induced electromotive force induced by Lamb wave at receiving terminal in EMAT

Lamb wave at the receiving terminal can be approximated as a plane wave front Lamb wave propagating along the x-direction and gradually decreasing in the propagation process.

As far-field approximation does not change the dispersion characteristics of Lamb wave, the analytic expression of Lamb wave displacement in combination (2.71) to Formula (2.74), and the dispersion equation of Lamb wave given by Formula (2.38) and (2.48), the analytic model of the Lamb wave sound field of the receiver is obtained.

The principle of induced electromotive force in EMAT at the receiving terminal Lamb wave field is shown in Fig. 2.7. As mentioned before, the plane wave front Lamb wave at the receiver only contains in-plane displacement u_x (t, x, y, z) along the x-direction and the out-of-plane displacement u_z (t, x, y, z) along the z-direction. The analytic expression of u_x (t, x, y, z) and u_z (t, x, y, z) is given by Formulas (2.71)–(2.74). By combining the displacement expression of the A mode wave to the S mode wave, the expression can be obtained:

$$\begin{aligned} u_x(t,x,y,z) = & \left\{ \frac{F_{\mathrm{L}}[2k^2q\cos(qh)\cos(pz) - q(k^2-q^2)\cos(ph)\cos(qz)]}{2G_{\mathrm{a}}[(k^2-q^2)^2\sin(qh)\cos(ph) + 4pqk^2\cos(qh)\sin(ph)]} \right. \\ & \left. + \frac{F_{\mathrm{L}}\left[2k^2q\sin(qh)\sin(pz) - q\left(k^2-q^2\right)\sin(ph)\sin(qz)\right]}{2G_{\mathrm{a}}\left[\left(k^2-q^2\right)^2\sin(qh)\cos(ph) + 4pqk^2\cos(qh)\sin(ph)\right]} \right\} \\ & + \sqrt{\frac{1}{2\pi kx}}\mathrm{e}^{-\mathrm{j}(kx-\omega t)+\mathrm{j}\left(kr_c+\frac{3}{4}\pi\right)} \end{aligned} \tag{2.75}$$

$$u_z(t,x,y,z) = \left\{ \frac{F_L\left[2kqp\cos(qh)\sin(pz) + k\left(k^2-q^2\right)\cos(ph)\sin(qz)\right]}{2G_a\left[\left(k^2-q^2\right)^2\sin(qh)\cos(ph) - 4pqk^2\cos(qh)\sin(ph)\right]} + \frac{F_L\left[-2kqp\sin(qh)\cos(pz) - k\left(k^2-q^2\right)\sin(ph)\cos(qz)\right]}{2G_a\left[\left(k^2-q^2\right)^2\sin(qh)\cos(ph) + 4pqk^2\cos(qh)\sin(ph)\right]} \right\} \times \sqrt{\frac{1}{2\pi kx}}e^{-j(kx-\omega t)+j\left(kr_c+\frac{1}{4}\pi\right)} \tag{2.76}$$

The vibration of the aluminum plate caused by the Lamb wave causes the motion of charged particles in aluminum plates. The velocity is as follows:

$$\boldsymbol{v} = \frac{\partial \boldsymbol{u}}{\partial t} = \frac{\partial u_x(t,x,y,z)}{\partial t}\boldsymbol{x} + \frac{\partial u_z(t,x,y,z)}{\partial t}\boldsymbol{z} \tag{2.77}$$

The charged particles moving in the aluminum plate generate dynamic current density under the static magnetic field $\boldsymbol{B}_p$.

$$\boldsymbol{J}_L = \sigma \boldsymbol{v} \times \boldsymbol{B}_p \tag{2.78}$$

$\boldsymbol{B}_p$ has only the component of the z-direction, then:

$$\begin{aligned} \boldsymbol{J}_L &= \sigma\left[\frac{\partial u_x(t,x,y,z)}{\partial t}\boldsymbol{x} + \frac{\partial u_z(t,x,y,z)}{\partial t}\boldsymbol{z}\right] \times \left(-B_p\boldsymbol{z}\right) \\ &= -\sigma\frac{\partial u_x(t,x,y,z)}{\partial t}B_p\boldsymbol{y} \end{aligned} \tag{2.79}$$

According to Formula (2.79), the dynamic current density $\boldsymbol{J}_L$ has only y-direction components and distributes uniformly along the y-direction.

Dynamic electric current density $\boldsymbol{J}_L$ will produce a dynamic electric field and the magnetic field around and around the aluminum plate. In this process, the receiving EMAT single turn coil is in the open state, its total current is zero, and the control equation of the magnetic vector potential in the aluminum plate, air gap and coil, and other regions should be satisfied.

$$\frac{1}{\mu}\nabla^2\boldsymbol{A} - \sigma\frac{\partial \boldsymbol{A}}{\partial t} - \varepsilon\frac{\partial^2 \boldsymbol{A}}{\partial t^2} + \frac{\sigma}{S_c}\frac{\partial}{\partial t}\iint_{S_c} \boldsymbol{A}\mathrm{d}S = -\boldsymbol{J}_L \tag{2.80}$$

The vector magnetic potential $\boldsymbol{A}$ generated is the same as the direction of $\boldsymbol{J}_L$, that is, only the y-direction component. As EMAT excites the omnidirectional Lamb wave model, ignoring the width of the single turn loop, the output voltage of the coil is:

$$\boldsymbol{V}_{out} = \int -\frac{\partial \boldsymbol{A}}{\partial t}\cdot \mathrm{d}\boldsymbol{l} \tag{2.81}$$

2.1.2 Finite Element Simulation of EMAT Based on COMSOL Multiphysics

Using COMSOL Multiphysics to model and analyze the problems described by the PDE equation, two ways can be chosen. One way is to model the PDE equation directly. According to the expression of PDE equation, it is divided into three types: coefficient, universal, and weak form. Under the known parameters and boundary conditions of various forms of PDE expression, these parameters and boundary condition values can be input to complete the modeling of the physical model. Another way is to use some of the commonly used physical field analysis modules built-in COMSOL Multiphysics itself. In the COMSOL Multiphysics 3.5A version, it includes eight modules, such as AC/DC module, RF module, acoustic module, chemical engineering module, earth science module, a heat transfer module, MEMS module, and structural mechanics module. In any way, the COMSOL Multiphysics is used to solve the finite element solution and the essence of the multifield coupling problem, which is the partial differential equation and the boundary and boundary conditions corresponding to the actual physical problems, and the equations are converted into weak form, and then the weak form is solved. The advantage of this method is that it can reduce the continuity requirement of integral variables and improve the solving ability of nonlinear and multiphysical field problems. Therefore, the weak form of the multi physical field coupling modeling is better universality, which can not only clarify the essence of the finite element calculation for various physical fields but also solve the problems that cannot be solved in the COMSOL Multiphysics built-in module.

COMSOL Multiphysics is used to carry out the numerical simulation analysis of EMAT based on Lorenz force mechanism. The method and the concrete steps of the numerical simulation analysis are:

(1) Select the required solution coordinate system. If the two-dimensional analysis is adopted, 2D Cartesian coordinate system or axisymmetric coordinate system can be selected. It should be noted that the 2D axisymmetric coordinate system does not support weak form modeling itself. In the weak form modeling, the axis coordinate system can be transformed to the direct, coordinate system by simple coordinate transformation.
(2) The solution form of various fields is selected as weak form. In this way, four groups of equations should be chosen.
(3) Establish the geometric model of each component of EMAT. In the numerical calculation of electromagnetic field, a solution area is needed. Therefore, a solution area must be delineated outside the EMAT model.
(4) According to the weak form and boundary conditions of various field equations mentioned before, the solution area and boundary are set, respectively. For the EMAT based on the Lorenz force mechanism, the calculation of the static magnetic field and the mechanical field can be accomplished by the built-in module of the COMSOL Multiphysics, except for the weak form of the skin effect and the adjacent utility. Because of the coupling of various fields, it is

necessary to set the coupling variable. The static bias magnetic field is used to solve the product of the eddy density of the static magnetic induction intensity and the pulsed eddy current field. The static bias magnetic field is used to solve the static magnetic induction intensity and the tested permeability and the particle. The product of velocity and the pulsed eddy current field is used to calculate the current source density when the coil receives ultrasonic waves. Due to the calculation of the induction electromotive force of the coil conductor when the coil receives the ultrasonic signal, the integral coupling variable is required to calculate the result of the induction electromotive force.

(5) Mesh subdivision of the solution area. In mesh generation, it is necessary to pay attention to two points in order to improve the accuracy of calculation: One is that more than 2 grid units should be set in the skin depth of the surface of the tested specimen; and two, within the specimen, more than 7 grid units should be guaranteed in the wavelength of the ultrasonic excitation.

(6) In order to solve the model, the static bias magnetic field is solved by a steady state, and the rest of the fields are solved by a transient state. The transient solution requires setting relative error and absolute error of step size and solution. Since the pulse excitation signal is a high-frequency tone-burst signal, a smaller step should be set to ensure the smoothness and stability of the obtained solution. When solving the displacement of the internal particle of the specimen, the displacement value is small, so the absolute error of the solution should be small enough to ensure the correctness of the solution.

(7) Post-processing of the results, including variables such as cloud and transient waveforms.

The Lamb wave is excited by the back-folding coil on the aluminum plate. Because of the dispersion characteristic of the Lamb wave, the phase velocity and the group velocity dispersion curve of the Lamb wave should be calculated according to the elastic modulus of the aluminum plate, the Poisson's ratio and the thickness of the plate before making the EMAT, and the suitable working point is selected according to the frequency dispersion curve, and the design of the EMAT coil is carried out according to the working point. The parameters of the tested aluminum plates used in the calculation and test are as follows (Table 2.1).

Table 2.1 Size and parameters of the tested aluminum plate

Parameter	Value
Length	500 mm
Width	350 mm
Thickness	3 mm
Conductivity	$3.5 \times e^{-7}$ S/m
Modulus of elasticity	70 Gpa
Poisson ratio	0.33

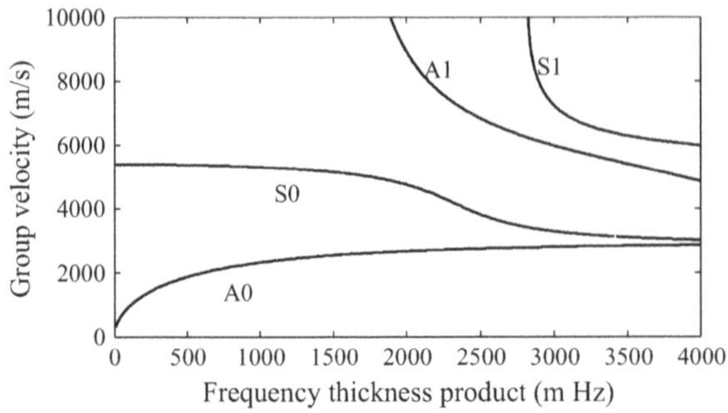

(a) Phase velocity dispersion curve.

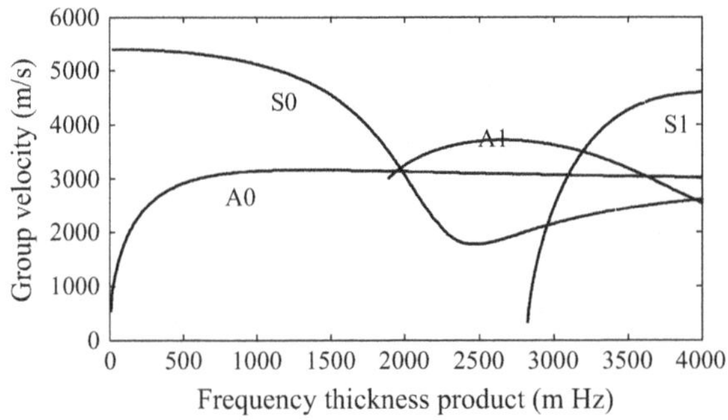

(b) Group velocity dispersion curve.

Fig. 2.8 Phase velocity and group velocity dispersion curves of Lamb waves in the aluminum plate

By using the software of dispersion curve calculation, the phase velocity and group velocity dispersion curves of the tested aluminum plates under different frequency–thickness are calculated (Fig. 2.8).

The corresponding frequency–thickness product of the selected 3 mm aluminum plate is 1100; that is, the selective excitation frequency is 366.7 kHz. At this time, there are two kinds of wave modes corresponding to the Lamb wave, namely A0 and S0 mode, in which the phase velocity and group velocity corresponding to the A0 mode Lamb wave are 2314 and 3148 m/s, respectively; the phase velocity and group velocity corresponding to the S0 mode Lamb wave are 5287 m/s and group velocities, respectively. The double-layer double-split folded coil with a center of 6.5 mm distance is shown in Fig. 2.9.

The folded coil is manufactured by PCB technology, and its dimensions and material parameters are given in Table 2.2.

In the example, the excitation and reception of ultrasonic wave are realized by using two probes. The probe is placed at 90 mm from the left side of the tested aluminum plate, and the receiving probe is placed at 120 mm from the left side of the aluminum plate. The center distance between the two probes is 140 mm. The distance between the two-probe coil and the measured aluminum plate is 1 mm,

Fig. 2.9 Double-split back-folding winding coil

Table 2.2 Coil size and material parameters

Parameter	Value
Substrate thickness	0.500 mm
Copper platinum width	0.720 mm
The thickness of copper and platinum	0.035 mm
Line spacing	0.905 mm
Fold spacing	3.25 mm
The magnetic permeability of copper foil	$4\pi \times 10^{-7}$ H/m
The electrical conductivity of copper foil	2.667×10^{7} S/m

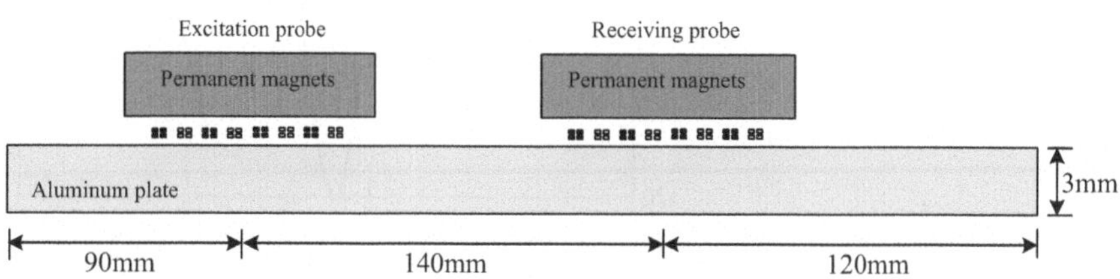

Fig. 2.10 Lamb wave probe layout

the permanent magnet with the residual magnetic induction intensity is 1T above the refolding coil, and the distance from the refolding coil is 0.5 mm. The probe is arranged as shown in Fig. 2.10.

In order to verify the effectiveness of the numerical simulation results, the EPR-4000 pulse occurrence and receiver produced by RITEC company are used as the excitation source of the EMAT probe to produce RF tone burst signal. The amplitude and frequency of the signal can be adjusted according to demand. Between the EPR-4000 and the excitation coils, the impedance matching between the coil impedance and the excitation source output impedance is achieved through the impedance matching device. EPR-4000 can also filter and amplify the receiving signal of the coil, and the bandwidth and magnification of the filter can also be adjusted. The receiving

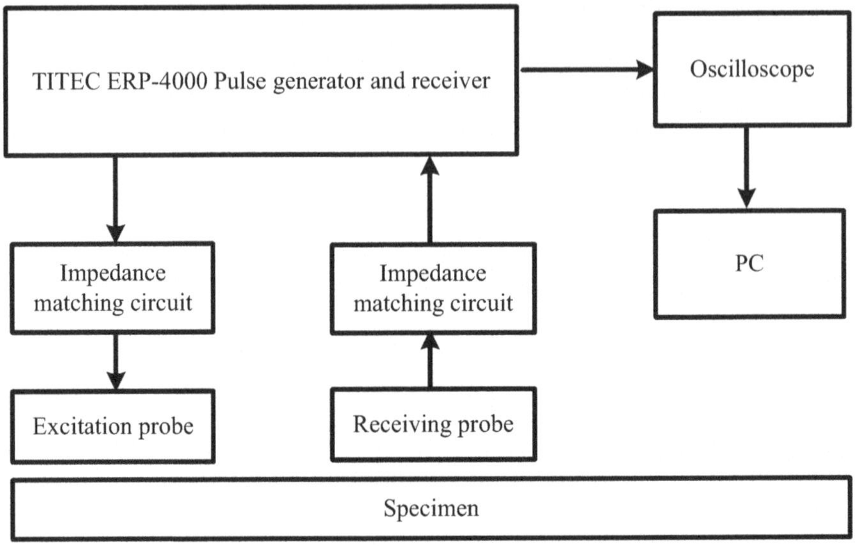

Fig. 2.11 Connection diagram of test equipment

Fig. 2.12 Waveform of the excitation signal

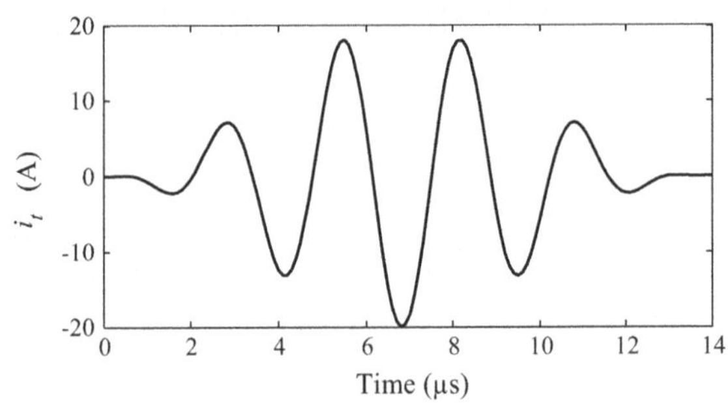

coil is connected with the impedance matching device to match the coil impedance and the EPR-4000 input impedance, to achieve a larger power output. The signals received by EPR-4000 are displayed, and output and the data acquisition and waveform display on the PC machine are realized by the data acquisition software Wave Star for Oscilloscopes, which is matched with the oscilloscope. The connection of the test equipment is shown in Fig. 2.11.

The waveform of the tone-burst current excitation signal with the frequency 366.7 kHz, and the number of cycles is shown in Fig. 2.12.

The excitation current is replaced by COMSOL Multiphysics for modeling and calculation. The maximum unit size of the specimen is set to 1/8 Lamb wavelength, which is 0.8125 mm, and the step size is 0.00000002 s. To observe the results obtained, the observation points A and B are selected within the specimen, the A point is located below the left conductor group of the excitation coil and below the

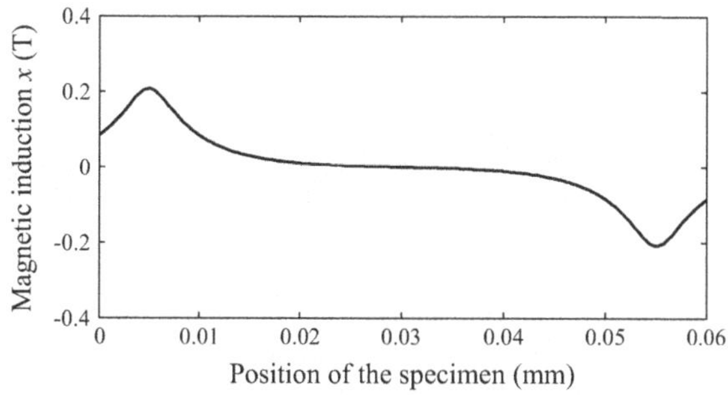

Fig. 2.13 *x* component of static magnetic induction intensity

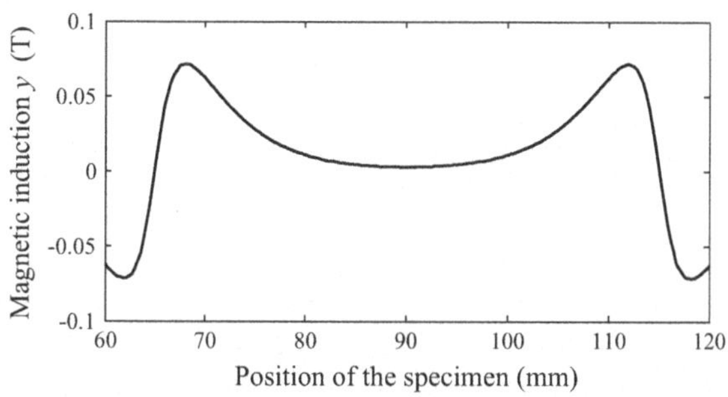

Fig. 2.14 *y* component of static magnetic induction intensity

surface of the specimen on the surface of the 0.01 mm; the B point is located below the center of the receiving coil and below the surface 0.01 mm at the surface of the specimen. An observation area C is selected, which is a line segment at the bottom of the probe, 0.01 mm below the surface of the specimen along the width of the surface of the specimen. The center of the line corresponds to the winding coil, and the center of the permanent magnet is shown in the abscissa.

Figures 2.13 and 2.14 show the *x* and *y* components of the static magnetic induction produced by the permanent magnet in the region C, respectively. The *x* component of magnetic induction intensity distributes symmetrically along the center of the region, and the value increases gradually from the center to the sides and reaches the maximum at the edge of the magnet. The *y* component of magnetic induction intensity distributes in saddle shape along the region, and near the center of the region, the *y* component of magnetic induction intensity is a constant, extending to the two sides of the region, the value becomes larger, and the maximum is reached at the edge of the permanent magnet. It can be seen that when the design of the permanent magnet is suitable, the *y* component of the bias magnetic field in the coil area can be approximated to be a constant.

Figure 2.15 shows the vector potential position of the coil generated by the tone-burst pulse excitation at 10 s time. Under the action of the pulsed magnetic field, the induced eddy current and eddy current distribution in the skin depth of the specimen

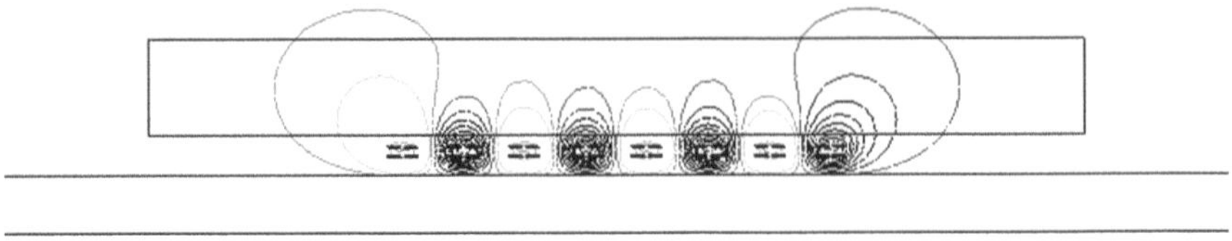

Fig. 2.15 Vector magnetic potential equipotential line at 10 μs

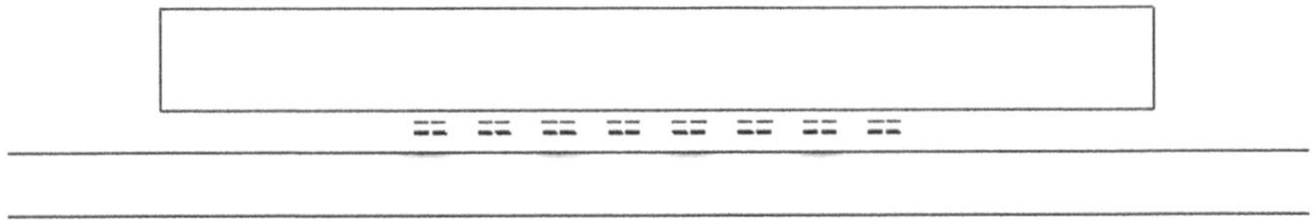

Fig. 2.16 Eddy current distribution in a tested specimen at 10 μs

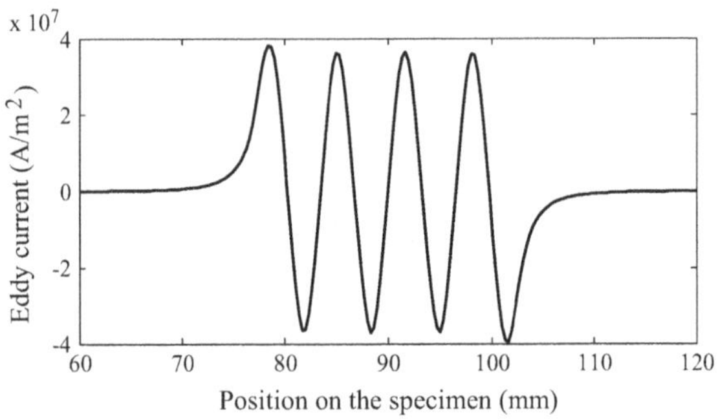

Fig. 2.17 Eddy current distribution on C region at 10 μs

are shown in Figs. 2.16 and 2.17. It can be seen that the direction of the eddy current in the specimen under the refolding coil is opposite because of the opposite direction of the 80% off line segments adjacent to the refolding coil, and the numerical value of the eddy density under each line section of the coil is approximately equal. Under the action of bias magnetic field and pulsed eddy current, Lorenz force is generated inside the specimen and acted on the inner point of the specimen (Figs. 2.18 and 2.19).

Figures 2.20 and 2.21 show the characteristics of the x and y components of the Lorenz force at the observation point A over time. Under the effect of pulsed eddy current and bias magnetic field, the duration of Lorenz force produced and acted on the specimen is the same as the duration of the coil impulse excitation. Under the effect of Lorenz force, the particle in the specimen is vibrated, excited ultrasonic, and propagated along the specimen of the measured aluminum plate.

Figure 2.22 shows the cloud figures of the internal particle displacement x components of the internal particles of the specimens at 0, 10, 30, and 50 μs, reflecting the situation of ultrasonic propagation in the specimen. At the time of 0 μs, the coil

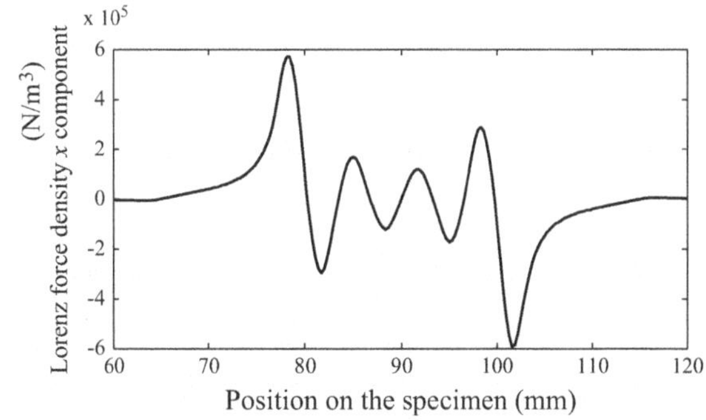

Fig. 2.18 *x* component of Lorenz force on *C* region at 10 μs

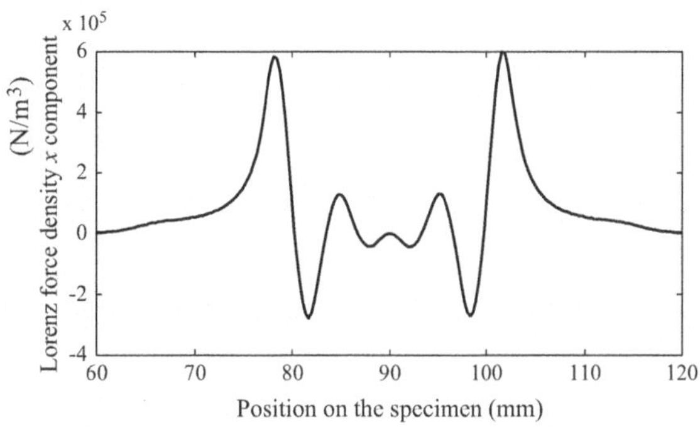

Fig. 2.19 *y* component of Lorenz force on *C* region at the moment 10 μs

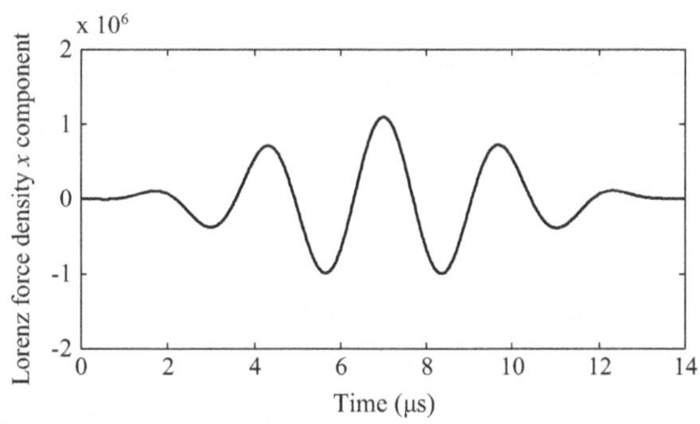

Fig. 2.20 Curves of Lorenz force *x* component varying with time at *A* point

has not been stimulated, and the ultrasonic is not produced; at the time of 10 μs, ultrasonic wave has been produced, and it has begun to spread to the left and right two directions; at the time of 30 μs, the coil excitation has ended; and the ultrasonic wave continues to spread to the two directions. At this time, it can be seen that the sound waves of A0 and S0 modes are excited in the specimen. At 50 μs, the ultrasonic wave of the S0 mode that propagates to the right has passed the receiving probe, and the ultrasonic wave of the A0 mode has also been propagated to the receiving probe,

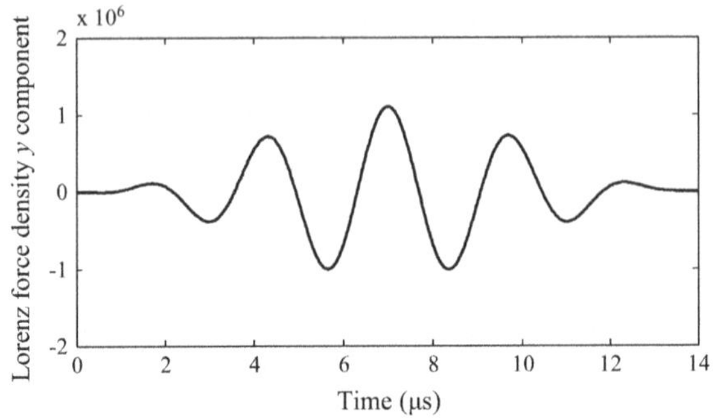

Fig. 2.21 Curves of Lorenz force y component varying with time at A point

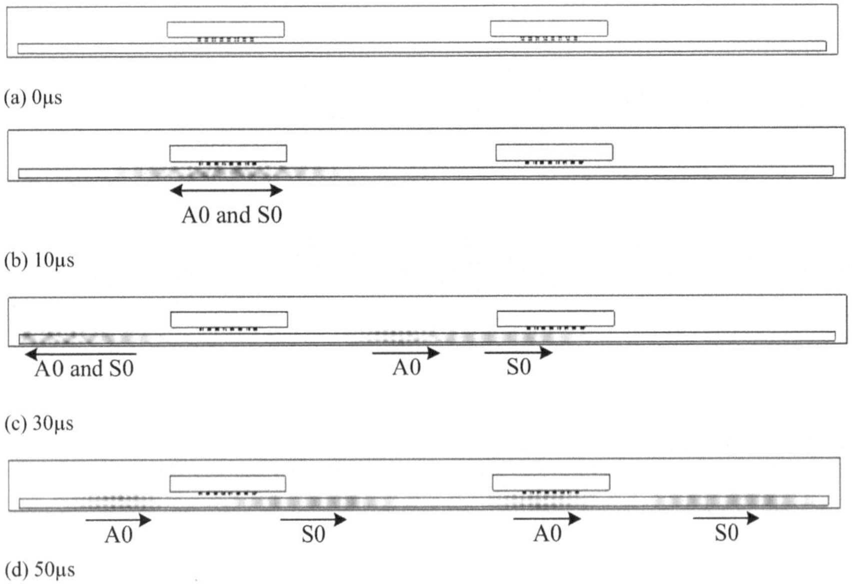

Fig. 2.22 Cloud distribution of x component of particle displacement

while the A0 and S0 mode ultrasonic wave propagating to the left has been reflected by the left end of the specimen and changed to propagate at the right side of the tested specimen.

When the ultrasonic wave reached the receiving probe, the motion of the specimen's endoplasmic reticulum produces a dynamic magnetic field under the bias magnetic field. The induced electromotive force is generated in the receiving coil to realize the signal reception. At the time of 50 μs, the equipotential line of the magnetic vector potential produced by the movement of the particle is shown in Fig. 2.23.

Figure 2.24 shows the normalized waveform of the signal received by the receiving coil within 120 μs after the excitation of the excitation probe. It can be seen

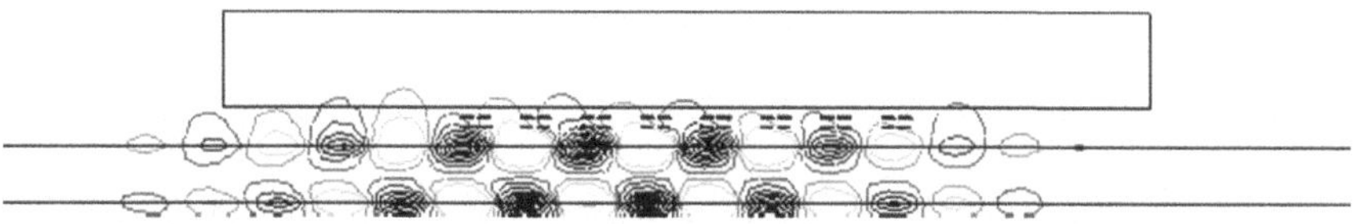

Fig. 2.23 Equipotential line of the magnetic vector potential produced by the movement of the particle in the specimen

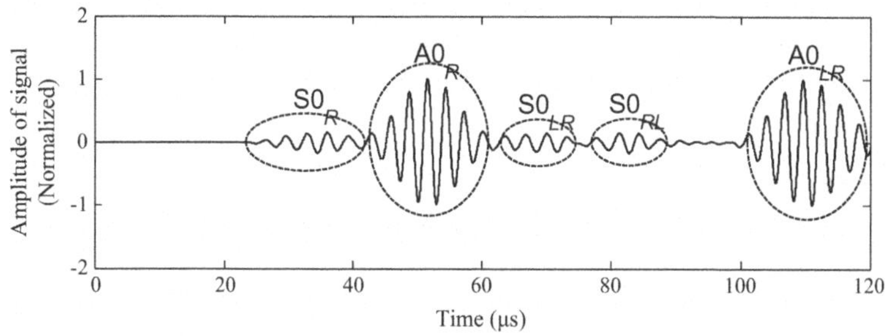

Fig. 2.24 Induction electromotive force of the receiving coil

from Fig. 2.24 that the receiving probe has received five packets at 120 μs. For the convenience of analysis, each wave packet is named according to the mode and direction of the ultrasonic wave: S_{0R}, A_{0R}, S_{0LR}, S_{0RL}, A_{0LR}, respectively. S_{0R} is the S0 mode Lamb wave that is propagating to the right side of the specimen; A_{0R} is the Lamb wave of the S0 mode that is propagating to the right side of the specimen; S_{0LR} represents the S0 mode Lamb wave that starts to propagate to the left side of the specimen and is reflected by the left end face and propagates to the right side of the specimen; S_{0RL} is a Lamb wave of S0 mode that starts to propagate to the right side of the specimen and is reflected by the right end face to the left side of the specimen., and A_{0LR} is the Lamb wave of the A0 mode that propagates to the left side of the tested specimen, and propagates to the right side of the specimen after the left side is reflected.

In physical experiments, parameters are carried out using EMAT, including the model parameters, and impulse excitation with the same frequency and number of cycles. In the experiment, the normalized waveform of the received signal is shown in Fig. 2.25. The ultrasonic wave of the physical test and the corresponding time of the wave packet are in agreement with the simulation results. This proves the correctness and validity of the numerical simulation analysis method based on the Lorenz force mechanism EMAT complete energy exchange process [5, 6].

According to the distance of the ultrasonic wave propagating in a specific time, the group velocity of the A0 mode and the Lamb wave in the S0 mode obtained by the numerical simulation and the physical test can be calculated, and the results are compared with the theoretical design values. The results are given in Table 2.3.

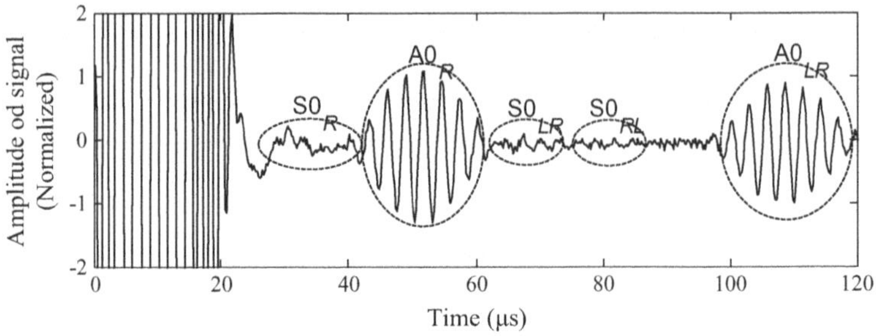

Fig. 2.25 Measured results of the receiving probe

Table 2.3 Comparison of wave velocity

Mode	Theoretical velocity (m/s)	Simulation velocity (m/s)	Experimental velocity (m/s)
S_0	3148	3102	3147
A_0	5044	5291	5263

As can be seen from Table 2.3, the theoretical, numerical simulation and the measured wave velocity have little difference, which further demonstrates the correctness and reliability of the numerical simulation results and experimental results.

2.1.3 Analytical Modeling and Calculation of Spiral Coil EMAT

A spiral coil is used to excite and receive ultrasonic body waves in the specimen. The structure of a typical body wave EMAT is shown in Fig. 2.19. Permanent magnets or electromagnets provide a static bias magnetic field perpendicular to the surface of the coil and the specimen, and the spiral coil exerts a suitable central frequency pulse excitation so that the eddy current of the specimen will be in the skin depth. The eddy current is induced to reverse the current of the coil. Under the bias magnetic field, the specimen is affected by the radial Lorenz force along the circle, and the ultrasonic body wave is propagating perpendicular to the surface of the specimen which is stimulated inside the specimen. The reception of the echo signal is contrary to the process mentioned above (Fig. 2.26).

In practical applications, spiral coils are made of the printed circuit board (PCB). The coils made by this kind of coils have many advantages, such as precise size, compact structure, and easy to use. At design time, the coil can be designed as a single- or double-layer structure as required, as shown in Figs. 2.27 and 2.28.

In solving the analytical solution, the following assumptions should be satisfied with the derivation of the model and formula.

(1) All the media in the solution domain are linear and isotropic homogeneous media.
(2) Ignore the skin effect and proximity effect in the copper foil coil; that is, the current density distribution in the copper foil coil is uniform everywhere.
(3) Neglect the effect of displacement current.
(4) The spiral coil is equivalent to the superposition of concentric circular coils.

Using the cylindrical coordinate system to calculate the frequency-domain solution of the magnetic field, the magnetic vector potential in each solution area has only circumferential components. The geometric model of the double-helix coil EMAT in cylindrical coordinates is shown in Fig. 2.29. The spiral coil can be regarded as a combined array of N concentric copper foil coils, whose radii are r_{11}, r_{21}, r_{12}, $r_{22}...r_{1N}$, r_{2N}. The coordinates of the upper and lower edges of the spiral conductor are labeled l_1, l_2, l_3, l_4, respectively. The thickness of the specimen is c. When the spiral coil is a single coil, remove the l_3, l_4 in the analytical expression. According to Fig. 2.29, the solution area can be divided into seven parts along the z-axis, marked as $a = 1, 2 \ldots 7$. Because the coil substrate has the same permeability and conductivity as air, it can be regarded as air treatment in the course of the derivation.

First, the magnetic vector potential (MVP) of the δ function coil is calculated, and the $\delta\,(r - r_0)\,(z - z_0)$ coil above the specimen is considered, as shown in Fig. 2.30; at this time, the number of the solution area is 4.

The frequency-domain differential equation satisfying the MVP of the δ function coil above the specimen can be expressed as

$$\left(\frac{\partial^2}{\partial r^2} + \frac{1}{r}\frac{\partial}{\partial r} + \frac{\partial^2}{\partial z^2} - \frac{1}{r^2} - j\omega\mu_a\sigma_a\right)A_a(\omega, r, z)$$
$$= -\mu_a i(\omega)\delta(r - r_0)\delta(z - z_0) \tag{2.82}$$

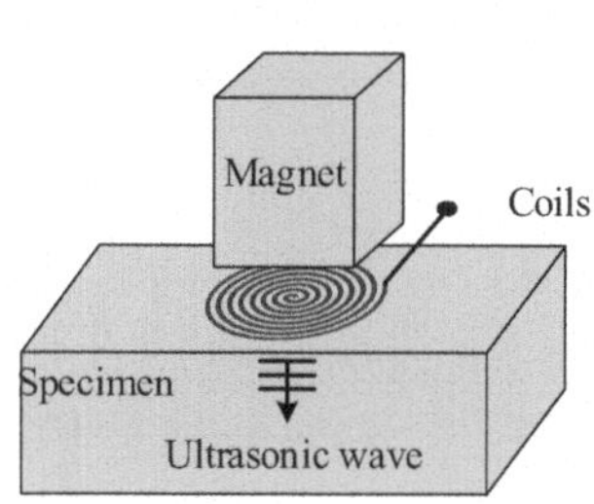

Fig. 2.26 EMAT structure: permanent magnet, spiral coil, and the specimen

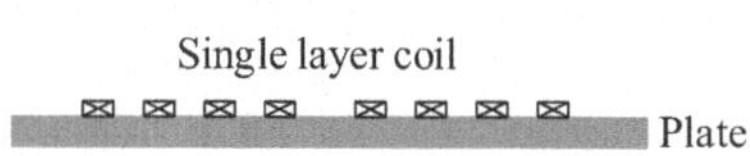

Fig. 2.27 Cross-sectional diagram of single-layer spiral coil

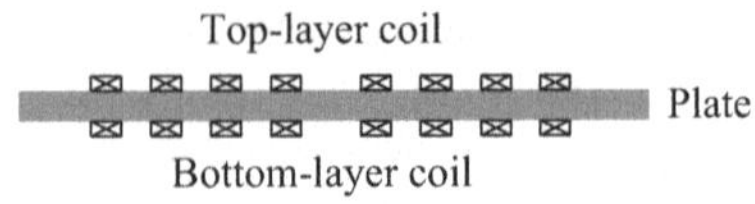

Fig. 2.28 Cross-sectional diagram of the double spiral coil

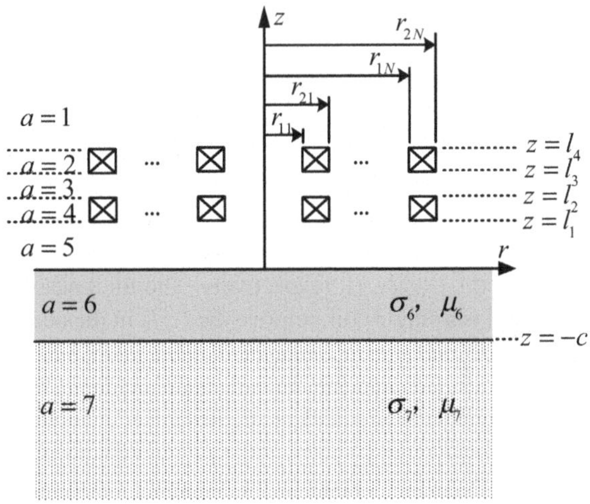

Fig. 2.29 Double spiral coil placed above the specimen

where A_a (ω, r, z) is MVP; i (ω) is the excitation current density; μ_a and σ_a are permeability and conductivity, respectively; a = 1, 2, 3, 4 represents four solution regions.

At the boundary of the adjacent solution region, the following boundary conditions are satisfied.

$$A_a(\omega, r, z_a) = A_{a+1}(\omega, r, z_a) \tag{2.83}$$

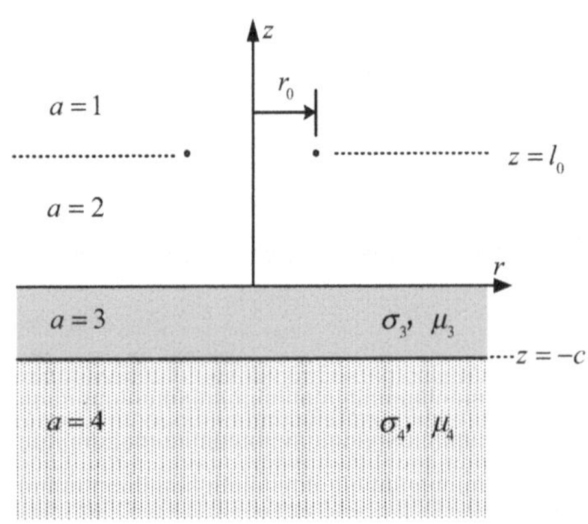

Fig. 2.30 δ function coil locates above the specimen

$$\left.\frac{1}{\mu_a}\frac{\partial A_a}{\partial z}\right|_{z=z_a} = \left.\frac{1}{\mu_{a+1}}\frac{\partial A_{a+1}}{\partial z}\right|_{z=z_a} - i(\omega)\delta(r-r_0)\delta(z-z_0) \tag{2.84}$$

By using the separation of variables method, the solution of each solution area is obtained.

$$A_1(r,z) = \frac{1}{2}\mu_0 i(\omega) r_0 \int_0^\infty J_1(kr_0)J_1(kr)\mathrm{e}^{-kl-kz}\left(\mathrm{e}^{2kl}+P_1\right)\mathrm{d}k \tag{2.85}$$

$$A_2(r,z) = \frac{1}{2}\mu_0 i(\omega) r_0 \int_0^\infty J_1(kr_0)J_1(kr)\mathrm{e}^{-kl}\left(\mathrm{e}^{kz}+P_1\mathrm{e}^{-kz}\right)\mathrm{d}k \tag{2.86}$$

$$A_3(r,z) = \mu_0 i(\omega) r_0 \int_0^\infty J_1(kr_0)J_1(kr)\mathrm{e}^{-kl}\left(P_2\mathrm{e}^{K_3 z}+P_3\mathrm{e}^{-K_3 z}\right)\mathrm{d}k \tag{2.87}$$

$$A_4(r,z) = \mu_0 i(\omega) r_0 \int_0^\infty J_1(kr_0)J_1(kr)\mathrm{e}^{-kl}P_4\mathrm{e}^{-K_4 z}\mathrm{d}k \tag{2.88}$$

where

$$K_a = \sqrt{k^2 + j\omega\mu_a\sigma_a}$$
$$P_1 = \frac{(k+K_3)(K_3-K_4)+(k-K_3)(K_4+K_3)\mathrm{e}^{2K_3c}}{(k-K_3)(K_3-K_4)+(k+K_3)(K_4+K_3)\mathrm{e}^{2K_3c}}$$
$$P_2 = \frac{k(K_3+K_4)\mathrm{e}^{2K_3c}}{(k-K_3)(K_3-K_4)+(k+K_3)(K_4+K_3)\mathrm{e}^{2K_3c}}$$
$$P_3 = \frac{k(K_3-K_4)}{(k-K_3)(K_3-K_4)+(k+K_3)(K_4+K_3)\mathrm{e}^{2K_3c}}$$
$$P_4 = \frac{2K_3k\mathrm{e}^{(K_3+K_4)c}}{(k-K_3)(K_3-K_4)+(k+K_3)(K_4+K_3)\mathrm{e}^{2K_3c}}$$

$J_1(x)$ is the first-class, first-order Bessel functions.

First, consider the situation of a single ring rectangular section. The MVP of the whole coil can be obtained by superposition method. For the single ring rectangular cross-sectional coils with a radius of r_{11}, r_{21} and height of l_1, l_2, the MVP in each region of the solution can be obtained by the MVP integral of the δ function coil.

$$\begin{aligned} A_\mathrm{a}(\omega,r,z) &= \int_S A_\mathrm{a}(\omega,r,z,r_0,l_0)\mathrm{d}s \\ &= \int_{r_1}^{r_2}\int_{l_1}^{l_2} A_\mathrm{a}(\omega,r,z,r_0,l_0)\mathrm{d}r_0\mathrm{d}l \end{aligned} \tag{2.89}$$

where A_a (ω, r, z, r_0, l_0) is the MVP of the δ function coil.

By calculating the integral, the MVP of the single ring rectangular section coil can be obtained.

$$A_{1,2,3}^{c}(\omega, r, z) = \frac{1}{2}\mu_0 i(\omega) \int_0^{\infty} \frac{1}{k^3} I(kr_1, kr_2) J_1(kr) \mathrm{e}^{-kz} \left[\mathrm{e}^{kl_2} - \mathrm{e}^{kl_1} - P_1\left(\mathrm{e}^{-kl_2} - \mathrm{e}^{-kl_1}\right)\right]\mathrm{d}k \tag{2.90}$$

$$A_{5}^{c}(\omega, r, z) = \frac{1}{2}\mu_0 i(\omega) \int_0^{\infty} \frac{1}{k^3} I(kr_1, kr_2) J_1(kr) \left(\mathrm{e}^{-kl_1} - \mathrm{e}^{-kl_2}\right)\left(\mathrm{e}^{kz} + P_1 \mathrm{e}^{-kz}\right)\mathrm{d}k \tag{2.91}$$

$$A_{6}^{c}(\omega, r, z) = \mu_0 i(\omega) \int_0^{\infty} \frac{1}{k^3} I(kr_1, kr_2) J_1(kr) \left(\mathrm{e}^{-kl_1} - \mathrm{e}^{-kl_2}\right) \left(P_2 \mathrm{e}^{K_1 z} + P_3 \mathrm{e}^{-K_1 z}\right)\mathrm{d}k \tag{2.92}$$

$$A_{7}^{c}(\omega, r, z) = \mu_0 i(\omega) \int_0^{\infty} \frac{1}{k^3} J(kr_1, kr_2) J_1(kr) \left(\mathrm{e}^{-kl_1} - \mathrm{e}^{-kl_2}\right) P_4 \mathrm{e}^{K_2 z}\mathrm{d}k \tag{2.93}$$

Superscript c represents the single ring rectangular cross section coil.

$$\begin{aligned} I(x_1, x_2) &= \int_{x_1}^{x_2} x J_1(x)\mathrm{d}x \\ &= \frac{\pi}{2}\{x_2[J_0(x_2)H_1(x_2) - J_1(x_2)H_0(x_2)] \\ &\quad - x_1[J_0(x_1)H_1(x_1) - J_1(x_1)H_0(x_1)] \end{aligned} \tag{2.94}$$

H_n represents the Struve function.

The MVP in zone 4 can be obtained by substituting $l_2 = z$ and $l_1 = z$ to (2.64) and (2.65) and adding them together.

$$A_{4}^{c}(\omega, r, z) = \frac{1}{2}\mu_0 i(\omega) \int_0^{\infty} \frac{1}{k^3} I(kr_1, kr_2) J_1(kr) \left[2 - \mathrm{e}^{k(z-l_2)} - \mathrm{e}^{-k(z-l_1)} + P_1\left(\mathrm{e}^{-kl_1} - \mathrm{e}^{-kl_2}\right)\mathrm{e}^{-kz}\right]\mathrm{d}k \tag{2.95}$$

By the MVP of the single loop rectangular section coil, the MVP of the other $n - 1$ single ring rectangular section coils can also be obtained similarly. By adding all the MVP of the single ring rectangle section coil, the MVP of the lower coil can be obtained.

$$A_{\mathrm{a}}^{l}(\omega, r, z) = \sum_{i}^{N} A_{\mathrm{a}}^{lci}(\omega, r, z) \tag{2.96}$$

Superscript l represents of the lower coil. In the same way, l_1 and $l\,2$ are replaced by $l\,3$ and $l\,4$, and the MVP of the upper coil can be obtained.

$$A_{\mathrm{a}}^{u}(\omega, r, z) = \sum_{i}^{N} A_{\mathrm{a}}^{uci}(\omega, r, z) \tag{2.97}$$

The superscript u represents the upper coil.

So, the MVP of the double coils is

$$A_{a}^{d}(\omega, r, z) = A_{a}^{l}(\omega, r, z) + A_{a}^{u}(\omega, r, z) \tag{2.98}$$

The superscript d represents a double-layer coil.

The derivation can be obtained

$$\begin{aligned} A_{6}^{d}(\omega, r, z) = \mu_0 i(\omega) \int_0^{\infty} \frac{1}{k^3} \sum_{i}^{N} I(kr_{1i}, kr_{2i}) J_1(kr) \big(\mathrm{e}^{-kl_1} - \mathrm{e}^{-kl_2} \\ + e^{-kl_3} - \mathrm{e}^{-kl_4}\big)\big(P_2 \mathrm{e}^{K_1 z} + P_3 \mathrm{e}^{-K_1 z}\big) \mathrm{d}k \end{aligned} \tag{2.99}$$

$$\begin{aligned} A_{2}^{d}(\omega, r, z) = \frac{1}{2}\mu_0 i(\omega) \int_0^{\infty} \frac{1}{k^3} \sum_{i}^{N} I(kr_{1i}, kr_{2i}) J_1(kr) [2 \\ - \mathrm{e}^{k(z-l_4)} - \mathrm{e}^{-k(z-l_3)} + \mathrm{e}^{-k(z-l_2)} - \mathrm{e}^{-k(z-l_1)} \\ + P_1 \big(\mathrm{e}^{-kl_1} - \mathrm{e}^{-kl_2} + \mathrm{e}^{-kl_3} - \mathrm{e}^{-kl_4}\big) \mathrm{e}^{-kz} \big] \mathrm{d}k \end{aligned} \tag{2.100}$$

$$\begin{aligned} A_{4}^{d}(\omega, r, z) = \frac{1}{2}\mu_0 i(\omega) \int_0^{\infty} \frac{1}{k^3} \sum_{i}^{N} I(kr_{1i}, kr_{2i}) J_1(kr) [2 \\ - \mathrm{e}^{k(z-l_2)} - \mathrm{e}^{-k(z-l_1)} + \mathrm{e}^{k(z-l_3)} - \mathrm{e}^{k(z-l_4)} \\ + P_1 \big(\mathrm{e}^{-kl_1} - \mathrm{e}^{-kl_2} + \mathrm{e}^{-kl_3} - \mathrm{e}^{-kl_4}\big) \mathrm{e}^{-kz}] \mathrm{d}k \end{aligned}$$

The dynamic magnetic induction intensity in the specimen can be calculated by:

$$\mathbf{B} = \nabla \times \boldsymbol{A} \tag{2.101}$$

Because A contains only the φ component, B will contain r components and z components. By derivation, there are

$$B_r(\omega, r, z) = -\mu_0 i(\omega) \int_0^{\infty} \frac{1}{k^3} \sum_i^N I(kr_{1i}, kr_{2i}) J_1(kr) \left(e^{-kl_1} - e^{-kl_2}\right.$$
$$\left. + e^{-kl_3} - e^{-kl_4}\right)\left(P_2 K_1 e^{K_1 z} - P_3 K_1 e^{-K_1 z}\right) dk \tag{2.102}$$

$$B_z(\omega, r, z) = \mu_0 i(\omega) \int_0^{\infty} \frac{1}{k^2} \sum_i^N I(kr_{1i}, kr_{2i}) J_0(kr) \left(e^{-kl_1} - e^{-kl_2}\right.$$
$$\left. + e^{-kl_3} - e^{-kl_4}\right)\left(P_2 e^{K_1 z} + P_3 e^{-K_1 z}\right) dk \tag{2.103}$$

In the skin depth of the specimen, the pulsed eddy current is induced. According to the calculation relation between the eddy current and the MVP, that is, $\boldsymbol{J} = -j\omega\sigma\boldsymbol{A}$, Formula (2.72) is replaced by the upper formula, and the pulsed eddy current can be obtained.

$$J_e(\omega, r, z) = -j\omega\sigma_6\mu_0 i(\omega) \int_0^{\infty} \frac{1}{k^3} \sum_i^N I(kr_{1i}, kr_{2i}) J_1(kr) \left(e^{-kl_1}\right.$$
$$\left. - e^{-kl_2} + e^{-kl_3} - e^{-kl_4}\right)\left(P_2 e^{K_1 z} + P_3 e^{-K_1 z}\right) dk \tag{2.104}$$

Before calculating the input impedance, the induction electromotive force of the coil must first be calculated. Induction electromotive force in a rectangular cross-sectional coil is

$$V(\omega) = \frac{j\omega 2\pi}{\text{coil cross section}} \iint_{\text{coil cross section}} rA(\omega, r, z) \tag{2.105}$$

The induction electromotive force of the double coil can be obtained by accumulating the induction electromotive force of the single ring rectangle section coil.

$$V^d(\omega) = \sum_i^N \left[\frac{j\omega 2\pi}{(l_2 - l_1)(r_{2i} - r_{1i})} \int_{l_1}^{l_2} \int_{r_{1i}}^{r_{2i}} rA_2^d(\omega, r, z)\, dr dz \right.$$
$$\left. + \frac{j\omega 2\pi}{(l_4 - l_3)(r_{2i} - r_{1i})} \int_{l_3}^{l_4} \int_{r_{1i}}^{r_{2i}} rA_4^d(\omega, r, z) \right] dr dz \tag{2.106}$$

According to Ohm's law, the expression of the input impedance of the coil can be obtained.

$$
\begin{aligned}
Z(\omega) = Z_0 + \sum_{i=1}^{N} \Bigg\{ & \frac{j\omega\mu_0 2\pi}{(l_2 - l_1)^2 (r_{2i} - r_{1i})^2} \int_0^\infty \frac{1}{k^6} I(kr_{1i}, kr_{2i}) \sum_{i=1}^{N} I(r_{1i}, r_{2i}) \\
& \left[2k(l_2 - l_1) + 2\left(e^{-k(l_2 - l_1)} - 1\right) + \left(e^{-kl_1} - e^{-kl_2}\right)\left(e^{kl_4} - e^{kl_3}\right)\right. \\
& \left. + P_1\left(e^{-kl_1} - e^{-kl_2} + e^{-kl_3} - e^{-kl_4}\right)\left(e^{-kl_1} - e^{-kl_2}\right)\right] dk \\
& + \frac{j\omega\mu_0 2\pi}{(l_4 - l_3)^2 (r_{2i} - r_{1i})^2} \int_0^\infty \frac{1}{k^6} I(r_{1i}, r_{2i}) \sum_{i=1}^{N} I(r_{1i}, r_{2i}) [2k(l_4 - l_3) \\
& + 2\left(e^{-k(l_4 - l_3)} - 1\right) + \left(e^{-kl_1} - e^{-kl_2}\right)\left(e^{kl_4} - e^{kl_3}\right) \\
& + P_1\left(e^{-kl_3} - e^{-kl_4} + e^{-kl_1} - e^{-kl_2}\right)\left(e^{-kl_3} - e^{-kl_4}\right)] dk \Bigg\}
\end{aligned} \tag{2.107}
$$

DC impedance Z_0 is

$$
Z_0 = \frac{1}{\sigma_6} \sum_{i=1}^{N} \left[\frac{\pi(r_{2i} + r_{1i})}{(r_{2i} - r_{1i})(l_2 - l_1)} + \frac{\pi(r_{2i} + r_{1i})}{(r_{2i} - r_{1i})(l_4 - l_3)} \right] \tag{2.108}
$$

In the analytical expressions of magnetic induction intensity, pulsed eddy current, and input impedance, the calculation is quite difficult because of the existence of the double infinite integral of the first-order Bessel function. In order to reduce the calculation difficulty and reduce the calculation time, the expression of the magnetic induction intensity, the pulse eddy current, and the input impedance integral form is approximated by the regional feature function truncation (TREE). This truncation processing will cause some calculation errors, but the error range is easier to control.

By choosing the solution region $0 \le r \le R$ instead of $0 \le r \le \infty$, the infinite integral problem is converted to the summation problem of finite series. When R is large enough, the result will be close to the true value. According to the physical meaning of MVP, it can be satisfied at the boundary.

$$
|A|_{r<R}| < \infty \tag{2.109}
$$

$$
A|_{r=R} = 0 \tag{2.110}
$$

Then, the general solution of MVP can be expressed as

$$
A_n(\omega, r, z) = \sum_{m=1}^{\infty} \left(A_m(k_m) e^{K_n z} + B_m(k_m) e^{-K_n z}\right) C_m(k_m) J_1(k_m r) \tag{2.111}
$$

And

$$
J_1(k_m R) = 0 \tag{2.112}
$$

It can be seen that $k_m R$ is the m solution x_m of J1 $J_1(x)$,

$$k_m = \frac{x_m}{R} \tag{2.113}$$

The expressions of magnetic induction intensity, pulsed eddy current, and input impedance series are obtained.

$$\begin{aligned} B_{\mathrm{r}} = -2\mu_0 i(\omega) \sum_{i=1}^{N} \sum_{m=1}^{\infty} \frac{I(kr_{1i}, kr_{2i}) J_1(k_m r)}{k_m^4 R^2 J_0^2(k_m R)} \big(\mathrm{e}^{-k_m l_1} \\ - \mathrm{e}^{-k_m l_2} + \mathrm{e}^{-k_m l_3} - \mathrm{e}^{-k_m l_4}\big)\big(P_{2m} K_{1m} \mathrm{e}^{K_{1m} z} - P_{3m} K_{1m} \mathrm{e}^{-K_{1m} z}\big) \end{aligned} \tag{2.114}$$

$$\begin{aligned} B_z(\omega, r, z) = 2\mu_0 i(\omega) \sum_{i=1}^{N} \sum_{m=1}^{\infty} \frac{I(kr_{1i}, kr_{2i}) J_0(k_m r)}{k_m^3 R^2 J_0^2(k_m R)} \big(\mathrm{e}^{-kl_1} \\ -\mathrm{e}^{-kl_2} + \mathrm{e}^{-kl_3} - \mathrm{e}^{-kl_4}\big)\big(P_{2m} \mathrm{e}^{K_1 z} + P_{3m} \mathrm{e}^{-K_1 z}\big) \end{aligned} \tag{2.115}$$

$$\begin{aligned} J_e(\omega, r, z) = -2j\omega\sigma_6\mu_0 i(\omega) \sum_{i=1}^{N} \sum_{m=1}^{\infty} \frac{I(kr_{1i}, kr_{2i}) J_1(k_m r)}{k_m^4 R^2 J_0^2(k_m R)} \big(\mathrm{e}^{-k_m l_1} \\ -\mathrm{e}^{-k_m l_2} + \mathrm{e}^{-k_m l_3} - \mathrm{e}^{-k_m l_4}\big)\big(P_{2m} \mathrm{e}^{K_{1m} z} + P_{3m} \mathrm{e}^{-K_{1m} z}\big) \end{aligned} \tag{2.116}$$

$$\begin{aligned} Z(\omega) = Z_0 + \sum_{i=1}^{N} \sum_{m=1}^{\infty} \Bigg\{ \frac{j2\pi\omega\mu_0 I(r_{1i}, r_{2i}) \sum_{i=1}^{N} I(kr_{1i}, kr_{2i})}{(l_2 - l_1)^2 (r_{2i} - r_{1i})^2 k_m^7 R^2 J_0^2(k_m R)} [2(l_2 - l_1) \\ + 2\big(\mathrm{e}^{-k(l_2 - l_1)} - 1\big) + \big(\mathrm{e}^{-kl_1} - \mathrm{e}^{-kl_2}\big)\big(\mathrm{e}^{kl_4} - \mathrm{e}^{kl_3}\big) \\ + P_{1m}\big(\mathrm{e}^{-kl_1} - \mathrm{e}^{-kl_2} + \mathrm{e}^{-kl_3} - \mathrm{e}^{-kl_4}\big)\big(\mathrm{e}^{-kl_1} - \mathrm{e}^{-kl_2}\big)] \\ + \frac{j2\pi\omega\mu_0 I(r_{1i}, r_{2i}) \sum_{i=1}^{N} I(kr_{1i}, kr_{2i})}{(l_4 - l_3)^2 (r_{2i} - r_{1i})^2 k_m^7 R^2 J_0^2(k_m R)} [2(l_4 - l_3) \\ + 2\big(\mathrm{e}^{-k(l_4 - l_3)} - 1\big) + \big(\mathrm{e}^{-kl_1} - \mathrm{e}^{-kl_2}\big)\big(\mathrm{e}^{kl_4} - \mathrm{e}^{kl_3}\big) \\ + P_{1m}\big(\mathrm{e}^{-kl_3} - \mathrm{e}^{-kl_4} + \mathrm{e}^{-kl_1} - \mathrm{e}^{-kl_2}\big)\big(\mathrm{e}^{-kl_3} - \mathrm{e}^{-kl_4}\big)]\Bigg\} \end{aligned}$$

The impulse excitation and response can be expressed as a series of sinusoidal signals with different frequencies. On the premise of known magnetic induction intensity and pulse current frequency-domain analytic expression, Fourier inverse transform can be used to obtain the time-domain response.

Time-domain expression is

$$B_{\mathrm{r}}(t, k, z) = \frac{1}{2\pi} \int_{-\infty}^{\infty} B_{\mathrm{r}}(\omega, k, z) \mathrm{e}^{\mathrm{j}\omega t} \mathrm{d}\omega \tag{2.117}$$

$$B_z(t, r, z) = \frac{1}{2\pi} \int_{-\infty}^{\infty} B_z(\omega, r, z) \mathrm{e}^{\mathrm{j}\omega t} \mathrm{d}\omega \tag{2.118}$$

$$J_e(t, r, z) = \frac{1}{2\pi} \int_{-\infty}^{\infty} J_e(\omega, r, z) e^{j\omega t} d\omega \tag{2.119}$$

It is difficult to obtain B_r (ω, r, z), B_z (ω, r, z) and J_e (ω, r, z) by directly solving analytic integral. In this regard, the adoption of FFT-IFFT method is an effective choice. The concrete steps of this method are as follows: First, the time-domain current excitation signal is converted to the frequency domain by FFT, and the magnetic induction intensity and the pulsed eddy current value at different frequencies are calculated. Finally, the time-domain signals of magnetic induction intensity and pulsed eddy current are calculated by using IFFT.

When the pulsed eddy current is known in the skin depth of the specimen, the Lorenz force in the specimen can be calculated by the following function.

$$\boldsymbol{f}_L(t, r, z) = \boldsymbol{B}_0 \times \boldsymbol{J}_e(t, r, z) \tag{2.120}$$

$\boldsymbol{f}_L(t, r, z)$ represents Lorenz force; $\boldsymbol{B}_0$ is the static magnetic induction intensity.

In order to verify the effectiveness of the analytical and computational methods, a calculation example is given, and the results are compared with the TSFEM method and the experimental results.

Figure 2.31 presents a test model for double PCB spiral coils, measured aluminum plates, and 0.5 mm. The size and material parameters of the coil and the tested aluminum plate are shown in Tables 2.4 and 2.5.

Generally speaking, the excitation signal of EMAT is a narrowband or modulated tone-burst signal, which is generated by the signal generator and amplified by the power amplifier. Modulation of tone-burst signals can be characterized

$$i(t) = \begin{cases} I_0\left(1 - \cos\frac{\omega t}{n}\right)\cos\omega t & 0 \le t \le \frac{2n\pi}{\omega} \\ 0 & t \ge \frac{2n\pi}{\omega} \end{cases} \tag{2.121}$$

Here, $\omega = 2\pi f$, select $I_0 = 20$ A, $n = 2$, $f = 1$ MHz; the excitation signal waveform is shown in Fig. 2.32.

Fig. 2.31 Double PCB spiral coils located above the aluminum plate specimen

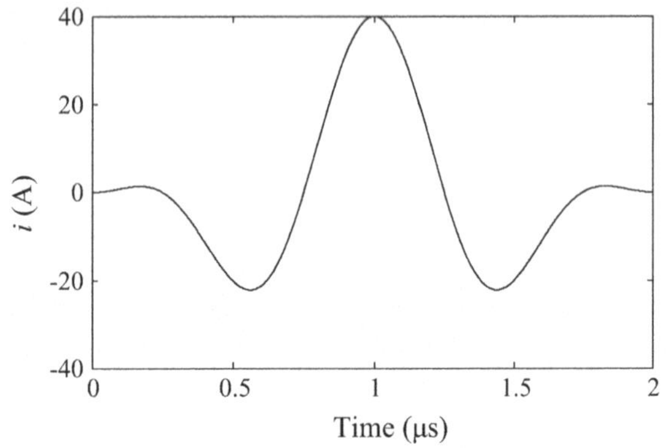

Fig. 2.32 Waveform of tone-burst current signal applied to the spiral coil

The derived analytical expressions of magnetic induction intensity, pulsed eddy current and input impedance and the calculation method proposed are calculated. When calculating, the cutoff radius is $R = 696$ mm, and the summation times are 50. The sampling frequency of FFT-IFFT is 32 MHz, and the number of data is 64. Subsequently, impedance analyzer is used to measure the input impedance of the coil to verify the correctness of the impedance calculation results. The time-domain response of magnetic induction intensity and a pulsed eddy current is verified by the time-step finite element method (TSFEM). In order to improve the computational efficiency, the mesh area of the ring section needs to mesh adequately. The number of grids is 13,193.

Table 2.4 Width and material parameters of the double-helix coil

Parameter	Value
Substrate thickness	0.500 mm
Copper foil width	0.254 mm
Copper foil thickness	0.050 mm
r_{11}	2.921 mm
r_{21}	3.175 mm
Conductor ring spacing	0.254 mm
The number of turns	29
The magnetic permeability of copper foil	$4\pi \times 10^{-7}$ H/m
The electrical conductivity of copper foil	2.667×10^{7} S/m

Table 2.5 Size and material parameters of the tested aluminum sheet

Parameter	Value
Thickness of aluminum plate	5.00 mm
Magnetic permeability	$4\pi \times 10^{-7}$ H/m
Conductivity	3.571×10^{-7} S/m

In order to achieve more accurate circuit matching to improve the transmission efficiency of ultrasonic power, it is necessary to calculate the input impedance of the coil in the working state. The impedance values of coils at 0.5, 1, and 1.5 MHz were calculated, and physical tests were carried out.

The specific results obtained are given in Table 2.6. It is clear that the calculated results are in good agreement with the experimental results, which validates the validity and correctness of the impedance analytical expressions and the TREE method.

The transient time-domain waveforms of B_r, B_z, and J_e are calculated by the analytical method and TSFEM method presented in this paper (Figs. 2.33, 2.34 and 2.35).

The calculated results of a given method are in agreement with the results of the TEFEM method, and the specific difference lies in the following two points: (1) The

Table 2.6 Numerical calculation and experimental measurement of input impedance

Frequency (MHz)	Calculated value (Ω)	Measurement value (Ω)
0.5	12.51 + j30.64	13.03 + j31.34
1	13.19 + j60.34	13.89 + j63.25
1.5	13.71 + j89.87	14.75 + j93.80

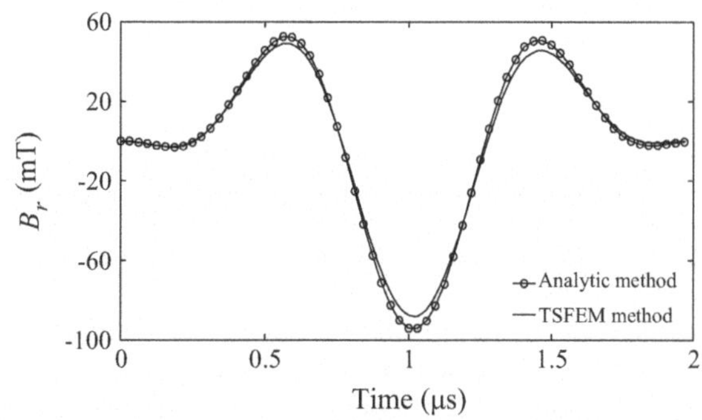

Fig. 2.33 Magnetic induction intensity in the *r*-direction near the surface of the specimen (3.048, −0.01 mm)

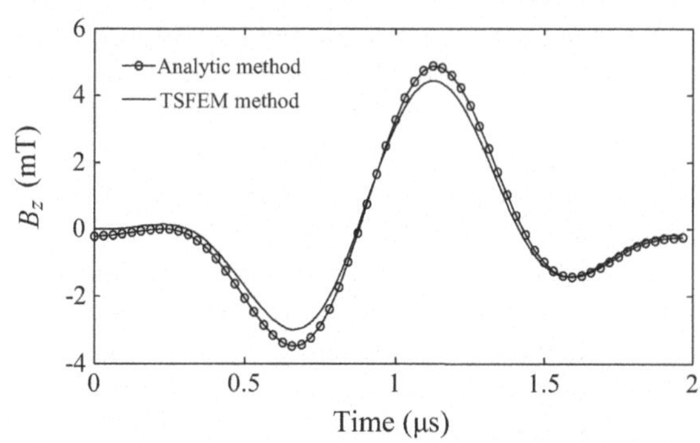

Fig. 2.34 Magnetic induction intensity in the *z*-direction near the surface of the specimen (3.048, −0.01 mm)

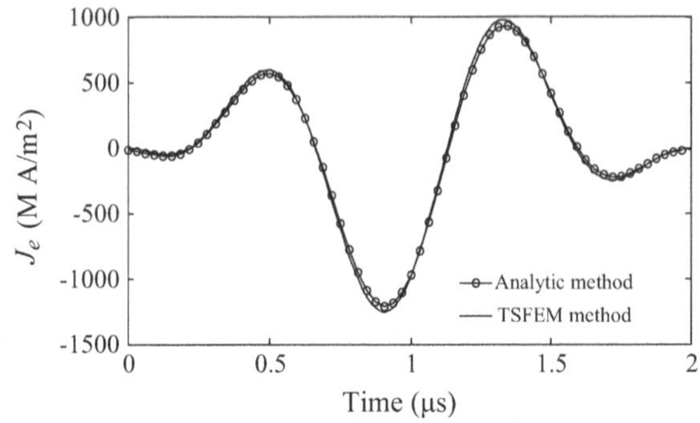

Fig. 2.35 Eddy density at the coordinates near the surface of the specimen (3.048, −0.01 mm)

Table 2.7 Error of peak and peak time under two methods

Error	B_r	B_z	J_e
Peak value (%)	7.4	9.7	3.1
Peak time (%)	0.78	1.3	0.41

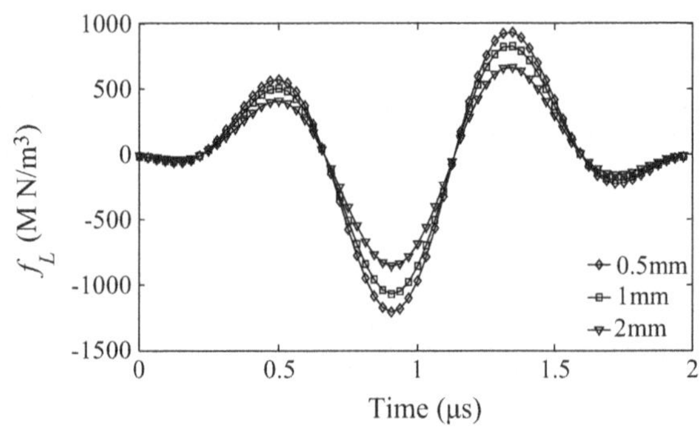

Fig. 2.36 Lorenz force density at the coordinates near the test surface (3.048, −0.01) at 0.5, 1, and 2 mm, respectively

calculated B_r and B_z given by a given method are slightly larger than those calculated by the TSFEM method, while J_e is the opposite; (2) the peak time of the waveform has a deviation. In order to evaluate the error of peak and peak time, the results are shown in Table 2.7.

Assuming that the static bias field is uniformly distributed and the value is 1T, the Lorenz force in the specimen is calculated under different liftoff values.

From Fig. 2.36, we find that the Lorenz force decreases with the increase of liftoff value. In the case of known Lorenz force, the displacement of the particle in the specimen can be calculated according to the elastic theory, and then the ultrasonic wave excited by the tested sample is obtained.

Although the analytical expressions of magnetic induction intensity, pulsed eddy current, and input impedance are more complex, the TREE method and the FFT-IFFT method can be easily and accurately calculated. Compared with the TEFEM method, the computation method is simple, the computation time is short, and the physical meaning is clear.

The following reasons can lead to a deviation between the calculated results and the test results: The skin effect and the proximity effect of the coils are ignored in the analytical model, and in this way, the error of the high frequency is greater, and the TREE method itself also brings a certain amount of calculation error.

2.1.4 Analytical Modeling and Calculation of Meander-Line-Coil EMAT

In the use of EMAT's nondestructive testing, with different static bias magnetic fields, the meander line coil can be used to stimulate and receive Lamb wave [7–10], SH wave [11], Rayleigh wave, and so on in the specimen. This section analyzes the modeling and calculation of the EMAT of the meander-line-coil structure.

The meander line coil used to make EMAT is mainly made by PCB and FPC technology to improve the manufacturing accuracy of coil size and facilitate practical application. The typical structure of the meander line coil is shown in Fig. 2.37.

The purpose of multisplitting is to increase the number of turns per coil. In order to achieve the same goal, the coil is sometimes designed to be a sandwich-like structure. The isometric or non-equidistant structure is used to control the ultrasonic mode excited by controlling the distance between the coils.

When the coil is applied to the specimen, the coil must be placed on it. The cross-sectional model, which ignores the end connection of the coil, considering the multilayer, multi splitting, and un-equidistant structure over the test sample, is shown in Fig. 2.31. The coil is an M-layer N-split coil, and the total coil number is $2q$. The

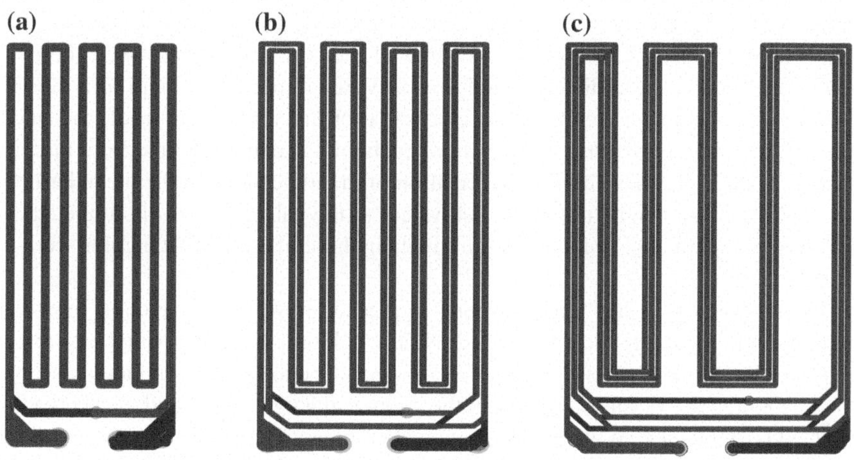

Fig. 2.37 Meander-line-coil structure: **a** equidistant single split coil; **b** equidistant multisplitting coil; **c** unequal distance multisplitting coil

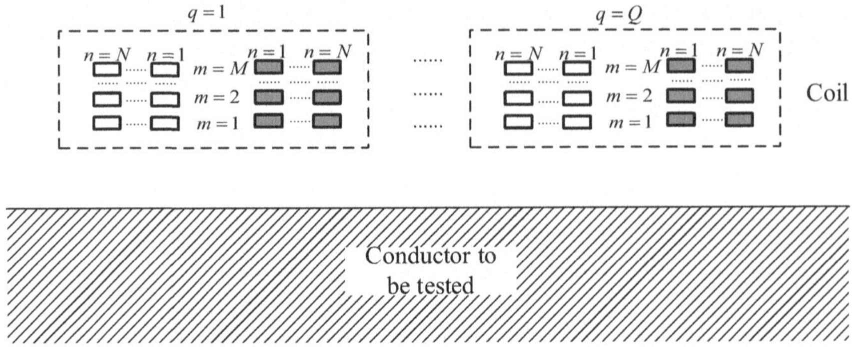

Fig. 2.38 General model of meander line coil

Fig. 2.39 Rectangle delta coil placed above infinite medium

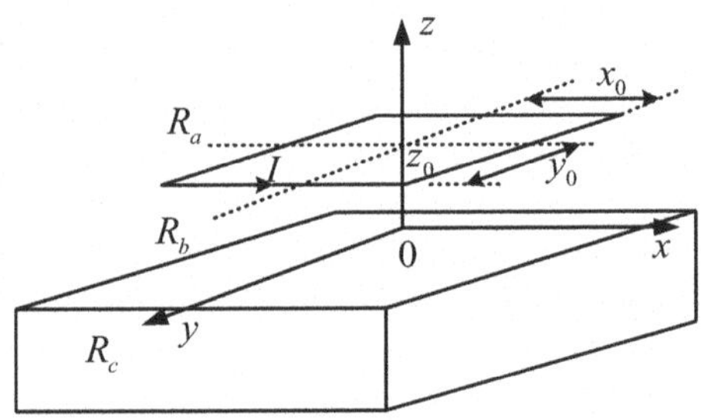

model is a general model of the meander line coil, and the equidistant structure coil is a special example of the model (Fig. 2.38).

According to the characteristics of the meander line structure, the current direction in the 80% off adjacent lines of the coil is opposite. Therefore, it is possible to approximate the two adjacent 80% off lines of the coil in the same layer, the two conductors on the same division, and a short wire connected to them to a closed loop, in which there will be Q similar returns in each layer. Since the whole meander line coil is a series structure, the current of each loop is equal after the excitation is applied to the coil. Therefore, the calculation of the impedance of the meander line coil and the pulse magnetic field is equivalent to the calculation of the impedance and pulse magnetic field of a single turn rectangular coil with a rectangular section of $M \times N \times Q$.

The rectangular δ coil with current I is placed above the conduction medium in the infinite region as shown in Fig. 2.39. The solution space is divided into three regions: The area above the coil R_a, the R_b between the coil and the conductor, and the conductor area R_c. x_0, and y_0 represent the width and length of the coil, respectively.

In NDT implemented by an electromagnetic ultrasonic method, the frequency of the excitation current is generally less than 10 MHz, and the displacement current and velocity effect can be neglected. In this way, the governing equation of time-harmonic electromagnetic field can be expressed as

$$\nabla^2 \boldsymbol{A} + \gamma^2 \boldsymbol{A} = -\mu \boldsymbol{J}_s \tag{2.122}$$

In the formula, $\boldsymbol{A}$ is the magnetic vector potential; J_s is the current source density; $\gamma_2 = -j\omega\mu\sigma$; ω is angular frequency; μ is medium permeability; σ is the medium conductivity; ∇^2 is the Laplasse operator.

According to the Coulomb criterion $\nabla \cdot \boldsymbol{A} = 0$, there is a two-order vector $\boldsymbol{W}$, so that $\boldsymbol{A} = \nabla \times \boldsymbol{W}$, and $\boldsymbol{W}$ can be decomposed into two vertical components; each component can be derived from one potential function; that is, $\boldsymbol{W}$ can be expressed as

$$\boldsymbol{W} = \boldsymbol{u} W_1 + \boldsymbol{u} \times \nabla W_2 \tag{2.123}$$

In the formula, $\boldsymbol{W}_1$ and $\boldsymbol{W}_2$ are the potential; $\boldsymbol{u}$ is the vector that satisfies the requirement.

The vector $\boldsymbol{W}$ is used to represent the magnetic vector potential $\boldsymbol{A}$

$$\boldsymbol{A} = \nabla \times \left(W_1 \hat{z} + \hat{z} \times \nabla W\right)_2 \tag{2.124}$$

$$\nabla \times \left[\hat{\mathbf{z}} \nabla^2 W_1 + \hat{\mathbf{z}} \times \nabla\left(\nabla^2 W_2\right) + \gamma^2\left(W_1 \hat{\mathbf{z}} + \hat{\mathbf{z}} \times \nabla W_2\right)\right] = -\mu \mathbf{J}_s \tag{2.125}$$

In the air region R_a and R_b, the source current and conductivity are 0, so there is

$$\nabla^2 W_{1(a,b)} = 0 \tag{2.126}$$

$$\nabla^2 W_{2(a,b)} = 0 \tag{2.127}$$

$$\boldsymbol{A}_{(a,b)} = \nabla \times \left(W_{1(a,b)} \hat{z}\right) - (\hat{z} \cdot \nabla) \nabla W_{2(a,b)} \tag{2.128}$$

In the formula, the subscript (a, b) represents the quantity in the region a and b, and the following is the same.

Therefore,

$$\boldsymbol{B}_{(a,b)} = \nabla \left(\frac{\partial W_{1(a,b)}}{\partial z} \right) \tag{2.129}$$

In the calculation, only the $\boldsymbol{W}_{1(a,b)}$ can be calculated. $\boldsymbol{W}_{1(a,b)}$ can be expressed as the superposition of the $\boldsymbol{W}_{1(a,b)s}$ generated by the coil and the $\boldsymbol{W}_{1(a,b)e}$ generated by the eddy in the conductor.

$$W_{1(a,b)} = W_{1(a,b)s} + W_{1(a,b)e} \tag{2.130}$$

Then

$$\nabla^2 \left(W_{1(a,b)s} + W_{1(a,b)e} \right) = 0 \tag{2.131}$$

In the conductor region R_c, according to the boundary condition of the interface, $W_2(c)$ must be satisfied at the $z = 0$ interface.

$$\begin{cases} \frac{\partial W_{2(c)}}{\partial z} = 0 \\ W_{2(c)} = 0 \end{cases} \tag{2.132}$$

In the formula, the subscript (c) represents the variables in the c region.
Because there is no source in the conductor area, there must be

$$\boldsymbol{A}_{(c)} = \nabla \times \left(W_{1(c)} \hat{z} \right) \tag{2.133}$$

$$\nabla^2 W_{1(c)} + \gamma^2 W_{1(c)} = 0 \tag{2.134}$$

The general solution can be expressed as a form of double Fourier transform.

$$W_{1(b)s} = \int_{-\infty}^{+\infty} \int_{-\infty}^{+\infty} C(\alpha, \beta) \mathrm{e}^{kz} \mathrm{e}^{\mathrm{j}\alpha x} \mathrm{e}^{\mathrm{j}\beta y} \mathrm{d}\alpha \mathrm{d}\beta \tag{2.135}$$

$$W_{1(b)e} = \int_{-\infty}^{+\infty} \int_{-\infty}^{+\infty} C(\alpha, \beta) \frac{k\mu_{\mathrm{r}} - \lambda}{k\mu_{\mathrm{r}} + \lambda} \mathrm{e}^{-kz} \mathrm{e}^{\mathrm{j}\alpha x} \mathrm{e}^{\mathrm{j}\beta y} \mathrm{d}\alpha \mathrm{d}\beta \tag{2.136}$$

$$W_{1(c)} = \int_{-\infty}^{+\infty} \int_{-\infty}^{+\infty} C(\alpha, \beta) \frac{2k\mu_{\mathrm{r}}}{k\mu_{\mathrm{r}} + \lambda} \mathrm{e}^{\lambda z} \mathrm{e}^{\mathrm{j}\alpha x} \mathrm{e}^{\mathrm{j}\beta y} \mathrm{d}\alpha \mathrm{d}\beta \tag{2.137}$$

α and β are integral variables; $C(\alpha, \beta)$ is the undetermined coefficients; $k = \sqrt{\alpha^2 + \beta^2}$; $\lambda = \sqrt{k^2 + \gamma^2}$. From the above three formulas, we can see that when the coefficient $C(\alpha, \beta)$ is determined, the W_1 of the whole solution area can be calculated.

According to Biot–Savart's law, there is

$$B_{(a,b)s} = \nabla \times \frac{\mu_0 I}{4\pi} \oint_l \frac{\mathrm{d}l}{r} \tag{2.138}$$

$r = \sqrt{(x - x_0)^2 + (y - y_0)^2 + (z - z_0)^2}$ is the distance from the field point (x, y, z) to the source point (x_0, y_0, z_0). μ_0 is a vacuum permeability, and l represents the integral path.

The Stokes theorem is applied to the formula:

$$B_{(a,b)s} = \nabla \left[-\frac{\mu_0 I}{4\pi} \iint_{\Omega} \mathrm{d}s \cdot \nabla \left(\frac{1}{r} \right) \right] \tag{2.139}$$

In the formula, Ω represents any curved surface bounded by curve l.

$$W_{1(a,b)s} = -\frac{\mu_0 I}{4\pi} \int\int_S \mathrm{d}s \cdot \nabla\left(\int \frac{\mathrm{d}z}{r}\right) \tag{2.140}$$

And

$$\frac{1}{r} = \frac{1}{2\pi} \int_{-\infty}^{+\infty}\int_{-\infty}^{+\infty} \frac{\mathrm{e}^{\mp k(z-z_0)}}{k} \mathrm{e}^{j[\alpha(x-x_0)+\beta(y-y_0)]} \mathrm{d}\alpha \mathrm{d}\beta \tag{2.141}$$

$$W_{1(a)s} = \frac{\mu I}{2\pi^2} \int_{-\infty}^{+\infty}\int_{-\infty}^{+\infty} \frac{\mathrm{e}^{-k(z-z_0)}}{k} \frac{\sin(\alpha y_0)}{\alpha} \times \frac{\sin(\beta y_0)}{\beta} \mathrm{e}^{\mathrm{j}\alpha x} \mathrm{e}^{\mathrm{j}\beta y} \mathrm{d}\alpha \mathrm{d}\beta \tag{2.142}$$

$$W_{1(b)s} = \frac{\mu I}{2\pi^2} \int_{-\infty}^{+\infty}\int_{-\infty}^{+\infty} \frac{\mathrm{e}^{k(z-z_0)}}{k} \frac{\sin(\alpha y_0)}{\alpha} \times \frac{\sin(\beta y_0)}{\beta} \mathrm{e}^{\mathrm{j}\alpha x} \mathrm{e}^{\mathrm{j}\beta y} \mathrm{d}\alpha \mathrm{d}\beta \tag{2.143}$$

First, consider the situation of the single turn coil, and its structure is shown in Fig. 2.40. In Fig. 2.40, x_0 and y_0 represent the distance between the inner edge of the coil length and the width of the two directions to the z-axis, and w represents the width of the coil wire; h_1 and h_2 represent the distance from the coils to the upper edge to the xoy plane, respectively.

The potential of a single turn rectangular section coil can be regarded as the superposition of the potential of the δ coil. The continuous distribution of the current density of the section of the conductor has been set up.

$$W^S_{1(a,b)s} = \int_0^w \int_{h_1}^{h_2} W_{1(a,b)s} \mathrm{d}z_0 dw \tag{2.144}$$

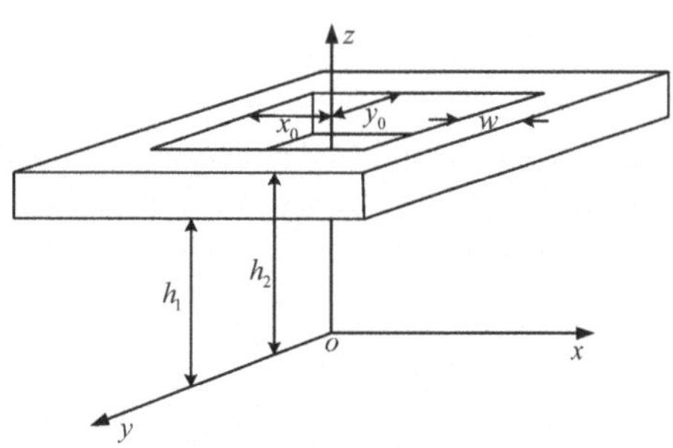

Fig. 2.40 Single turn coil of rectangular cross section

The superscript S represents the single turn coil.

$$W^S_{1(a)s} = \frac{\mu_0 I}{2\pi^2(h_2 - h_1)w} \int\limits_{-\infty}^{+\infty}\int\limits_{-\infty}^{+\infty} \frac{e^{-kz}}{k^2}(\mathrm{e}^{kh_2} - \mathrm{e}^{kh_1})\frac{P}{\alpha\beta}\mathrm{e}^{j\alpha x}\mathrm{e}^{j\beta y}\mathrm{d}\alpha\mathrm{d}\beta \quad (2.145)$$

$$W^S_{1(b)s} = \frac{\mu_0 I}{2\pi^2(h_2 - h_1)w} \int\limits_{-\infty}^{+\infty}\int\limits_{-\infty}^{+\infty} \frac{\mathrm{e}^{kz}}{-k^2}(\mathrm{e}^{-kh_2} - \mathrm{e}^{-kh_1})\frac{P}{\alpha\beta}\mathrm{e}^{j\alpha x}\mathrm{e}^{j\beta y}\mathrm{d}\alpha\mathrm{d}\beta \quad (2.146)$$

where

$$P = \begin{cases} \frac{\sin[(\alpha-\beta)w+\alpha x_0-\beta y_0]-\sin(\alpha x_0-\beta y_0)}{2(\alpha-\beta)} - \frac{\sin[(\alpha+\beta)w+\alpha x_0+\beta y_0]-\sin(\alpha x_0+\beta y_0)}{2(\alpha+\beta)} & \alpha \neq \beta \\ \frac{w}{2}\cos[\alpha(x_0 - y_0)] - \frac{\sin[\alpha(2w+x_0+y_0)]-\sin[\alpha(x_0+y_0)]}{4\alpha} & \alpha = \beta \end{cases}$$

The region of the coil section is treated as follows: The potential of any point in the region (x, y, z) can be regarded as the superposition of the potential of the coils and the coils above the point, then $h_2 = z$ and $h_1 = z$ are replaced, respectively, and the two expressions are added to the results.

$$W_{1\text{coil section}} = \frac{\mu_0 I}{2\pi^2(h_2 - h_1)w} \int\limits_{-\infty}^{+\infty}\int\limits_{-\infty}^{+\infty} \frac{1}{k^2}[2 - \mathrm{e}^{-k(z-h_1)} - \mathrm{e}^{k(z-h_2)})\frac{P}{\alpha\beta}\mathrm{e}^{j\alpha x}\mathrm{e}^{j\beta y}\mathrm{d}\alpha\mathrm{d}\beta \quad (2.147)$$

Under the premise of known single rectangular section coil potential, the potential generated by the meander line coil can be calculated using coordinate transformation and superposition.

For the meander-line-coil structure, the center coordinates of the $q = 1$ coil set are $(0, 0, z_m)$, and the center coordinates of the $q = 2$~Q coil set are $(x'_{\mathrm{q}}, 0, z_m)$. The scalar potential of the meander line coil is calculated by layering. The m layer contains $N \times Q$ independent single turn rectangular coils, and the potential of the m coil itself is generated.

$$W^m_{1(a)s} = \frac{\mu_0 I}{2\pi^2(h_2 - h_1)w} \int\limits_{-\infty}^{+\infty}\int\limits_{-\infty}^{+\infty} \frac{\mathrm{e}^{-kz}}{k^2}(\mathrm{e}^{kh_{2m}}$$

$$- e^{kh_{1m}}) \frac{\sum_{n=1}^{N} P_n}{\alpha\beta} \sum_{q=1}^{Q} e^{-j\alpha x'_q} e^{j\alpha x} e^{j\beta y} d\alpha d\beta \tag{2.148}$$

$$W^m_{1(b)s} = -\frac{\mu_0 I}{2\pi^2(h_2 - h_1)w} \int_{-\infty}^{+\infty}\int_{-\infty}^{+\infty} \frac{e^{kz}}{k^2} (e^{-kh_{2m}}$$

$$- e^{-kh_{1m}}) \frac{\sum_{n=1}^{N} P_n}{\alpha\beta} \sum_{q=1}^{Q} e^{-j\alpha x'_q} e^{j\alpha x} e^{j\beta y} d\alpha d\beta \tag{2.149}$$

$$W^m_{1\text{coil section}} = \frac{\mu_0 I}{2\pi^2(h_2 - h_1)w} \int_{-\infty}^{+\infty}\int_{-\infty}^{+\infty} \frac{1}{k^2} [2 - e^{-k(z-h_{2m})}$$

$$-e^{-k(z-h_{1m})}) \frac{\sum_{n=1}^{N} P_n}{\alpha\beta} \sum_{q=1}^{Q} e^{-j\alpha x'_q} e^{j\alpha x} e^{j\beta y} d\alpha d\beta \tag{2.150}$$

where

$$P_n = \begin{cases} \frac{\sin[(\alpha-\beta)w + \alpha x_{0n} - \beta y_{0n}] - \sin(\alpha x_{0n} - \beta y_{0n})}{2(\alpha-\beta)} - \frac{\sin[(\alpha+\beta)w + \alpha x_{0n} + \beta y_{0n}] - \sin(\alpha x_{0n} + \beta y_{0n})}{2(\alpha+\beta)} & \alpha \neq \beta \\ \frac{w}{2}\cos[\alpha(x_{0n} - y_{0n})] - \frac{\sin[\alpha(2w + x_{0n} + y_{0n})] - \sin[\alpha(x_{0n} + y_{0n})]}{4\alpha} & \alpha = \beta \end{cases}$$

The subscript m denotes the m layer in the formula.

The scalar potential in the region R_b is the superposition of the scalar potential of each layer coil.

$$W_{1(b)s} = \frac{\mu_0 I}{2\pi^2(h_2 - h_1)w} \int_{-\infty}^{+\infty}\int_{-\infty}^{+\infty} \frac{e^{kz}}{-k^2} \sum_{m=1}^{M} (e^{-kh_{2m}} - e^{-kh_{1m}})$$

$$\times \frac{1}{\alpha\beta} \sum_{n=1}^{N} P_n \sum_{q=1}^{Q} e^{-j\alpha x'_q} e^{j\alpha x} e^{j\beta y} d\alpha d\beta \tag{2.151}$$

The obtained coefficient $C(\alpha, \beta)$ is

$$C(\alpha, \beta) = -\frac{\mu_0 I}{2\pi^2(h_2 - h_1)wk^2\alpha\beta} \sum_{m=1}^{M} (e^{-kh_{2m}}$$

$$-e^{-kh_{1m}}) \sum_{n=1}^{N} P_n \sum_{q=1}^{Q} e^{-j\alpha x'_q} \tag{2.152}$$

For the meander line coil, the magnetic field in each conductor coil area is also worth paying attention to. For each coil, the potential of the conductor region is equal to the sum of the potential of the layer itself and the sum of the potential of the other layer coils in the region, so that the coil area of the m' layer is available.

$$\begin{aligned} W_{1\,\text{coil section}}^{m'} = & \frac{\mu_0 I}{2\pi^2(h_2 - h_1)w} \int_{-\infty}^{+\infty}\int_{-\infty}^{+\infty} \left\{ \frac{1}{k^2}\left[2 - \mathrm{e}^{-k(z-h_{1m'})} \right.\right. \\ & \left. -\mathrm{e}^{k(z-h_{2m'})}\right] + \sum_{m=1}^{m'-1} \frac{\mathrm{e}^{-kz}}{k^2}\left(\mathrm{e}^{kh_{2m}} - \mathrm{e}^{kh_{1m}}\right) \\ & \left. - \sum_{m=m'+1}^{M} \frac{\mathrm{e}^{kz}}{k^2}\left(\mathrm{e}^{-kh_{2m}} - \mathrm{e}^{-kh_{1m}}\right)\right\} \\ & \times \frac{1}{\alpha\beta}\sum_{n=1}^{N} P_n \sum_{q=1}^{Q} \mathrm{e}^{-j\alpha x_q'} \mathrm{e}^{j\alpha x} \mathrm{e}^{j\beta y}\,\mathrm{d}\alpha\,\mathrm{d}\beta \end{aligned} \tag{2.153}$$

In the formula, superscript m' represents the m' layer.

The eddy current induced by the conductor is a necessary condition for electromagnetic ultrasonic testing. According to the electromagnetic field theory, the eddy generated by the folded coil on the surface of the conductor specimen can be expressed as

$$\boldsymbol{J}_e = -j\omega\sigma \boldsymbol{A}_{(c)} \tag{2.154}$$

In the formula, $\boldsymbol{J}_\mathrm{e}$ represents the eddy current in the conductor.

Then

$$\begin{aligned} \mathbf{J}_\mathrm{e} = & \frac{j\omega\sigma\mu_0 I}{\pi^2(h_2 - h_1)w} \int_{-\infty}^{+\infty}\int_{-\infty}^{+\infty} \frac{\mathrm{e}^{\lambda z}}{k} \times \frac{\mu_\mathrm{r}}{k\mu_\mathrm{r} + \lambda} \\ & \times (j\beta\hat{\mathbf{x}} - j\alpha\hat{\mathbf{y}}) \sum_{m=1}^{M} \left(\mathrm{e}^{-kh_{2m}} - \mathrm{e}^{-kh_{1m}}\right) \\ & \times \frac{1}{\alpha\beta}\sum_{n=1}^{N} P_n \sum_{q=1}^{Q} \mathrm{e}^{-j\alpha x_q'} \mathrm{e}^{j\alpha x} \mathrm{e}^{j\beta y}\,\mathrm{d}\alpha\,\mathrm{d}\beta \end{aligned} \tag{2.155}$$

$\hat{\boldsymbol{x}}$ and $\hat{\boldsymbol{y}}$ are the direction vectors.

The magnetic induction intensity in the available region R_b and R_c is

$$
\begin{aligned}
\mathbf{B}_{(b)} = {} & \frac{\mu_0 I}{2\pi^2(h_2 - h_1)w} \int\limits_{-\infty}^{+\infty}\int\limits_{-\infty}^{+\infty} \left[\frac{1}{-k}\left(\mathrm{e}^{kz} - \frac{k\mu_r - \lambda}{k\mu_r + \lambda}\mathrm{e}^{-kz} \right) \right. \\
& \left. \times (j\alpha\hat{\mathbf{x}} + j\beta\hat{\mathbf{y}}) - \left(\mathrm{e}^{kz} + \frac{k\mu_r - \lambda}{k\mu_r + \lambda}\mathrm{e}^{-kz} \right)\hat{\mathbf{z}} \right] \sum_{m=1}^{M} (\mathrm{e}^{-kh_{2m}} \\
& - \mathrm{e}^{-kh_{1m}}) \times \frac{1}{\alpha\beta} \sum_{n=1}^{N} P_n \sum_{q=1}^{Q} \mathrm{e}^{-j\alpha x'_q} \mathrm{e}^{j\alpha x} \mathrm{e}^{j\beta y} \mathrm{d}\alpha \mathrm{d}\beta
\end{aligned}
\tag{2.156}
$$

$$
\begin{aligned}
\mathbf{B}_{(c)} = {} & \frac{\mu_0 I}{2\pi^2(h_2 - h_1)w} \int\limits_{-\infty}^{+\infty}\int\limits_{-\infty}^{+\infty} \left[\frac{\mathrm{e}^{\lambda z}}{-k^2} \times \frac{2k\mu_r\lambda}{k\mu_r + \lambda} \right. \\
& \left. \times \left(j\alpha\hat{\mathbf{x}} + j\beta\hat{\mathbf{y}} + \lambda\hat{\mathbf{z}} \right) \right] \sum_{m=1}^{M} \left(\mathrm{e}^{-kh_{2m}} - \mathrm{e}^{-kh_{1m}} \right) \\
& \times \frac{1}{\alpha\beta} \sum_{n=1}^{N} P_n \sum_{q=1}^{Q} \mathrm{e}^{-j\alpha x'_q} \mathrm{e}^{j\alpha x} \mathrm{e}^{j\beta y} \mathrm{d}\alpha \mathrm{d}\beta
\end{aligned}
\tag{2.157}
$$

The impedance of the coil can be expressed as the sum of the DC impedance of the coil, the impedance of the coil in the free space, and the impedance variation induced by the induced eddy current of the conductor.

$$
Z = Z_d + Z_0 + \Delta Z \tag{2.158}
$$

Z_d is DC impedance; Z_0 is the impedance of free space; ΔZ is the impedance variation caused by the eddy current.

The impedance Z_0 can be calculated by calculating the induced electromotive force on the coil divided by the current of the coil.

$$
Z_0 = \frac{V}{I} \tag{2.159}
$$

Therefore, to calculate impedance Z_0, the induced electromotive force of the coil must be calculated first. According to the calculation of the scalar potential, we first calculate the induced electromotive force of the single turn coil, which can be expressed as

$$
V^S = \frac{j\omega}{(h_2 - h_1)w} \int\limits_{\text{coil cross section}} \left(\int\limits_S \boldsymbol{B} \cdot \mathrm{ds} \right) \mathrm{d\,Area} \tag{2.160}
$$

The induced electromotive force of the meander line coil is equal to the superposition of all equivalent rectangular coil EMF.

$$
\begin{aligned}
V = {} & \frac{2j\omega\mu_0 I}{\pi^2(h_2-h_1)^2 w^2}\sum_{m'=1}^{M}\int_{-\infty}^{+\infty}\int_{-\infty}^{+\infty}\{2(h_{2m'}-h_{1m'}) \\
& +\frac{2}{k}\left[e^{-k(h_{2m'}-h_{1m'})}-1\right]+\sum_{m=1}^{m'-1}\frac{1}{k}\left(e^{kh_{2m}}-e^{kh_{1m}}\right)\left(e^{-kh_{1m'}}\right. \\
& \left.-e^{-kh_{2m'}}\right)+\sum_{m=m'+1}^{M}\frac{1}{k}\left(e^{-kh_{2m}}-e^{-kh_{1m}}\right)\left(e^{kh_{1m'}}-e^{kh_{2m'}}\right)\Bigg\} \\
& \frac{1}{\times(\alpha\beta)^2}\left(\sum_{n=1}^{N}P_n\right)^2\left(\sum_{q=1}^{Q}e^{-j\alpha x'_q}+\sum_{q=1}^{Q}e^{j\alpha x'_q}\right)\mathrm{d}\alpha\mathrm{d}\beta
\end{aligned} \tag{2.161}
$$

Then

$$
\begin{aligned}
Z_0 = {} & \frac{2j\omega\mu_0}{\pi^2(h_2-h_1)^2 w^2}\sum_{m'=1}^{M}\int_{-\infty}^{+\infty}\int_{-\infty}^{+\infty}\{2(h_{2m'}-h_{1m'}) \\
& +\frac{2}{k}\left[e^{-k(h_{2m'}-h_{1m'})}-1\right]+\sum_{m=1}^{m'-1}\frac{1}{k}\left(e^{kh_{2m}}-e^{kh_{1m}}\right)\left(e^{-kh_{1m'}}\right. \\
& \left.-e^{-kh_{2m'}}\right)+\sum_{m=m'+1}^{M}\frac{1}{k}\left(e^{-kh_{2m}}-e^{-kh_{1m}}\right)\left(e^{kh_{1m'}}-e^{kh_{2m'}}\right)\Bigg\} \\
& \times\frac{1}{(\alpha\beta)^2}\left(\sum_{n=1}^{N}P_n\right)^2\left(\sum_{q=1}^{Q}e^{-j\alpha x'_q}+\sum_{q=1}^{Q}e^{j\alpha x'_q}\right)\mathrm{d}\alpha\mathrm{d}\beta
\end{aligned} \tag{2.162}
$$

For the impedance variation ΔZ induced by the induced eddy current, the following formula can be used.

$$
\begin{aligned}
\Delta Z = \Delta R + j\Delta X = {} & \frac{j8\pi^2\omega}{\mu_0 i^2}\int_{-\infty}^{+\infty}\int_{-\infty}^{+\infty}k^3 C(\alpha,\beta) \\
& \times C(-\alpha,-\beta)\frac{k\mu_r-\lambda}{k\mu_r+\lambda}\mathrm{d}\alpha\mathrm{d}\beta
\end{aligned} \tag{2.163}
$$

$$\Delta Z = \frac{j2\omega\mu_0}{\pi^2(h_2-h_1)^2w^2}\int\limits_{-\infty}^{+\infty}\int\limits_{-\infty}^{+\infty}\frac{1}{k(\alpha\beta)^2}\left[\sum_{m=1}^{M}(\mathrm{e}^{-kh_{2m}}\right.$$

$$-\mathrm{e}^{-kh_{1m}})\Big]^2\left(\sum_{n=1}^{N}P_n\right)^2\left(\sum_{q=1}^{Q}\mathrm{e}^{-\mathrm{j}\alpha x'_q}\right.$$

$$\left.\left.\times\sum_{q=1}^{Q}\mathrm{e}^{j\alpha x'_q}\right)\right]\times\frac{k\mu_r-\lambda}{k\mu_r+\lambda}\mathrm{d}\alpha\mathrm{d}\beta \tag{2.164}$$

The DC impedance Z_0 of the meander line coil is the DC resistance, which can be calculated according to the coil size and conductivity.

$$Z_d = \frac{L}{\sigma_c(h_2-h_1)w} \tag{2.165}$$

L is the total length of the meander line conductor, and σ_c is the conductivity of the coil conductor.

In electromagnetic ultrasonic testing, the applied excitation is a pulse signal. It is necessary to calculate the pulse magnetic field produced by the coil and the pulsed eddy in the specimen; that is, the impulse response should be solved.

The impulse excitation and the corresponding response can be represented as a series of superposition of sinusoidal signals with different frequencies. When the magnetic induction intensity and the frequency-domain expression of the pulsed eddy current are known, the inverse Fourier transform (IFT) can be used to obtain the time-domain response of the magnetic induction intensity and the pulsed eddy current; that is to say, the time-domain response of the magnetic induction intensity and the pulsed eddy current can be obtained.

$$\boldsymbol{J}_e(t) = \frac{1}{2\pi}\int\limits_{-\infty}^{\infty}\boldsymbol{J}_e\mathrm{e}^{\mathrm{j}\omega t}\mathrm{d}\omega \tag{2.166}$$

$$\boldsymbol{B}_{(b)}(t) = \frac{1}{2\pi}\int\limits_{-\infty}^{\infty}\boldsymbol{B}_{(b)}\mathrm{e}^{\mathrm{j}\omega t}\mathrm{d}\omega \tag{2.167}$$

$$\boldsymbol{B}_{(c)}(t) = \frac{1}{2\pi}\int\limits_{-\infty}^{\infty}\boldsymbol{B}_{(c)}\mathrm{e}^{\mathrm{j}\omega t}\mathrm{d}\omega \tag{2.168}$$

However, it is difficult to get the explicit expression of magnetic induction intensity and pulsed eddy current in the time domain by directly obtaining IFT. When solving the problem, FFT-IFFT method is adopted.

In order to verify the analytical expression and the effectiveness of the calculation method, a 20% off equal interval double-layer, double-split meander line coil is taken

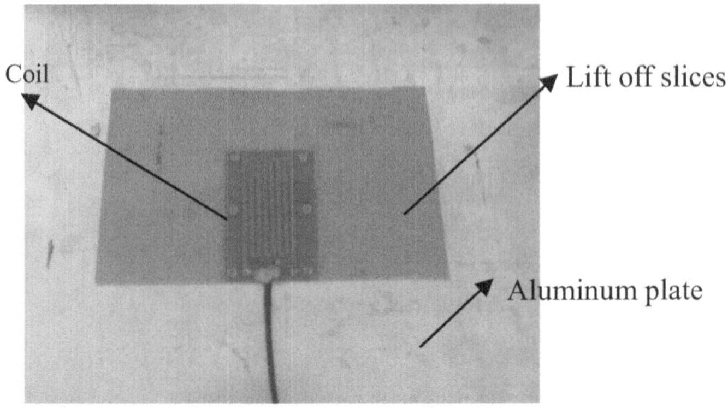

Fig. 2.41 Experiment setting

as an example. The coil is placed above the sample of the measured aluminum plate, and the coil and the aluminum plate are removed by the non-conductive stripping thin slices. The liftoff value can be adjusted, as shown in Fig. 2.41.

In physical tests, the coil is provided by the high-power RF power amplifier (AG1024) to provide excitation signals. The power amplifier can produce continuous sinusoidal signals with continuous and adjustable frequency and amplitude and can also produce tone-burst signals with adjustable frequency, amplitude and periodic number. In the experiment, the voltage signal of the coil port is measured by an oscilloscope which can record the waveform data, and the current signal of the coil is measured by measuring the voltage signal on the 1 Ω resistor in series with the coil. The amplitude and phase angle of the coil impedance can be calculated from the voltage and current waveforms of the measured sinusoidal signals.

When calculating the pulse magnetic field of the coil, pulse excitation is applied to the coil. The tone-burst signal with a frequency of 500 kHz and a cycle number of 4 is applied to the coil when the loop is lifted from 1 mm.

In order to compare with the numerical calculation method, the TSFEM method is used to calculate the above calculation models. Considering that the calculation of the three-dimensional model is too large, the end effect of the coil is ignored, and the length of the coil and the conductor in the length direction are assumed to be infinitely long, so the two-dimensional model of the coil is established. In the two-dimensional model, the y component of the eddy current, the x component, and the z component of the magnetic induction intensity can only be obtained. The finite element analysis software COMSOL Multiphysics is used for calculation.

The simulation uses the current signal as the coil excitation. In order to ensure the accuracy of the calculation results, a denser grid must be divided into the skin depth of the conductor and the coil conductor because of the higher frequency of the excitation signal, the smaller skin depth of the conductor and the small size of the conductor. The total number of grid cells divided into TSFEM is 15,348 as shown in Fig. 2.42.

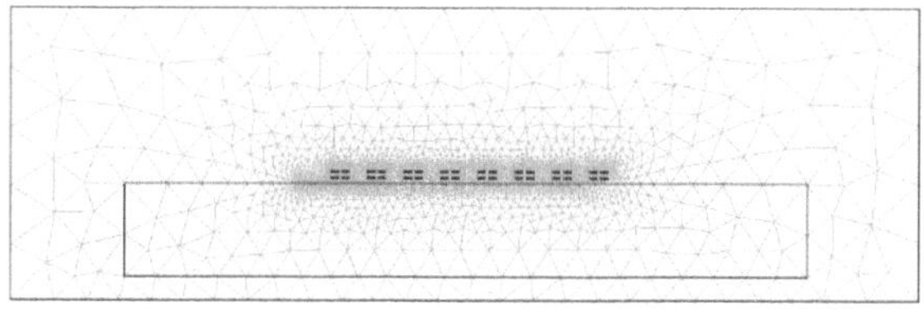

Fig. 2.42 TSFEM simulation model

First, the impedance of the coil at different liftoff values is calculated. The recursive adaptive Simpson method with high computation accuracy is used to calculate the numerical integration. Because the calculation formula only contains trigonometric function and does not have the Bessel function in the calculation formula of the spiral coil, it has good convergence, which makes the calculation simpler.

The FFT-IFFT method is used to calculate the pulse magnetic field of the coil and the induced eddy current in the tested sample. For the current signal in Fig. 2.36, 1024 data points are taken for FFT transformation, and the corresponding Simpson method is used to calculate the eddy current and magnetic induction intensity of the corresponding frequency. Finally, the time-domain signals of eddy current and magnetic induction intensity are obtained by IFFT.

Figures 2.43, 2.44 and 2.45 show an analytical model of the y component of the eddy density, the x and z components of the magnetic induction intensity at (1.1725, 11.665, −0.1 mm), and the calculated values of the TSFEM method. Figures 2.46 and 2.47 show the calculated values of the analytical model of the x and z components of the magnetic induction intensity at (1.1725, 11.665, −0.1 mm) and the calculated values of the TSFEM method. It can be seen from these diagrams that although the calculated values of the analytical model and the calculated values of the TSFEM method have some errors in time and amplitude, they are basically consistent.

Due to the neglect of the skin effect and proximity effect in the derivation of the analytical formula, the current density in the conductor is all equal, which is bound to cause a deviation between the analytical formula and the actual physical model. Because it is difficult to consider the skin effect and proximity effect in the analytical model of the coil, it is a common practice to deduce the analysis model of the coil,

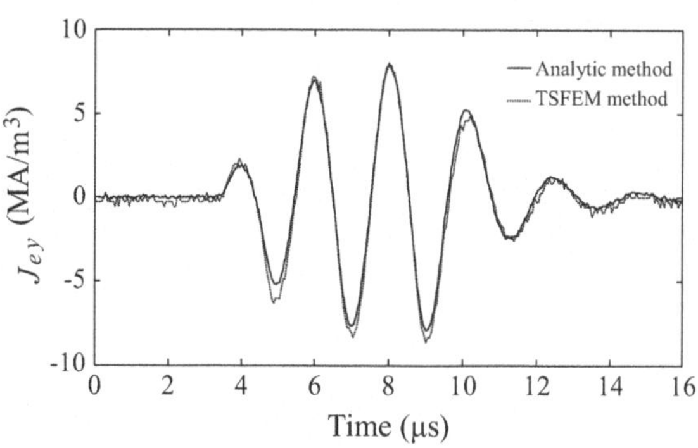

Fig. 2.43 y component of the eddy current density (1.1725, 11.665, −0.1 mm)

and the calculation error is often in the acceptable range of the practical engineering application. Also, the idea that the meander line coil is equivalent to the superposition of the single turn closed loop will also bring some calculation errors.

Compared with numerical calculation, the advantage of analytical calculation is that the analytical formula has clear physical meaning and fast computation speed.

2.1.5 Omnidirectional Mode-Controlled Lamb Wave EMAT

The single turn loop coil EMAT can be used to receive omnidirectional Lamb waves in the aluminum plate. However, because of the low energy exchange efficiency of EMAT and the gradual attenuation of the omnidirectional Lamb wave during the propagation process, the Lamb wave detection signal obtained by the single turn coil EMAT is usually relatively weak. At present, the traditional EMAT uses spiral wound coils with relatively large turns to improve the intensity of the detection signal. However, the traditional EMAT based on the spiral wound coil does not possess a good omnidirectional Lamb wave mode control function. Given this, based on the EMAT energy exchange model of the single turn coil, the mode control method for

Fig. 2.44 *x* component of the magnetic induction intensity (1.1725, 11.665, −0.1 mm)

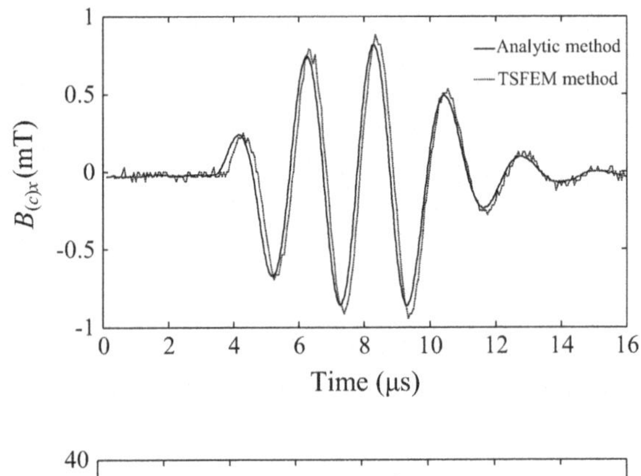

Fig. 2.45 *z* component of the magnetic induction intensity (1.1725, 11.665, −0.1 mm)

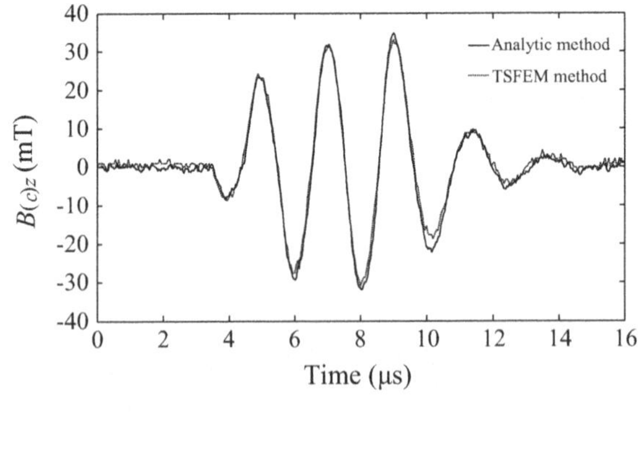

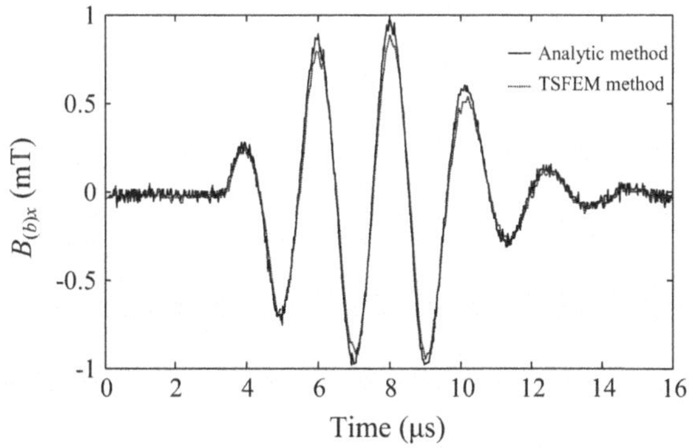

Fig. 2.46 x component of the magnetic induction intensity (1.1725, 11.665, −0.1 mm)

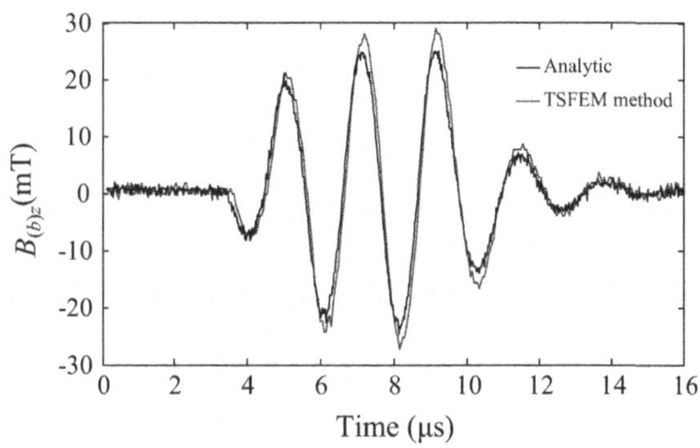

Fig. 2.47 z component of the magnetic induction intensity (1.1725, 11.665, −0.1 mm)

omnidirectional Lamb wave EMAT is studied, and a new design method of the new omnidirectional mode-controlled Lamb wave EMAT is put forward.

The new EMAT with excitation and reception has the same structure. The new EMAT consists of three parts: cylindrical permanent magnet, coil, and aluminum plate. Unlike the traditional EMAT, the new EMAT adopts the meander line coil, the loop wires in the coil are concentric wires, and the distance l_c of the adjacent 80% off ring wires is fixed. When the new EMAT is used to excite the omnidirectional Lamb wave, the folded coil has a pulse exciting current $\boldsymbol{I}_{exc}$. The winding structure of the coil determines that the currents in the 80% off adjacent circular conductors be opposite at any time. By the complete energy exchange model of the single turn loop EMAT, it is known that under the action of $\boldsymbol{I}_{exc}$, all 90% off ring wires will cooperate with permanent magnets to excite the omnidirectional Lamb wave propagating along the radial direction. For the convenience of discussion, a cylindrical coordinate system is built on the new EMAT: The original point is located on the middle of the aluminum plate and is located just below the center of the cylindrical permanent magnet; the direction of the z is perpendicular to the surface of the aluminum plate,

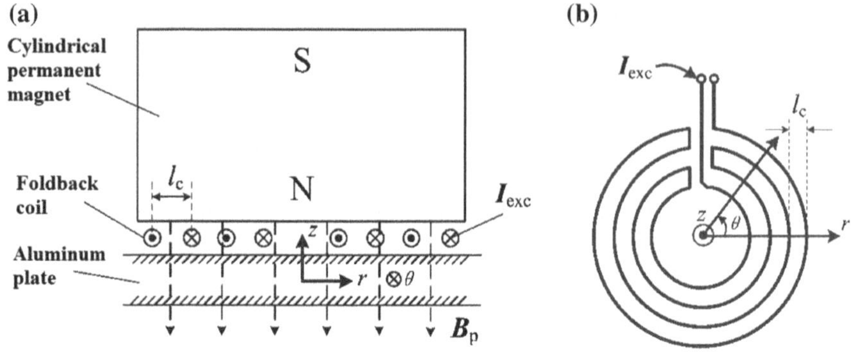

Fig. 2.48 Schematic diagram of the structural principle of the new EMAT: **a** EMAT structure principle (side view section); **b** meander coil (top view)

and the direction of the r and θ is radial and circumferential of the coil, respectively (Fig. 2.48).

According to the dispersion relation of Lamb wave, the wavelength λ of different modes Lamb wave is different at any working frequency f. Therefore, the mode control of the omnidirectional Lamb wave can be realized through the implementation of "wavelength control"; that is, through the structure design of the back coil, the new EMAT can only excite the omnidirectional Lamb wave with specific wavelengths at the working frequency f, so that it can only excite the corresponding Lamb wave in the specific mode under the f. The design principle of the meander coil is as follows.

The internal wave component of the omnidirectional Lamb wave is similar to that of the external traveling wave component. For any 80% off adjacent circular wires W_1 and W_2 in the coil, their radius is r_1 and r_2 ($r_1 < r_2$), respectively, and then, there is $r_2 - r_1 = l_c$. Under a given working frequency f_0, the omnidirectional Lamb wave s_1 and s_2 of a specific mode with wavelengths of λ_0 generated by W_1 and W_2 are considered.

The analytical expression of the omnidirectional Lamb wave vibration displacement along the r-direction (the z-direction vibration displacement does not produce induction voltage and thus neglecting it) shows that the phase of the r-direction vibration displacement caused by s_1 and s_2 at the receiver at any position (r_a, θ_a, z_a) ($r_a > r_2$) is φ_1 and φ_2, respectively:

$$
\begin{aligned}
\varphi_1 &= -(kr_a - \omega t) + \left(kr_1 + \frac{3}{4}\pi\right) + \varphi_{10} \\
&= -\left(\frac{2\pi}{\lambda_0}r_a - \omega t\right) + \left(\frac{2\pi}{\lambda_0}r_1 + \frac{3}{4}\pi\right) + \varphi_{10} \\
\varphi_2 &= -(kr_a - \omega t) + \left(kr_2 + \frac{3}{4}\pi\right) + \varphi_{20}
\end{aligned}
\tag{2.169}
$$

$$= -\left(\frac{2\pi}{\lambda_0}r_a - \omega t\right) + \left(\frac{2\pi}{\lambda_0}r_2 + \frac{3}{4}\pi\right) + \varphi_{20} \tag{2.170}$$

Among them, φ_{10} and φ_{20} are the initial phase. The reverse structure of the coil determines the phase difference of the current in the adjacent loop wire W_1 and W_2 is π. Therefore, the initial phase difference of the corresponding vibration displacement $(\varphi_{20} - \varphi_{10})$ is also π. The difference between φ_2 and φ_1 can be calculated.

$$\begin{aligned}\Delta\varphi = \varphi_2 - \varphi_1 &= \left[-\left(\frac{2\pi}{\lambda_0}r_a - \omega t\right) + \left(\frac{2\pi}{\lambda_0}r_2 + \frac{3}{4}\pi\right) + \varphi_{20}\right] \\ k4 &- \left[-\left(\frac{2\pi}{\lambda_0}r_a - \omega t\right) + \left(\frac{2\pi}{\lambda_0}r_1 + \frac{3}{4}\pi\right) + \varphi_{10}\right] \\ &= \frac{2\pi}{\lambda_0}(r_2 - r_1) + (\varphi_{20} - \varphi_{10}) \\ &= \frac{2\pi l_c}{\lambda_0} + \pi\end{aligned} \tag{2.171}$$

If the wavelength λ_0 of s_1 and s_2 satisfies the following relationship:

$$\lambda_0 = \frac{2l_c}{2P_{int} + 1} \tag{2.172}$$

where P_{int} is any integer and $P_{int} \geq 0$, and the phase difference $\Delta\varphi$ is $2\,(P_{int} + 1)\pi$. This indicates that the vibration displacement caused by s_1 and s_2 at any end of the receiver is always the same phase. Therefore, the vibration caused by s_1 and s_2 will always enhance each other after stacking. By analogy, the omnidirectional Lamb wave with a wavelength of λ_0 produced by the meander coil leads to each other after superposition, which will make the overall wavelength of the new EMAT excited by the omnidirectional Lamb wave of λ_0 can be effectively enhanced.

Similarly, in the working frequency f_0, we consider the other mode omnidirectional Lamb wave s_1' and s_2' generated by W_1 and W_2, which are wavelengths of λ_0'. If the λ_0' is not satisfied (2.172), the vibration caused by s_1' and s_2' at any location of the receiver will weaken each other after superposition. By analogy, the omnidirectional Lamb wave of wavelength of λ_0' produced by the folded loop wire will weaken each other after superposition, which will make the overall Lamb wave amplitude of the new EMAT excited by the new type of λ_0' is suppressed effectively.

It is concluded that, under the working frequency f_0, if the wavelength of a new type of EMAT is designed by the design of the coil parameter l_c of a new type of EMAT (2.172), the new EMAT will be able to effectively stimulate the omnidirectional Lamb wave of the mode and effectively suppress the appearance of the other mode omnidirectional Lamb waves. Based on this, the design method of the new EMAT meander line coil is proposed. The specific steps are as follows:

(1) According to the demand of tomography, we choose the mode of Lamb wave to be excited.

(2) Select the working frequency f_0 of the new EMAT.
(3) Using the phase velocity dispersion curve of Lamb wave in the aluminum plate, the phase velocity of each mode Lamb wave in the aluminum plate under the frequency thick product f_0d_0 is calculated according to the working frequency f_0 and the thickness of the measured aluminum plate thickness d_0.
(4) Using the formula $\lambda = c_p/f_0$, the wavelength λ of each mode Lamb wave in the aluminum plate under the frequency thick product f_0d_0 is calculated. Among them, the phase velocity c_p of the Lamb waves of each mode is calculated in step (3).
(5) Using the $l_c = 0.5\lambda_0$, the EMAT coil parameter l_c is calculated, so that the l_c and the λ_0 are satisfied (2.172), of which, the λ_0 is the wavelength of the selected modal Lamb wave under the frequency thick product f_0d_0.
(6) According to the l_c calculated in step (5), it is judged whether there is any other mode Lamb wave of the wavelength satisfying (2.172) under the frequency thick product f_0d_0. If it does not exist, the design is finished; if there is $f_0 = f_0 + \Delta f_0$, and return to the step (3), the Δf_0 is the working frequency adjustment step and the $\Delta f_0 = 5$ kHz.

Through the above design, the new EMAT can only effectively stimulate the omnidirectional Lamb wave of the selected mode under the working frequency f_0 and effectively suppress the appearance of the other mode omnidirectional Lamb wave, so that only the pure single modal Lamb wave signal is contained in the detection waveform received under f_0; that is, the ideal Lamb wave mode is realized.

2.1.6 Experimental Verification

In order to verify the modal control capability of the new EMAT, a series of confirmatory tests were carried out on 3-mm-thick aluminum plate. The structure principle of the test system is shown in Fig. 2.49. The upper computer is used to control the high-power pulse excitation power, and the excitation signal is provided for EMAT excitation. The peak value of the burst excitation signal used is about 300 V, and the number of cycles is 7. The Lamb wave direct wave signal and the Lamb wave reflected by the aluminum plate edge are received and detected by EMAT. The detection signal is filtered and amplified by the signal conditioning circuit, and then, it is collected by the high-speed data acquisition card and sent to the upper computer for analysis and processing.

The parameters of the EMAT used in the experiment are shown in Table 2.8. The EMAT coils are fabricated by PCB Technology (the following EMAT coils are fabricated by this technology). From the phase velocity of Lamb wave in aluminum plate and the dispersion curve of group velocity, it is known that only A0 and S0 mode Lamb waves exist in the 3-mm-thick aluminum plate used for the test when the working frequency is 80 kHz. In addition, using phase velocity and group velocity dispersion curve, the theoretical phase velocity, group velocity, and wavelength of

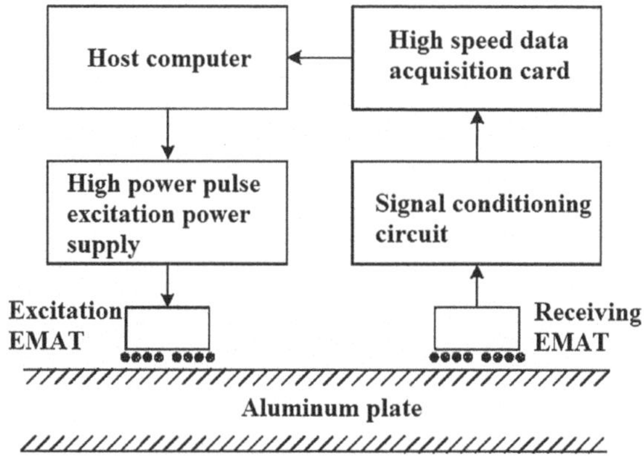

Fig. 2.49 Schematic diagram of the structure of an experimental system

Table 2.8 Parameters for EMAT used in the experiment

	Traditional EMAT	New EMAT
Type of coil structure	Wound spiral coil	Meander line coil
Lamb wave modes	–	A0
Working frequency (kHz)	80	80
l_c (mm)	–	9
The outer diameter of coil (mm)	53	63
The inner diameter of the coil (mm)	32	27
Total coil number of coils	26	30

the A0 mode Lamb wave of 80 kHz can be calculated. The values are 1446, 2519 m/s, and 18 mm, respectively.

The comparison test on the aluminum plates of 3 mm × 1000 mm × 1000 mm is shown in Fig. 2.50: Use traditional EMAT and new EMAT to stimulate and receive 80 kHz Lamb waves, and compare their ability to receive and transmit pure single A0 mode Lamb waves.

The specific test process is as follows:

(1) Use the new EMAT to excite and receive omnidirectional Lamb waves. One excitation is placed on the corner of the aluminum plate, and the three receiving EMAT is placed in three different positions along the edge of the aluminum plate. An omnidirectional Lamb wave is launched by using EMAT, and three receiving EMAT are used to detect the direct and reflected waves, respectively.
(2) Use conventional EMAT to excite and receive omnidirectional Lamb waves. Four traditional EMAT are used to replace the four new EMAT in the previous step. An omnidirectional Lamb wave is launched using EMAT, and three receivers are used for EMAT detection.

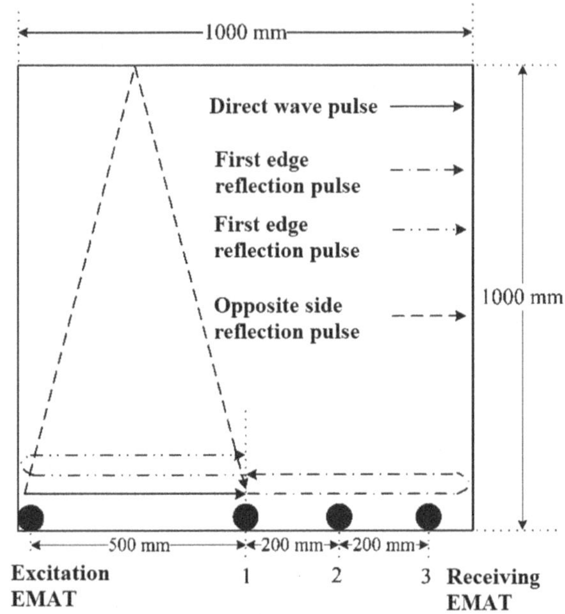

Fig. 2.50 EMAT location and A0 wave propagation path schematic diagram

When the omnidirectional Lamb wave is excited, it first reaches the first four Lamb wave propagation paths of position 1. In Fig. 2.51a, the signal *a* is the first pulse, which is the excitation pulse that the receiving EMAT is induced in the space; the signal *b* is a direct wave pulse; the signal *c* and *e* are, respectively, the first and second edge reflection pulses, and the signal *d* is the opposite side reflection pulse. The meaning of signal *a~e* in Fig. 2.11b–f is the same as that of the corresponding signal. Below, the test results are analyzed and discussed.

First of all, a quantitative analysis of the detection waveform is performed to identify the mode of the omnidirectional Lamb wave excited and received by the new EMAT. The detection waveforms are analyzed. (1) The propagation time difference Δt_{prop} of each edge reflection pulse signal (*c, d* and *e*) and the direct wave pulse signal *b* are extracted from the waveform, respectively. (2) The geometric relationship is used to calculate the actual propagation path difference Δd_{prop} of each edge reflection pulse signal (*c, d* and *e*) and the direct wave pulse signal *b*, respectively; (3) using the calculated propagation distance difference Δd_{prop} divided by the corresponding propagation time difference Δt_{prop}, to estimate the group velocity of the omnidirectional Lamb wave, so as to verify the mode of the omnidirectional Lamb wave. Similarly, the group velocity of omnidirectional Lamb waves is estimated. The relative error of the estimated group velocity $c_{\text{g\_est}}$ E_{err} can be calculated by $E_{\text{err}}=|c_{\text{g\_est}}- c_{\text{g\_the}}|/c_{\text{g\_the}}$, in which the $c_{\text{g\_the}} = 2519$ m/s is the theoretical group velocity of the A0 wave of the 80 kHz in the 3-mm-thick aluminum plate. The result of group velocity estimation is shown in Table 2.9. The results in Table 2.9 show that the group velocity of the estimated omnidirectional Lamb wave is in good agreement

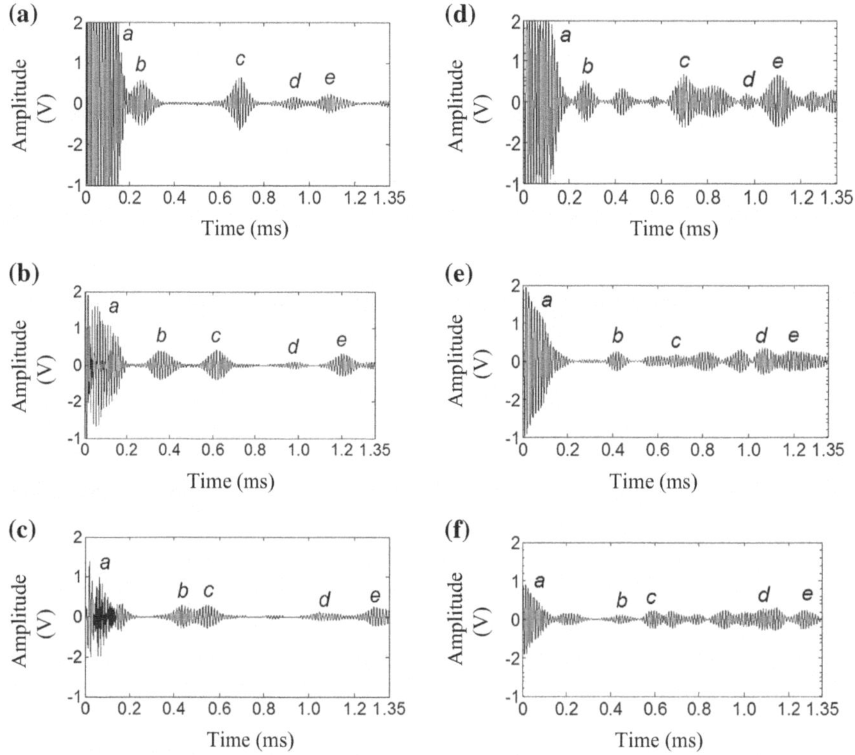

Fig. 2.51 Detection waveform of A0 wave transceiver test: **a** new EMAT, receiving position 1; **b** new EMAT, receiving position 2; **c** new EMAT, receiving position 3; **d** traditional EMAT, receiving position 1; **e** traditional EMAT, receiving position 2; **f** tradition, receiving position 3

Table 2.9 Estimation of the group velocity of Lamb wave produced by new EMAT

	Figure 2.11a			Figure 2.11b			Figure 2.11c		
	Signal	Signal *d*	Signal *e*	Signal *c*	Signal *d*	Signal *e*	Signal *c*	Signal *d*	Signal *e*
Δt_{prop} (μs)	420	645	810	247	595	833	84	541	825
Δd_{prop} (m)	1.00	1.56	2.00	0.60	1.42	2.00	0.20	1.29	2.00
c_{g_est} (m/s)	2381	2419	2469	2429	2387	2401	2381	2384	2424
E_{err} (%)	5.5	4.0	2.0	3.6	5.2	4.7	5.5	5.4	3.8

with the theoretical group velocity of the A0 wave, and the maximum relative error of the estimated value is 5.5%. The results show that the new EMAT developed in this work can effectively stimulate and receive the omnidirectional A0 modal Lamb wave.

Secondly, in order to verify the mode control capability of the new EMAT, the new EMAT detection waveform is compared with the traditional EMAT detection waveform. In the experiments corresponding to the above two detection waveforms, the location of sending and receiving EMAT is the same. The comparison results show that there are only A0 mode Lamb wave signals, that is, the signal *b~e*, and there are many non-A0 mode Lamb wave signals in the waveform shown in (*d*), except for the A0 wave. Through group velocity estimation, it is not difficult to find that these multiple signals are direct wave pulses and edge reflection pulses of S0 mode Lamb waves. The appearance of this S0 mode (the interference mode) Lamb wave signal will seriously affect the identification and feature extraction of the A0 mode (the useful mode) Lamb wave signal. This effect is particularly prominent when the S0 wave signal and the A0 wave signal overlap. For example, the amplitude of the signal *b* in the two detection waveforms is similar, and the characteristics of the signal *c* and the signal *d* in the two waveforms are also the same; however, the amplitude of the CITIC number *e* of Fig. 2.51d is far larger than the amplitude of the CITIC number *e* of Fig. 2.51a. The reasons for this phenomenon are as follows: By using the Lamb dispersion curve, the group velocity of A0 wave and S0 wave in the test is 2519 and 5612 m/s, respectively, while the propagation path of the A0 wave second edge reflection pulse and the S0 wave fifth marginal reflection pulse is 2.5 and 5.5 m, respectively. Thus, the propagation of the two pulses can be obtained by calculation. The time is 992 and 980 s, respectively, and the two are very close, which means that the signal E in Fig. 2.51d is the superposition of the two pulses. The overlap between the S0 wave signal and the A0 wave signal increases the difficulty of the recognition of the A0 wave signal and will seriously affect the accuracy of the extraction of the characteristic information of the A0 wave signal. Similarly, by drawing a comparison between Fig. 2.51b, c and e, f, we can get the same conclusion.

In conclusion, compared to the traditional EMAT, the new EMAT can stimulate and receive a purer single-mode omnidirectional Lamb wave; that is, it has more ideal Lamb wave mode control ability, which is beneficial to the accurate extraction of the travel time projection data required for tomography. Compared with the use of traditional EMAT, the improvement of Lamb wave tomography performance with the new EMAT will be studied in the next chapter.

2.2 Magnetostrictive Guided SH Wave Direction Controllable EMAT in Steel Plate

Due to its unique characteristics and advantages, the SH guided wave in the metal plate is suitable for the study of the detection of guided wave scattering defects, especially the SH0 mode guided wave. There is no dispersion phenomenon in the propagation of the plate in the steel plate; that is, the group velocity is a constant value, it does not change with the frequency, and there is only one SH0 mode in the low-frequency working region of the SH guide wave. The guided wave makes the guided wave mode in the steel plate pure and single. In addition, when the SH0 guided wave propagated in the steel plate acts with the defect, due to the limitation of the working region of the SH0 guided wave frequency, the restriction of the SH0 guided wave reception of the EMAT structure and the narrowband filter of the signal processing module, there will be little of the guided wave of the mode conversion into the signal data of the guided wave detection, so the guided wave detection letter is obtained. In the number data, the intensity of the guided wave signal of the mode conversion is negligible relative to the pure single SH0 mode guided wave signal intensity in the steel plate, which is beneficial to the establishment of a more accurate model of the defect contour reconstruction of the guided wave scattering defect. Because the magnetostrictive effect is in the dominant position in the electromagnetic energy transfer mechanism of ferromagnetic materials, it is necessary to develop a SH guided wave electromagnetic transducer based on a magnetostrictive steel plate.

At present, the emission direction of the SH guided wave EMAT of the steel plate is generally omnidirectional or directional, and once the position of EMAT is fixed on the structure, the direction of the transmitting or receiving guided wave cannot be changed. On the one hand, under the same excitation signal intensity, the guided wave energy emitted by the omnidirectional emission EMAT is distributed in each direction. The intensity of the guided wave signal intensity in the specific emission direction is weaker than that of the directed emission, which is not conducive to the detection of the defect and the enhancement of the signal-to-noise ratio of the guided wave detection signal. The direction of the directional emission EMAT is single. If the receiving EMAT is still directed, it can only detect the defects in the specific propagation path of the guided wave, and the other regions of the structural parts will be detected blind area, which reduces the range of guided wave detection and reduces the detection efficiency of the guided wave. Given the above problems, it is urgent to develop a new type of ultrasonic guided wave EMAT, which can cover the direction of the transmitting or receiving ultrasonic guided waves as much as possible and can accurately control the direction of its launching or receiving. It not only makes the signal energy of the guided wave concentrated in a certain controlled detection direction, thus improving the aggregation of the guided wave but also can improve the conductivity of the guided wave. Changing the direction of emission or reception makes it possible to detect defects in almost all directions by using approximately uniform guided wave signal intensity. Therefore, guided wave emission or receiving direction controlled EMAT is of great significance for improving the controllability and intelligence of the defect guided wave detection process and improving the accuracy and reliability of defect detection.

In this chapter, in view of the demand of guided wave scattering detection for guided wave mode and characteristics, this chapter studies the propagation characteristics of SH guided wave in steel plate, solves the dispersion equation, draws the dispersion curve of SH guided wave, selects the appropriate guided wave mode and its working point, and secondly, analyses the theoretical basis of SH guided wave EMAT based on magnetostrictive steel plate. The EMAT structure based on the nickel strip and the winding coil is established, and the influence of key dimension parameters and design on the SH guided wave in the EMAT structure is analyzed. Through the study of the propagation characteristics of the SH guided wave and the working mechanism of the SH guided magnetostrictive EMAT of the steel plate, a new SH guided EMAT based on the magnetostrictive steel plate is proposed. The design idea, the expected performance, the composition structure, and the size parameters of the

EMAT are detailed and detailed, and the working principle and feasibility of this specific direction controllable EMAT are provided Working principle and feasibility. Then, the key performance of the direction of SH guided wave excitation and reception in the direction of EMAT is studied in the direction of the steel plate in a certain direction, the amplitude of the SH guided wave and the consistency of the signal intensity in all directions are carried out in all directions, and the key performance of the above parameters is evaluated by defining the relevant parameters. Finally, the direction controlled EMAT performance verification and the testing platform are set up, and the experimental method is used to study the calculation of the direction controlled EMAT to excite the focusing angle of the guided wave in a certain angle and to excite the consistency error of the SH guided wave in different directions.

2.2.1 Magnetization and Magnetostrictive Properties of Ferromagnetic Materials

The magnetization characteristics of ferromagnetic materials can be described by magnetization curves. The magnetization curves of typical ferromagnetic materials are shown in Fig. 3.1. From the magnetization curve, permeability $\mu = B/H$ is a nonlinear function of magnetic field H. According to the different types of applications, permeability can be defined differently (Fig. 2.52.).

When the value of the magnetic field H is very small, the magnetization process is reversible. The slope of the B-H curve at the origin can be approximated as the permeability in the low-value area of the magnetic field, called the initial permeability, which is expressed as the initial permeability.

$$\mu_{in} = \lim_{\Delta H \to 0} \left(\frac{\Delta B}{\Delta H} \right) \Big|_{H=0, B=0} \tag{2.173}$$

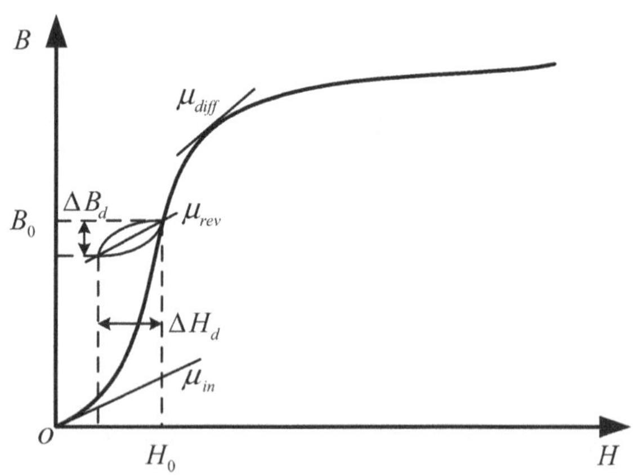

Fig. 2.52 Magnetization curves of typical ferromagnetic materials

The slope at every point other than the initial part of the magnetization curve is called differential permeability, which is expressed as

$$\mu_{\mathrm{diff}} = \lim_{\Delta H \to 0} \left(\frac{\Delta B}{\Delta H} \right) \Big|_{H,B} = \frac{\mathrm{d}B}{\mathrm{d}H} \tag{2.174}$$

In some fields of application, a small periodic component H_d is superimposed on a large static magnetic field, which will produce a small hysteresis on the *B-H* curve, and the large static magnetic field determines the working point on the *B-H* curve. At this time, the slope of the small hysteresis loop at the running point is defined as a reversible permeability.

$$\mu_{rev} = \lim_{\Delta H_d \to 0} \left(\frac{\Delta B_d}{\Delta H_d} \right) \Big|_{H=0,B=0} \tag{2.175}$$

The magnitude of reversible permeability is related to the magnitude of the static magnetic field and the frequency of the dynamic magnetic field.

In EMAT, the magnetization of ferromagnetic materials is corresponding to a small periodic component H_d superposed on a large static magnetic field. Therefore, it is necessary to use reversible permeability when analyzing and calculating. For an isotropic ferromagnetic material, its mechanical properties and permeability will be transversely isotropic under the bias magnetic field, which is similar to the magnetostrictive properties of polycrystalline ferromagnetic materials under the bias magnetic field and the piezoelectric properties of 6/m piezoelectric materials. Thus, the reversible permeability is the same at the interface perpendicular to the bias magnetic field. If the bias magnetic field is in the direction of x_3, the permeability matrix is

$$[\mu] = \begin{bmatrix} \mu_{11} & 0 & 0 \\ 0 & \mu_{11} & 0 \\ 0 & 0 & \mu_{33} \end{bmatrix} \tag{2.176}$$

Ferromagnetic material has a crystalline structure similar to that of iron ions in the center of positive ions of iron. In ferromagnetic materials, the adjacent atoms, due to the electron spin, produce the magnetic moment, and there are interaction forces between the elements and magnetic moments. It drives the adjacent element magnetic moments parallel in the same direction to form a magnetic domain, and the interaction between the magnetic domains is very small. When there is no external magnetic field, the magnetic domains are mutually balanced, and the total magnetization of materials is zero. When an external magnetic field is applied, the equilibrium is destroyed, and the magnetization vector of the domain is turned to the direction of the outer magnetic field, parallel to the external magnetic field. When the magnetization state of the ferromagnetic material changes, the magnetic domain will rotate, making the length or volume of the ferromagnetic material change slightly; this phenomenon is called

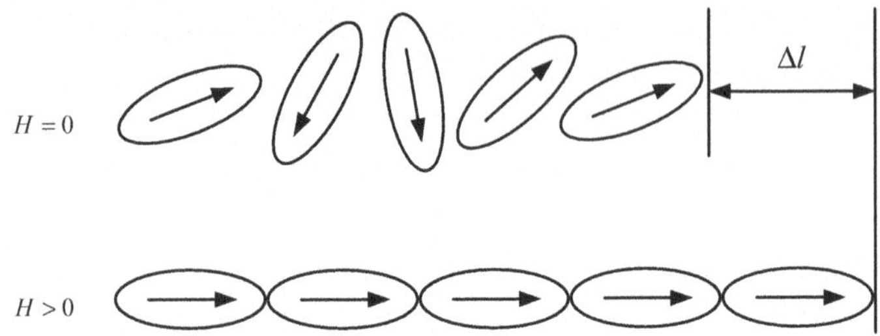

Fig. 2.53 Magnetic domain motion in magnetostriction of ferromagnetic materials

magnetostrictive effect, as shown in Fig. 3.2. There are two forms of magnetostrictive effect: linear magnetostriction and volume magnetostriction (Fig. 2.53).

When a ferromagnetic object is magnetized, it will be accompanied by the spontaneous deformation of the lattice, that is, elongation or shortening along the direction of magnetization, which is called linear magnetostriction. When the magnet is subjected to linear magnetostriction, the volume is almost unchanged. The main length of the magnet varies when the magnetization does not reach saturation. The line magnetostrictive is further divided into two types: longitudinal magnetostriction and transverse magnetostriction. Magnetostriction, which produces relative changes of size along the direction of the external magnetic field, is called longitudinal magnetostriction, and the magnetostriction in the direction of the vertical external magnetic field which produces relative changes in size is called transverse magnetostriction. The linear magnetostriction coefficient is defined as $\gamma = \Delta l / l$, and the length of material is l. γ is the positive value; it is shown that with the enhancement of the external magnetic field, the strain of the material is elongated, which is called positive magnetostriction; conversely, the strain of the material is shortened with the enhancement of the external magnetic field, which is called the negative magnetostrictive. The relationship between magnetostriction and magnetic field can be represented by the hysteresis curve. The magnetostrictive curve of the single crystal iron is shown in Fig. 3.2.

Volume magnetostriction refers to the expansion or contraction of the volume of magnets when the magnetization state changes. Saturation magnetization is mainly caused by volume magnetostriction. In general, ferromagnetic materials, the volume magnetostriction is very small. It is seldom considered in the measurement and research.

Magnetostrictive properties of ferromagnetic materials are similar to those of piezoelectric crystals, so they are also known as piezomagnetic properties of ferromagnetic materials. Unlike the piezoelectric properties, the strain produced by magnetostrictive material has nothing to do with the polarity of the applied magnetic field, that is, the two square relations between the magnetostrictive and the magnetic field, which is similar to the two effects in the electrostriction. When the magnetic field is applied to a small dynamic magnetic field superimposed on a large static bias

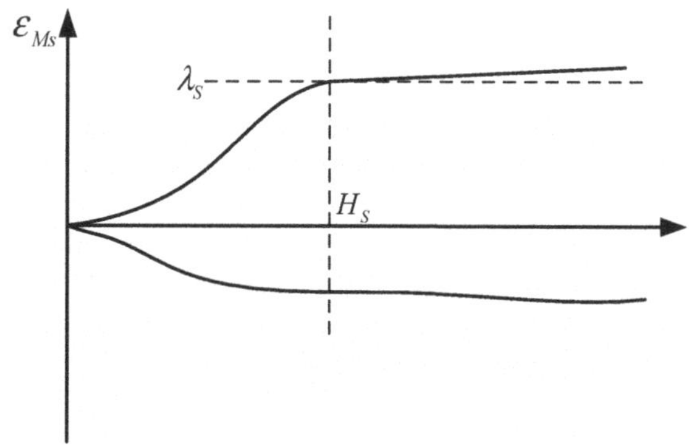

Fig. 2.54 Magnetostrictive curve of single crystal iron

magnetic field, the relationship between hysteresis and the magnetic field will be approximately linear (Fig. 2.54).

Contrary to the hysteresis of magnetic hysteresis, when the size of the ferromagnetic material changes, it will cause the rotation or movement of the magnetic domain and then produce the magnetic effect in the material. This phenomenon is called the reverse magnetostriction. The magnetostriction effect and inverse magnetostriction effect of ferromagnetic materials are the main principles for the excitation and reception of ultrasonic waves in the ferromagnetic specimen.

The relationship between the mechanical properties and magnetic properties of ferromagnetic materials can be represented by the following magneto-elastic constitutive equations.

$$S_i = s_{ij}^H \sigma_j + d_{ki} H_k \tag{2.177}$$

$$\sigma_i = c_{ij}^H S_j - e_{ki} H_k \tag{2.178}$$

$$B_m = d_{mj} \sigma_j + \mu_0 \mu_{mk}^T H_k \tag{2.179}$$

$i, j = 1, \ldots\ldots 6$; $m, k = 1, 2, 3$; Eqs. (2.177) and (2.178) describes the magnetostrictive effect of ferromagnetic materials, and Eq. (2.179) describes the inverse magnetostriction effect of materials. In the formula, σ_i, s_{ij}^H, c_{ij}^H are the stress matrix, the flexible coefficient matrix, and the stiffness coefficient matrix under a certain magnetic field, B_m and μ_{mk}^T is the total magnetic induction intensity and the anisotropic reversible permeability matrix under the constant static stress, and the piezomagnetic matrix and the inverse piezomagnetic matrix, respectively, d_{ki} are defined as the piezomagnetic matrix, and e_{ki} is the inverse piezomagnetic matrix, respectively.

$$d_{ki} = \left.\frac{\partial S_i}{\partial H_k}\right|_\sigma \tag{2.180}$$

$$e_{ki} = -\frac{\partial \sigma_i}{\partial H_k}\bigg|_{\varepsilon} \tag{2.181}$$

H_k is the intensity component of the dynamic magnetic field.

In ferromagnetic materials, the change of magnetic field can make the material produce elastic strain, and the elastic strain can produce a magnetic field inside the material, which is the key to realize the energy conversion based on the magnetostrictive mechanism of EMAT. Equations (3.8) and (3.9) show that the piezomagnetic coefficient d_{ki} and the inverse piezomagnetic matrix e_{ki} can be calculated from the hysteresis curve of the material.

In the theoretical analysis of magnetostrictive mechanism EMAT, the bias magnetic field is assumed to be a uniform magnetic field. The reason is that considering the inhomogeneous distribution of the bias magnetic field, the magnetization direction and the magnetization of each point in the ferromagnetic material are different. It is difficult to describe the magnetization and magnetostrictive properties of such a microcosmic magnetic field and apply it to the analysis and calculation of EMAT. Also, a reasonable magnet design can be used to obtain an approximately uniform magnetic field inside the specimen.

2.2.2 Finite Element Analysis Method Based on Magnetostrictive Mechanism EMAT

The calculation process based on the magnetostrictive mechanism EMAT is similar to the analysis process based on the Lorenz force mechanism EMAT. The difference lies in: (1) Due to the assumption that the bias magnetic field is unidirectional distribution, the magnet model is not needed to be reestablished for the bias magnetic field calculation; (2) the change of the material properties of the specimen material makes the dynamic magnetic field of the coil. Some changes have been made. In the EMAT based on the Lorenz force mechanism, the magnetic properties of the tested samples do not change with the changes in the external magnetic field, but in the magnetostrictive mechanism of EMAT, the ferromagnetic materials have magnetization and magnetostrictive properties. At this time, the calculation of magnetostrictive force and the calculation of magnetostrictive current density become the key to realize the multifield coupling of EMAT.

From the magnetization characteristics of ferromagnetic materials and the working mechanism of EMAT, it is known that the dynamic magnetic field produced by the excitation coil is relatively small under the effect of strong bias magnetic field, which makes the value of the superimposed dynamic magnetic field to fluctuate near the bias magnetic field, so that the permeability of the material corresponds to the reversible magnetic permeability at the paranoid magnetic field, which is a constant. The specific value can be measured by the experiment.

In the process of ultrasonic excitation, the current density produced by the inverse magnetostriction effect inside the material is much smaller than the eddy current density, so the influence is often neglected. In this way, the dynamic magnetic field generated by the coil satisfies the following equations.

$$\frac{1}{\boldsymbol{\mu}}\nabla^2 \boldsymbol{A} - \sigma\frac{\partial \boldsymbol{A}}{\partial t} + \frac{1}{S}\iint_S \sigma\frac{\partial \boldsymbol{A}}{\partial t}ds = -\frac{\boldsymbol{i}}{S} \tag{2.182}$$

The ferromagnetic material exhibits magnetic transversely isotropy under magnetization, and the specific permeability matrix depends on the direction of the bias magnetic field. Here, the general permeability matrix is used to represent the magnetic permeability matrix.

$$\boldsymbol{\mu} = \begin{bmatrix} \mu_{11} & 0 & 0 \\ 0 & \mu_{22} & 0 \\ 0 & 0 & \mu_{33} \end{bmatrix} \tag{2.183}$$

The force that causes the internal vibration of the material comes from the Lorenz force, magnetization force, and magnetostriction force of the material. The magnetization force is smaller than the other two forces and can be neglected. The magnitude of the magnetostrictive force is related to the dynamic magnetic field intensity $\boldsymbol{H}$ and the inverse piezomagnetic matrix $\boldsymbol{e}$. Under the condition of free stress, the magnetostrictive force can be expressed as

$$\boldsymbol{f}_{Ms} = -\nabla_t\left(\mathrm{e}^T \boldsymbol{H}\right) \tag{2.184}$$

where $\boldsymbol{e} = [e_{\mathrm{kj}}]$.

The total force of the ferromagnetic specimen is

$$f = f_{\mathrm{L}} + f_{Ms}$$

Under the joint action of Lorenz force and magnetostriction force, the equilibrium equation of the displacement of the specimen is calculated.

$$(\nabla \cdot c\nabla \boldsymbol{u}) + \boldsymbol{f} = \rho\frac{\partial^2 \boldsymbol{u}}{\partial t^2}$$

$$c = \begin{bmatrix} c_{11} & c_{12} & c_{13} & c_{14} & c_{15} & c_{16} \\ c_{21} & c_{22} & c_{23} & c_{24} & c_{25} & c_{26} \\ c_{31} & c_{32} & c_{33} & c_{34} & c_{35} & c_{36} \\ c_{41} & c_{42} & c_{43} & c_{44} & c_{45} & c_{46} \\ c_{51} & c_{52} & c_{53} & c_{54} & c_{55} & c_{56} \\ c_{61} & c_{62} & c_{63} & c_{64} & c_{65} & c_{66} \end{bmatrix}$$

$$\boldsymbol{u} = [\,u \;\; v \;\; w\,]^T \tag{2.185}$$

When the ultrasonic signal is received, the motion of the specimen's particles produces a dynamic magnetic field under the static bias magnetic field. The dynamic magnetic field induces the voltage in the receiving coil, thus realizing the reception of the ultrasonic signal.

Reverse magnetostrictive effect of materials is

$$\boldsymbol{B}_{Ms} = \boldsymbol{eS} \tag{2.186}$$

$\boldsymbol{B}_{Ms}$ is the magnetic induction intensity produced by the displacement of material particles.

The current density in the specimen is

$$\boldsymbol{J}_{Ms} = \nabla \times \boldsymbol{B}_{Ms} \tag{2.187}$$

$\boldsymbol{J}_{Ms}$ is the magnetostriction current density.

Lorenz current density equation caused by particle motion is the same as Eq. (2.22).

Therefore, the total current density in the specimen is

$$\boldsymbol{J} = \boldsymbol{J}_{\mathrm{L}} + \boldsymbol{J}_{Ms} \tag{2.188}$$

The control equations for each region of the receiving coil and the specimen are as follows.

$$-\frac{1}{\mu}\nabla^2 \boldsymbol{A} + \sigma\frac{\partial \boldsymbol{A}}{\partial t} - \frac{\sigma}{S}\frac{\partial}{\partial t}\iint\limits_{\Omega_c} \mathbf{A}\mathrm{d}s = \boldsymbol{J} \tag{2.189}$$

After obtaining the magnetic vector potential of each region, the induced electromotive force of the receiving coil can be calculated. The electric induction field in the coil conductor can be expressed as

$$\boldsymbol{E} = -\frac{\partial \boldsymbol{A}}{\partial t} \tag{2.190}$$

The electromotive force of a conductor in a coil can be obtained by the line integral of the electric field intensity.

$$V_{\mathrm{pout}} = \int\limits_{l} -\frac{\partial \boldsymbol{A}}{\partial t} \cdot \mathrm{d}\boldsymbol{l} \tag{2.191}$$

The output voltage of the coil can be obtained by averaging the electromotive force of the pointed conductor contained in the coil.

$$V_{\text{out}} = \frac{\int_{\Omega} V_{\text{pout}} \mathrm{d}\Omega}{\int_{\Omega} \mathrm{d}\Omega} \tag{2.192}$$

The two-dimensional finite element method is used to analyze and calculate the EMAT. The EMAT model is used to establish the mathematical model of the plate specimen. In a two-dimensional rectangular coordinate system, only considering the magnetic field and the x and y components of the mechanical field, the vector magnetic potential and eddy current have only z components. In order to realize the coupling between different fields, the formula for calculating the magnetostrictive force and the hysteresis current density must be deduced.

When the electromagnetic field of the coil is solved by the finite element method, the vector magnetic potential A_z is obtained, and the magnetostrictive force must be expressed by A_z.

According to Eq. (3.13), the inverse piezomagnetic matrix and differential operator are substituted. The hysteresis force in two-dimensional rectangular coordinates can be expressed as

$$f_{Ms} = \begin{bmatrix} f_{Msx} \\ f_{Msy} \\ f_{Msz} \end{bmatrix} = -\begin{bmatrix} \frac{\partial}{\partial x} & 0 & 0 & 0 & 0 & \frac{\partial}{\partial y} \\ 0 & \frac{\partial}{\partial y} & 0 & 0 & 0 & \frac{\partial}{\partial x} \\ 0 & 0 & 0 & \frac{\partial}{\partial y} & \frac{\partial}{\partial x} & 0 \end{bmatrix} \begin{bmatrix} e_{11} & e_{21} & e_{31} \\ e_{12} & e_{22} & e_{32} \\ e_{13} & e_{23} & e_{33} \\ e_{14} & e_{24} & e_{34} \\ e_{15} & e_{25} & e_{35} \\ e_{16} & e_{26} & e_{36} \end{bmatrix} \begin{bmatrix} H_x \\ H_y \\ 0 \end{bmatrix} \tag{2.193}$$

H_x and H_y are x and y components of dynamic magnetic intensity, respectively. Therefore,

$$f_{Ms} = \begin{bmatrix} f_{Msx} \\ f_{Msy} \\ f_{Msz} \end{bmatrix} = -\begin{bmatrix} e_{11} & e_{21} & e_{16} & e_{26} \\ e_{16} & e_{26} & e_{12} & e_{22} \\ e_{15} & e_{25} & e_{14} & e_{24} \end{bmatrix} \begin{bmatrix} \frac{\partial H_x}{\partial x} \\ \frac{\partial H_y}{\partial x} \\ \frac{\partial H_x}{\partial y} \\ \frac{\partial H_y}{\partial y} \end{bmatrix} \tag{2.194}$$

Then

$$\begin{bmatrix} H_x \\ H_y \end{bmatrix} = \begin{bmatrix} \frac{1}{\mu_{xx}} & 0 \\ 0 & \frac{1}{\mu_{yy}} \end{bmatrix} \begin{bmatrix} B_x \\ B_y \end{bmatrix} = \begin{bmatrix} \frac{1}{\mu_{xx}} & 0 \\ 0 & \frac{1}{\mu_{yy}} \end{bmatrix} \begin{bmatrix} \frac{\partial A_z}{\partial y} \\ -\frac{\partial A_z}{\partial x} \end{bmatrix} \tag{2.195}$$

$$\begin{bmatrix} \frac{\partial H_x}{\partial x} \\ \frac{\partial H_y}{\partial x} \\ \frac{\partial H_x}{\partial y} \\ \frac{\partial H_y}{\partial y} \end{bmatrix} = \begin{bmatrix} \frac{1}{\mu_{xx}} & 0 & 0 & 0 \\ 0 & -\frac{1}{\mu_{yy}} & 0 & 0 \\ 0 & 0 & \frac{1}{\mu_{xx}} & 0 \\ 0 & 0 & 0 & -\frac{1}{\mu_{yy}} \end{bmatrix} \begin{bmatrix} \frac{\partial^2 A_z}{\partial x \partial y} \\ \frac{\partial^2 A_z}{\partial x^2} \\ \frac{\partial^2 A_z}{\partial y^2} \\ \frac{\partial^2 A_z}{\partial y \partial x} \end{bmatrix} \tag{2.196}$$

$$f_{Ms} = \begin{bmatrix} f_{Msx} \\ f_{Msy} \\ f_{Msz} \end{bmatrix} = -\begin{bmatrix} e_{11}\ e_{21}\ e_{16}\ e_{26} \\ e_{16}\ e_{26}\ e_{12}\ e_{22} \\ e_{15}\ e_{25}\ e_{14}\ e_{24} \end{bmatrix} \begin{bmatrix} \frac{1}{\mu_{xx}} & 0 & 0 & 0 \\ 0 & -\frac{1}{\mu_{yy}} & 0 & 0 \\ 0 & 0 & \frac{1}{\mu_{xx}} & 0 \\ 0 & 0 & 0 & -\frac{1}{\mu_{yy}} \end{bmatrix} \begin{bmatrix} \frac{\partial^2 A_z}{\partial x \partial y} \\ \frac{\partial^2 A_z}{\partial x^2} \\ \frac{\partial^2 A_z}{\partial y^2} \\ \frac{\partial^2 A_z}{\partial y \partial x} \end{bmatrix} \tag{2.197}$$

Considering the relationship between the dynamic magnetic field intensity H and the magnetic induction intensity B in two-dimensional Cartesian coordinates, it is possible to obtain the relationship between the dynamic magnetic field intensity and the magnetic induction intensity.

$$\begin{bmatrix} H_{Msx} \\ H_{Msy} \end{bmatrix} = -\begin{bmatrix} \frac{1}{\mu_{xx}} & 0 \\ 0 & \frac{1}{\mu_{yy}} \end{bmatrix} \begin{bmatrix} e_{11}\ e_{12}\ e_{13}\ e_{14}\ e_{15}\ e_{16} \\ e_{21}\ e_{22}\ e_{23}\ e_{24}\ e_{25}\ e_{26} \end{bmatrix} \begin{bmatrix} S_1 \\ S_2 \\ S_3 \\ S_4 \\ S_5 \\ S_6 \end{bmatrix} \tag{2.198}$$

$\boldsymbol{S} = \begin{bmatrix} S_1 \\ S_2 \\ S_3 \\ S_4 \\ S_5 \\ S_6 \end{bmatrix}$ is the strain tensor matrix; H_{Msx} and H_{Msy} are the magnetostrictive magnetic field intensity.

The following geometric equations are satisfied between strain and displacement in the specimen.

$$\begin{bmatrix} S_1 \\ S_2 \\ S_3 \\ S_4 \\ S_5 \\ S_6 \end{bmatrix} = \begin{bmatrix} \frac{\partial}{\partial x} & 0 & 0 \\ 0 & \frac{\partial}{\partial y} & 0 \\ 0 & 0 & 0 \\ 0 & 0 & \frac{\partial}{\partial y} \\ 0 & 0 & \frac{\partial}{\partial x} \\ \frac{\partial}{\partial y} & \frac{\partial}{\partial x} & 0 \end{bmatrix} \begin{bmatrix} u \\ v \\ w \end{bmatrix} \tag{2.199}$$

The following derivation can be obtained

$$\begin{bmatrix} \frac{\partial H_{Msy}}{\partial x} \\ \frac{\partial H_{Msx}}{\partial y} \end{bmatrix} = \begin{bmatrix} \frac{1}{\mu_{xx}}\left[e_{11}\frac{\partial^2 u}{\partial y \partial x} + e_{12}\frac{\partial^2 v}{\partial y^2} + e_{14}\frac{\partial^2 w}{\partial y^2} + e_{15}\frac{\partial^2 w}{\partial y \partial x} + e_{16}\left(\frac{\partial^2 u}{\partial y^2} + \frac{\partial^2 v}{\partial y \partial x}\right)\right] \\ \frac{1}{\mu_{yy}}\left[e_{21}\frac{\partial^2 u}{\partial x^2} + e_{22}\frac{\partial^2 v}{\partial x \partial y} + e_{24}\frac{\partial^2 w}{\partial x \partial y} + e_{25}\frac{\partial^2 w}{\partial x^2} + e_{26}\left(\frac{\partial^2 v}{\partial x^2} + \frac{\partial^2 u}{\partial x \partial y}\right)\right] \end{bmatrix} \tag{2.200}$$

Magnetostriction current density is

$$\boldsymbol{J}_{Ms} = \nabla \times \boldsymbol{H}_{Ms} = \begin{vmatrix} \hat{x} & \hat{y} & \hat{z} \\ \frac{\partial}{\partial x} & \frac{\partial}{\partial y} & 0 \\ H_{Msx} & H_{Msy} & 0 \end{vmatrix} = \left(\frac{\partial H_{Msy}}{\partial x} - \frac{\partial H_{Msx}}{\partial y} \right) \hat{z} \tag{2.201}$$

The cylindrical coordinate system is suitable for analyzing the thickness of ferromagnetic materials and considering the excitation, propagation, and reception of guided waves propagating along the pipe axis. Considering the axisymmetric characteristics of spiral and circular pipes, the problem can be simplified to two-dimensional axisymmetric problems. In 2D axisymmetric coordinates, only the azimuth component of the vector potential A_θ must be considered.

According to the magnetostrictive property of ferromagnetic material, the stress in the specimen caused by the magnetic field is

$$\begin{bmatrix} \sigma_1 \\ \sigma_2 \\ \sigma_3 \\ \sigma_4 \\ \sigma_5 \\ \sigma_6 \end{bmatrix} = - \begin{bmatrix} e_{11} & e_{21} & e_{31} \\ e_{12} & e_{22} & e_{32} \\ e_{13} & e_{23} & e_{33} \\ e_{14} & e_{24} & e_{34} \\ e_{15} & e_{25} & e_{35} \\ e_{16} & e_{26} & e_{36} \end{bmatrix} \begin{bmatrix} H_{\mathrm{r}} \\ 0 \\ H_z \end{bmatrix} \tag{2.202}$$

$\boldsymbol{\sigma} = \begin{bmatrix} \sigma_1 \\ \sigma_2 \\ \sigma_3 \\ \sigma_4 \\ \sigma_5 \\ \sigma_6 \end{bmatrix}$ is the stress tensor matrix; H_{r} and H_z are the magnetic field intensity in the direction of R and Z, respectively.

The magnetostrictive force of two-dimensional axisymmetric coordinates can be expressed as

$$\boldsymbol{f}_{Ms} = \begin{bmatrix} f_{Msr} \\ f_{Ms\theta} \\ f_{Msz} \end{bmatrix} = - \begin{bmatrix} e_{11} & e_{31} & e_{16} & e_{36} \\ e_{14} & e_{34} & e_{15} & e_{35} \\ e_{16} & e_{36} & e_{13} & e_{33} \end{bmatrix} \begin{bmatrix} \frac{\partial H_{\mathrm{r}}}{\partial r} \\ \frac{\partial H_z}{\partial r} \\ \frac{\partial H_{\mathrm{r}}}{\partial z} \\ \frac{\partial H_z}{\partial z} \end{bmatrix} - \frac{1}{r} \begin{bmatrix} e_{11} - e_{12} & e_{31} - e_{32} \\ 2e_{14} & 2e_{34} \\ e_{16} & e_{36} \end{bmatrix} \begin{bmatrix} H_{\mathrm{r}} \\ H_z \end{bmatrix} \tag{2.203}$$

The relationship between the magnetic field intensity H and the magnetic induction intensity B is

$$\begin{bmatrix} H_{\mathrm{r}} \\ H_z \end{bmatrix} = \begin{bmatrix} \frac{1}{\mu_{rr}} & 0 \\ 0 & \frac{1}{\mu_{zz}} \end{bmatrix} \begin{bmatrix} B_{\mathrm{r}} \\ B_z \end{bmatrix} = \begin{bmatrix} \frac{1}{\mu_{rr}} & 0 \\ 0 & \frac{1}{\mu_{zz}} \end{bmatrix} \begin{bmatrix} -\frac{\partial A_\theta}{\partial z} \\ \frac{\partial A_\theta}{\partial r} \end{bmatrix} \tag{2.204}$$

μ_{rr} and μ_{zz} are the magnetic permeability in the direction of r and z, respectively. In Eq. (2.204), the partial differential is obtained for r and z:

$$\begin{bmatrix} \frac{\partial H_r}{\partial r} \\ \frac{\partial H_z}{\partial r} \\ \frac{\partial H_r}{\partial z} \\ \frac{\partial H_z}{\partial z} \end{bmatrix} = \begin{bmatrix} 0 & 0 & -\frac{1}{\mu_{rr}} \\ \frac{1}{\mu_{zz}} & 0 & 0 \\ 0 & -\frac{1}{\mu_{xx}} & 0 \\ 0 & 0 & \frac{1}{\mu_{zz}} \end{bmatrix} \begin{bmatrix} \frac{\partial^2 A_\theta}{\partial r^2} \\ \frac{\partial^2 A_\theta}{\partial z^2} \\ \frac{\partial^2 A_\theta}{\partial r \partial z} \end{bmatrix} \tag{2.205}$$

The magnetostrictive force is obtained by A_θ

$$f_{Ms} = \begin{bmatrix} f_{Msr} \\ f_{Ms\theta} \\ f_{Msz} \end{bmatrix} = -\begin{bmatrix} e_{11} & e_{31} & e_{16} & e_{36} \\ e_{14} & e_{34} & e_{15} & e_{35} \\ e_{16} & e_{36} & e_{13} & e_{33} \end{bmatrix} \begin{bmatrix} 0 & 0 & -\frac{1}{\mu_{rr}} \\ \frac{1}{\mu_{zz}} & 0 & 0 \\ 0 & -\frac{1}{\mu_{xx}} & 0 \\ 0 & 0 & \frac{1}{\mu_{zz}} \end{bmatrix} \begin{bmatrix} \frac{\partial^2 A_\theta}{\partial r^2} \\ \frac{\partial^2 A_\theta}{\partial z^2} \\ \frac{\partial^2 A_\theta}{\partial r \partial z} \end{bmatrix}$$
$$- \frac{1}{r} \begin{bmatrix} e_{11} - e_{12} & e_{31} - e_{32} \\ 2e_{14} & 2e_{34} \\ e_{16} & e_{36} \end{bmatrix} \begin{bmatrix} -\frac{1}{\mu_{rr}} & 0 \\ 0 & \frac{1}{\mu_{zz}} \end{bmatrix} \begin{bmatrix} \frac{\partial A_\theta}{\partial z} \\ \frac{\partial A_\theta}{\partial r} \end{bmatrix} \tag{2.206}$$

n the 2-D axisymmetric coordinate system, considering the relationship between the dynamic magnetic field intensity H and the magnetic induction intensity B, we can obtain

$$\begin{bmatrix} H_{Msr} \\ H_{Msz} \end{bmatrix} = \begin{bmatrix} \frac{1}{\mu_{rr}} & 0 \\ 0 & \frac{1}{\mu_{zz}} \end{bmatrix} \begin{bmatrix} e_{11} & e_{12} & e_{13} & e_{14} & e_{15} & e_{16} \\ e_{31} & e_{32} & e_{33} & e_{34} & e_{35} & e_{36} \end{bmatrix} \begin{bmatrix} S_1 \\ S_2 \\ S_3 \\ S_4 \\ S_5 \\ S_6 \end{bmatrix} \tag{2.207}$$

H_{Msr} and H_{Msz} are the magnetostrictive magnetic field.
Geometric equations between the strain and displacement are

$$\begin{bmatrix} S_1 \\ S_2 \\ S_3 \\ S_4 \\ S_5 \\ S_6 \end{bmatrix} = \begin{bmatrix} \frac{\partial}{\partial r} & 0 & 0 \\ 0 & 0 & 0 \\ 0 & 0 & \frac{\partial}{\partial z} \\ 0 & \frac{\partial}{\partial z} & 0 \\ \frac{\partial}{\partial z} & 0 & \frac{\partial}{\partial r} \\ 0 & \frac{\partial}{\partial r} - \frac{1}{r} & 0 \end{bmatrix} \begin{bmatrix} u \\ v \\ w \end{bmatrix} \tag{2.208}$$

Then, obtain the differential of magnetic field intensity:

$$\frac{\partial H_{Msr}}{\partial z}=\frac{1}{\mu_{rr}}\left[e_{11}\frac{\partial^2 u}{\partial r\partial z}+e_{13}\frac{\partial^2 w}{\partial z^2}+e_{14}\frac{\partial^2 v}{\partial z^2}+e_{15}\left(\frac{\partial^2 u}{\partial z^2}+\frac{\partial^2 w}{\partial r\partial z}\right)+e_{16}\left(\frac{\partial^2 v}{\partial r\partial z}-\frac{1}{r}\frac{\partial w}{\partial z}\right)\right] \tag{2.209}$$

$$\frac{\partial H_{Msz}}{\partial r}=\frac{1}{\mu_{zz}}\left[e_{31}\frac{\partial^2 u}{\partial r^2}+e_{33}\frac{\partial^2 w}{\partial z\partial r}+e_{34}\frac{\partial^2 v}{\partial z\partial r}+e_{35}\left(\frac{\partial^2 u}{\partial z\partial r}+\frac{\partial^2 w}{\partial r^2}\right)+e_{36}\left(\frac{\partial^2 v}{\partial r^2}-\frac{1}{r}\frac{\partial w}{\partial r}\right)\right] \tag{2.210}$$

The magnetostrictive current density in two-dimensional axisymmetric coordinates is

$$\boldsymbol{J}_{Ms}=\nabla\times\mathbf{H}_{Ms}=\begin{vmatrix}\frac{1}{r}\hat{r} & \hat{\theta} & \frac{1}{r}\hat{z}\\ \frac{\partial}{\partial r} & 0 & \frac{\partial}{\partial z}\\ H_{Msr} & 0 & H_{Mz}\end{vmatrix}=\frac{1}{r}\left(\frac{\partial H_{Msz}}{\partial r}-\frac{\partial H_{Msr}}{\partial z}\right)\hat{\theta} \tag{2.211}$$

The expression of magnetostrictive force and magnetostrictive current can be seen from the expression of magnetostrictive force and magnetostrictive current density. It is necessary to know the inverse piezomagnetic matrix $\boldsymbol{e}$. The elements in the inverse magnetostrictive matrix can be obtained by differentiating the intensity component of the magnetic field in all the strain conditions caused by the external magnetic field known in the tested sample. However, in reality, it is difficult to accurately measure the strain components in the tested samples caused by external magnetism. In the current simulation calculation based on the magnetostrictive mechanism EMAT, it is not necessary to fully know the components of the inverse piezomagnetic matrix, and according to the relation between the static magnetic field and the dynamic magnetic field in the numerical and the direction, the inverse piezomagnetic matrix component needed for the EMAT calculation can be obtained, which makes the realization of the magnetostrictive mechanism EMAT. The simulation analysis is possible.

In the free stress state, only stress and strain caused by magnetic field change are considered.

$$S_i^{(MS)}=d_{ki}^{(MS)}H_k \tag{2.212}$$

$$\sigma_i^{(MS)}=-e_{ki}^{(MS)}H_k \tag{2.213}$$

Then, according to the equation $\sigma_i^{(MS)}=-c_{ij}^H S_j^{(MS)}$

$$\sigma_i^{(MS)}=-c_{ij}^H d_{ki}^{(MS)}H_k=-e_{ik}^{(MS)}H_k \tag{2.214}$$

The inverse piezomagnetic matrix can be calculated by the piezomagnetic matrix

$$e_{ki}^{(MS)}=d_{ki}^{(MS)}c_{ij}^H \tag{2.215}$$

The calculation of the inverse magnetostriction coefficient must be classified according to the direction relationship between the paranoid magnetic field direction and the dynamic magnetic field. The following analysis is carried out for the

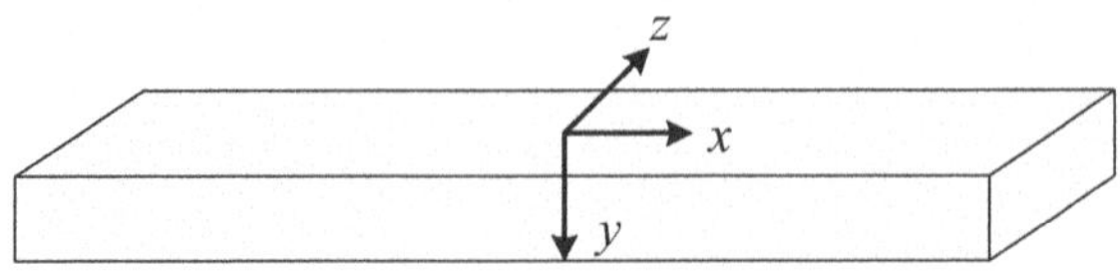

Fig. 2.55 Rectangular coordinate system for the specimen

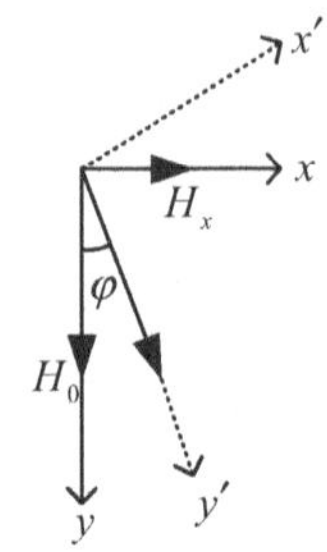

Fig. 2.56 Direction of the magnetic field

rectangular coordinate system of the plate specimen. The inverse magnetostrictive coefficient in the cylindrical coordinate system can be obtained by the r, θ, z analogies of x, y, z (Fig. 2.55).

When the static paranoid magnetic field is perpendicular to the tested sample surface along the y-direction, the corresponding EMAT is mainly composed of two types: body wave EMAT and Lamb wave EMAT. At this time, the magnetostrictive amount of the ferromagnetic sample in the free stress state in the direction of the static magnetic field is $\varepsilon\left(H_{0y}\right)$, the change of the magnetostrictive change is equal to the equal volume deformation, and then, the shape variable perpendicular to the static magnetic field direction is $-\varepsilon\left(H_{0y}\right)/2$.

The strain inside the specimen produced by a static magnetic field is (Fig. 2.56).

$$S_2^0 = \varepsilon\left(H_{0y}\right) \tag{2.216}$$

$$S_1^0 = S_3^0 = -\varepsilon\left(H_{0y}\right)/2 \tag{2.217}$$

$$S_4^0 = S_5^0 = S_6^0 = 0 \tag{2.218}$$

In two-dimensional Cartesian coordinates, the dynamic magnetic field has only H_x and H_y components. In this case, the magnetostriction of the ferromagnetic material will be in the H_t direction of the static magnetic field and the synthetic magnetic field of the dynamic magnetic field, and the angle φ of the H_t and H_0 will change with the change of the dynamic magnetic field in the direction of the synthetic magnetic field of the static magnetic field and the dynamic magnetic field. The synthetic magnetic field will cause the expansion and deformation of the ferromagnetic material, and the new coordinate system is rotated by the original coordinate system by counterclockwise. It is assumed that the object caused by magnetostriction is deformed into equal volume deformation. There are no shear strain components in S_1', S_2' and S_3'.

Then

$$S_2' = \varepsilon_{\mathrm{M}}(H_t) \tag{2.219}$$

$$S_1' = S_2' = -\frac{1}{2}\varepsilon_M(H_t) \tag{2.220}$$

$$S_4' = S_5' = S_6' = 0 \tag{2.221}$$

According to the coordinate transformation of the strain tensor,

$$\varepsilon_{ij}' = Q_{i'k}Q_{j'l}\varepsilon_{kl} \tag{2.222}$$

$$Q_{i'k} = \vec{e}_i' \cdot \vec{e}_k = Q_{ki'} \tag{2.223}$$

Obtain the expression of each strain in the original coordinate system

$$S_1 = S_1' \cos^2\varphi + S_2' \sin^2\varphi \tag{2.224}$$

$$S_2 = S_1' \sin^2\varphi + S_2' \cos^2\varphi \tag{2.225}$$

$$S_3 = S_3' \tag{2.226}$$

$$S_4 = S_5 = 0 \tag{2.227}$$

$$S_6 = \left(S_2' - S_1'\right)\sin 2\varphi \tag{2.228}$$

The elements of the piezomagnetic matrix are

$$\begin{aligned} d_{3z}^{(MS)} = d_{1x}^{(MS)} = \left.\frac{\partial S_1}{\partial H_x}\right|_\sigma &= \frac{3\varepsilon_{\mathrm{M}}}{H_0\cos\varphi}\cos^3\varphi\,\sin\varphi \\ &+ \lambda\,\sin\varphi\left(-\frac{1}{2}\cos^2\varphi + \sin^2\varphi\right) \end{aligned} \tag{2.229}$$

$$\begin{aligned} d_{2z}^{(MS)} = d_{2x}^{(MS)} = \left.\frac{\partial S_2}{\partial H_x}\right|_\sigma &= -\frac{3\varepsilon_{\mathrm{M}}}{H_0\cos\varphi}\cos^3\varphi\,\sin\varphi \\ &+ \lambda\,\sin\theta\left(-\frac{1}{2}\sin^2\varphi + \cos^2\varphi\right) \end{aligned} \tag{2.230}$$

$$d_{1z}^{(MS)} = d_{3x}^{(MS)} = \left.\frac{\partial S_3}{\partial H_x}\right|_\sigma = -\frac{1}{2}\lambda\,\sin\varphi \tag{2.231}$$

$$d_{4x}^{(MS)} = d_{5x}^{(MS)} = 0 \tag{2.232}$$

$$d_{4z}^{(MS)} = d_{6x}^{(MS)} = \left.\frac{\partial S_6}{\partial H_x}\right|_{\sigma} = \frac{3\varepsilon_M}{H_0 \cos\varphi}\cos^2\varphi \cos 2\varphi + \frac{3}{2}\lambda \sin 2\varphi \sin\varphi$$
$$d_{5z}^{(MS)} = d_{6z}^{(MS)} = 0 \tag{2.233}$$

λ represents the slope of the magnetostrictive curve of the specimen.

When the dynamic magnetic field is far less than the value of the static magnetic field,

$$S_2 = \left(\frac{\partial S_2}{\partial H_y}\right) H_y \tag{2.234}$$

$$S_1 = S_3 = -\frac{1}{2}\left(\frac{\partial S_2}{\partial H_y}\right) H_y \tag{2.235}$$

$$S_4 = S_5 = S_6 = 0 \tag{2.236}$$

The piezomagnetic coefficient of the y-direction is

$$d_{1y}^{(MS)} = d_{3y}^{(MS)} = -\frac{1}{2}\lambda \tag{2.237}$$

$$d_{2y}^{(MS)} = \lambda \tag{2.238}$$

$$d_{4y}^{(MS)} = d_{5y}^{(MS)} = d_{6y}^{(MS)} = 0 \tag{2.239}$$

The piezomagnetic matrix of the magnetic material can be expressed as

$$\left[d_{ik}^{(MS)}\right] = \begin{bmatrix} d_{1x}^{(MS)} & d_{2x}^{(MS)} & d_{3x}^{(MS)} & 0 & 0 & d_{6x}^{(MS)} \\ d_{1y}^{(MS)} & d_{2y}^{(MS)} & d_{1y}^{(MS)} & 0 & 0 & 0 \\ d_{3x}^{(MS)} & d_{2x}^{(MS)} & d_{1x}^{(MS)} & d_{6x}^{(MS)} & 0 & 0 \end{bmatrix}^T \tag{2.240}$$

When the static bias magnetic field is far larger than the dynamic magnetic field,

$$\left[d_{ik}^{(MS)}\right] = \begin{bmatrix} 0 & 0 & 0 & 0 & 0 & \frac{3\varepsilon_M}{H_0} \\ -\frac{\lambda}{2} & \lambda & -\frac{\lambda}{2} & 0 & 0 & 0 \\ 0 & 0 & 0 & \frac{3\varepsilon_M}{H_0} & 0 & 0 \end{bmatrix}^T \tag{2.241}$$

The mechanical properties and permeability of isotropic ferromagnetic materials will be transversely isotropic when the bias magnetic field is applied. Based on this, the magnetostriction behavior of polycrystalline ferromagnetic materials with a bias magnetic field is similar to that of hexagonal 6/m piezoelectric materials. When the bias field direction is x_2, the stiffness matrix of ferromagnetic material is obtained according to the symmetry of crystal structure.

$$
\left[c_{ij}\right] = \begin{bmatrix} c_{11} & c_{13} & c_{12} & 0 & 0 & 0 \\ c_{13} & c_{11} & c_{13} & 0 & 0 & 0 \\ c_{12} & c_{13} & c_{11} & 0 & 0 & 0 \\ 0 & 0 & 0 & c_{44} & 0 & 0 \\ 0 & 0 & 0 & 0 & c_{66} & 0 \\ 0 & 0 & 0 & 0 & 0 & c_{44} \end{bmatrix} \tag{2.242}
$$

Inverse piezomagnetic matrix is

$$
e_{ki}^{(MS)} = \begin{bmatrix} 0 & 0 & 0 & 0 & 0 & c_{44}\frac{3\varepsilon_M}{H_0} \\ \left[c_{13} - \frac{1}{2}(c_{11} + c_{22})\right]\lambda & (c_{11} - c_{13})\lambda & \left[c_{13} - \frac{1}{2}(c_{11} + c_{22})\right]\lambda & 0 & 0 & 0 \\ 0 & 0 & 0 & c_{44}\frac{3\varepsilon_M}{H_0} & 0 & 0 \end{bmatrix} \tag{2.243}
$$

When the static paranoid magnetic field is parallel to the sample surface along the z-direction, the corresponding EMAT is SH guided wave EMAT.

Taking into account the symmetric relationship of the coordinate system, the z-axis component of the piezomagnetic matrix is interchanged with the y-axis component, so that the piezomagnetic matrix, in this case, can be obtained.

$$
\left[d_{ik}^{(MS)}\right] = \begin{bmatrix} d_{1x}^{(MS)} & d_{2x}^{(MS)} & d_{3x}^{(MS)} & 0 & d_{6x}^{(MS)} & 0 \\ d_{3x}^{(MS)} & d_{2x}^{(MS)} & d_{1x}^{(MS)} & d_{6x}^{(MS)} & 0 & 0 \\ d_{1z}^{(MS)} & d_{2z}^{(MS)} & d_{1z}^{(MS)} & 0 & 0 & 0 \end{bmatrix}^T \tag{2.244}
$$

If the bias field direction is x_3, the ferromagnetic material is transversely isotropic along x_3, and the stiffness matrix under this condition is

$$
\left[c_{ij}\right] = \begin{bmatrix} c_{11} & c_{13} & c_{12} & 0 & 0 & 0 \\ c_{13} & c_{11} & c_{13} & 0 & 0 & 0 \\ c_{12} & c_{13} & c_{11} & 0 & 0 & 0 \\ 0 & 0 & 0 & c_{44} & 0 & 0 \\ 0 & 0 & 0 & 0 & c_{44} & 0 \\ 0 & 0 & 0 & 0 & 0 & c_{66} \end{bmatrix} \tag{2.245}
$$

The corresponding inverse piezomagnetic matrix is

$$
e_{ki}^{(MS)} = \begin{bmatrix} 0 & 0 & 0 & 0 & c_{44}\frac{3\varepsilon_M}{H_0} & 0 \\ 0 & 0 & 0 & c_{44}\frac{3\varepsilon_M}{H_0} & 0 & 0 \\ \left[c_{13} - \frac{1}{2}(c_{11} + c_{22})\right]\lambda & \left[c_{13} - \frac{1}{2}(c_{11} + c_{22})\right]\lambda & (c_{11} - c_{13})\lambda & 0 & 0 & 0 \end{bmatrix} \tag{2.246}
$$

When the bias magnetic field is parallel to the test sample surface along the x-direction, the corresponding EMAT is Lamb wave and surface wave EMAT.

In the same way as the two cases, the inverse piezomagnetic matrix under this condition can be obtained.

$$e_{ki}^{(MS)} = \begin{bmatrix} (c_{11}-c_{13})\lambda & \left[c_{13}-\frac{1}{2}(c_{11}+c_{22})\right]\lambda & \left[c_{13}-\frac{1}{2}(c_{11}+c_{22})\right]\lambda & 0 & 0 & 0 \\ 0 & 0 & 0 & 0 & 0 & c_{44}\frac{3\varepsilon_M}{H_0} \\ 0 & 0 & 0 & 0 & c_{44}\frac{3\varepsilon_M}{H_0} & 0 \end{bmatrix} \tag{2.247}$$

The finite element software COMSOL Multiphysics is used to realize the multi-field coupling numerical simulation based on magnetostriction mechanism EMAT.

The weak eddy currents of the coil in the rectangular coordinate system and the equilibrium equations of the sample are obtained, respectively.

$$\begin{aligned} &\int_\Omega \left(\frac{1}{\mu}\frac{\partial A_z}{\partial x}\frac{\partial \delta A_z}{\partial x} + \frac{1}{\mu}\frac{\partial A_z}{\partial y}\frac{\partial \delta A_z}{\partial y}\right)\mathrm{d}A \\ &= -\int_\Omega \sigma\frac{\partial A_z}{\partial t}\delta A\mathrm{d}A + \int_\Omega \frac{1}{S}\iint\limits_S \sigma\frac{\partial A_z}{\partial t}\mathrm{d}s\delta A_z\mathrm{d}A \\ &+ \int_\Omega J_{sz}\delta A_z\mathrm{d}A + \int_\Gamma \frac{1}{\mu}\frac{\partial A_z}{\partial n}\delta A_z\mathrm{d}l \end{aligned} \tag{2.248}$$

$$\begin{aligned} &\int\limits_\Omega (\sigma_x\delta\varepsilon_x + \sigma_y\delta\varepsilon_y + \sigma_{xy}\delta\varepsilon_{xy})\mathrm{d}V \\ &+ \int\limits_V \left(\rho\frac{\partial^2 u}{\partial t^2}\delta u + \rho\frac{\partial^2 v}{\partial t^2}\delta v\right)\mathrm{d}A \\ &= \int\limits_\Omega (f_x\delta u + f_y\delta v)\mathrm{d}A + \int\limits_\Gamma (T_x\delta u + T_y\delta v)\mathrm{d}l \end{aligned} \tag{2.249}$$

μ represents the permeability matrix of the sample.

In the axisymmetric coordinate system, the weak form of the eddy current field of the coil and the equilibrium motion equation of the sample are shown, respectively.

$$\begin{aligned} &\int\limits_\Omega \frac{1}{\mu}\left(r\frac{\partial A_\theta}{\partial r}\frac{\partial \delta A_\theta}{\partial r} + r\frac{\partial A_\theta}{\partial z}\frac{\partial \delta A_\theta}{\partial z}\right)\mathrm{d}A \\ &= -\int\limits_\Omega r\sigma\frac{\partial A_\theta}{\partial t}\delta A_\theta\mathrm{d}A + \int\limits_\Omega rJ_{s\theta}\delta A_\theta\mathrm{d}A \\ &+ \int\limits_\Omega r\frac{1}{S}\iint\limits_S \sigma\frac{\partial A_\theta}{\partial t}\mathrm{d}s\delta A_\theta\mathrm{d}A + \int\limits_\Gamma r\frac{1}{\mu}\frac{\partial A_\theta}{\partial n}\delta A_\theta\mathrm{d}l \end{aligned} \tag{2.250}$$

$$\begin{aligned} &\int_\Omega (\sigma_{rr}\delta\varepsilon_{rr} + \sigma_{zz}\delta\varepsilon_{zz} + \sigma_{rz}\delta\varepsilon_{rz})\mathrm{d}A \\ &+ \int_\Omega \left(\rho\frac{\partial^2 u}{\partial t^2}\delta u + \rho\frac{\partial^2 v}{\partial t^2}\delta v\right)\mathrm{d}A \\ &= \int_\Omega \left[\left(f_r + \frac{\sigma_{rr}-\sigma_{\theta\theta}}{r}\right)\delta u\right. \\ &+ \left.\left(\frac{\sigma_{rz}}{r} + f_z\right)\delta v\right)\mathrm{d}A + \int_\Gamma (T_r\delta u + T_z\delta v)\mathrm{d}l \end{aligned} \tag{2.251}$$

The finite element software COMSOL Multiphysics can be used to realize the numerical simulation of the whole process of ultrasonic excitation, propagation, and reception of the multifield coupling equation based on the magnetostriction mechanism EMAT, which is known as the weak form of the multifield coupling equation, the hysteresis force, and the magnetostriction current density. In the process of numerical simulation, the two quantities of magnetostriction force are realized from the electromagnetic field to mechanical field and coupling from the mechanical field to the electromagnetic field.

The calculation steps in COMSOL Multiphysics used for magnetostriction mechanism EMAT numerical simulation analysis are shown as follows:

(1) According to the magnetization curve and the magnetostriction curve of the ferromagnetic material, the values of the ε_M and λ the ferromagnetic samples under the set working state are calculated, the permeability of the tested samples and the values of the elements of the stiffness matrix of the sample are determined, and the values of the elements in the inverse piezomagnetic matrix are obtained.
(2) According to the specific type of EMAT, select 2D rectangular coordinate system or axisymmetric coordinate system.
(3) Choose the solution mode of different fields as the weak form and choose three sets of modes.
(4) Establish the geometric model of each component of EMAT. In the calculation of electromagnetic field, we need to give a solution area, so we need to delineate a solution area outside the EMAT model.
(5) According to the boundary conditions, the solution area and boundary are, respectively, set. The coupling variable between various fields is set, Lorenz force and magnetostriction force are set as the driving force of the vibration of the tested sample, and the current density of Lorenz current density and magnetostriction current density is set as the source current density in the tested sample when the ultrasonic receiving is received. Because it is necessary to calculate the induction electromotive force of the coil conductor when the coil receives the ultrasonic signal, and it is necessary to set the integral coupling variable to calculate the result of the induction electromotive force (EMF) for each step length.
(6) Grid subdivision of the solution area. When the grid is dissecting, it is necessary to pay attention to two points to improve the accuracy of calculation: One is that more than 2 grid units should be set in the skin depth on the surface of the tested sample, and two, within the sample, more than 7 grid elements should be guaranteed in the wavelength of the ultrasonic excitation.
(7) The transient solution of the model is carried out. Solving step size, relative error, and absolute error must be set. Since the pulse excitation signal is a high-frequency tone-burst signal, it is necessary to set a smaller step to ensure the smoothness and stability of the obtained solution. When solving the displacement of the internal particle of the test sample, the displacement value is smaller, and the absolute error of the solution is also small enough to ensure the correctness of the solution.

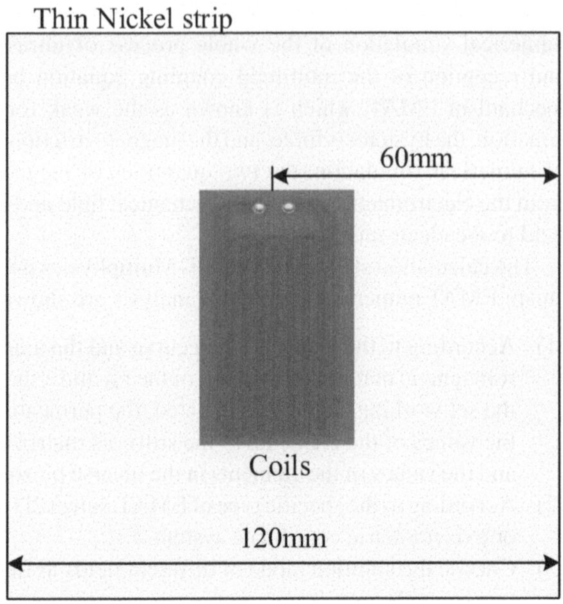

Fig. 2.57 Simulation and experiment model

Table 2.10 Coil size and material parameters

Parameter	Value
Substrate thickness	0.500 mm
Copper platinum width	1.625 mm
Copper platinum thickness	0.05 mm
Spacing	6.50 mm
Electrical conductivity of copper foil	2.667×10^7 S/m

(8) Post-processing of the results, including displaying the cloud and transient waveforms of each field variable.

In order to verify the deduced formula and the validity and correctness of the analytical method, the correctness of the result is verified by the calculation example and the method of physical experiment (Fig. 2.57).

The Lamb wave is excited and received on a thin nickel strip with a meander line coil. The coil is a single, unsplit loop, and the nickel band is pure nickel band. The size and material parameters of the coil and nickel band are given in Tables 2.10 and 2.11, respectively. The phase velocity and group velocity dispersion curves of the Lamb wave in the nickel strip are shown in Fig. 2.58.

According to the dispersion curve of the nickel strip, the Lamb wave excited by S0 mode in the nickel strip is selected. To meet the matching relationship between the distance between the folded coil and the Lamb wavelength, the excitation frequency is 783.3 kHz. At this frequency, the corresponding phase velocity and group velocity of the S0 mode Lamb wave are 5074 and 5067 m/s, respectively.

Table 2.11 Size and parameters of thin nickel plate

Parameter name	Value
Length	150 mm
Width	120 mm
Thickness	0.3 mm
Conductivity	1.43 × e7 S/m
Modulus of elasticity	206.9 GPa
density	8880 kg/m3
Poisson ratio	0.31

Fig. 2.58 Dispersion curve of nickel band

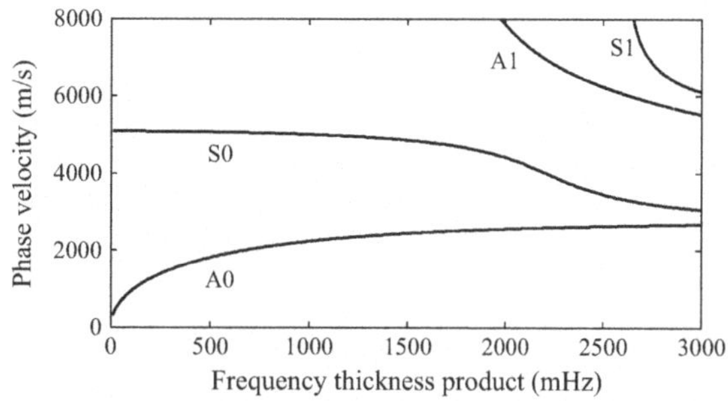

(a) Phase velocity dispersion curve of nickel band.

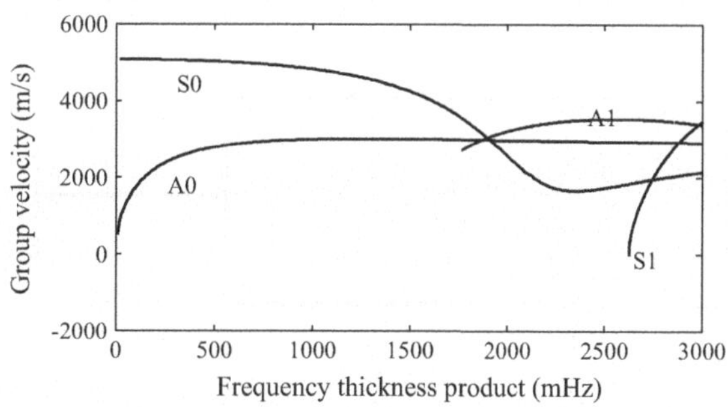

(b) Group velocity dispersion curve of nickel band.

The way EMAT works is spontaneous self-collection. The coil excites ultrasonic waves in the nickel strip and propagates in two directions. When the ultrasonic wave is reflected at the boundary field, the signal is received again when it reaches the coil. Because the distance between the two ends of the coil is equal, the reflection signal of the two boundaries will arrive at the coil at the same time.

In the experiment, the RF tone-burst signal is generated by the RITEC EPR-4000 pulse occurrence and the receiver as the excitation source of the EMAT probe. The amplitude and frequency of the signal can be adjusted according to the demand. EPR-4000 can also realize the filtering and amplification of the receiving signal of the coil, and the bandwidth and magnification of the filter can also be adjusted. Between the coil of EPR-4000 and EMAT, impedance matching device must be used to match the impedance of the coil and the output impedance of the excitation source, to achieve larger power output. The receiving signal of EPR-4000 can be output by the connection oscilloscope, and the data acquisition and waveform display on the PC machine can be realized by the data acquisition software Wave Star for Oscilloscopes, which is matched with the oscilloscope. The test equipment and its connection are shown in Fig. 2.59.

In this case, the direction of the static bias magnetic field is perpendicular to the surface of the nickel band, and the static bias magnetic field is provided by the remanence after the magnetization of the nickel band, and the residual magnetic induction intensity of the nickel band is 0.3325T. The magnetization curve and hysteresis curve of pure nickel are taken from references. In order to express the quantity relation of magnetostriction, magnetic induction intensity, and magnetic field intensity, the two sets of curves are fitted by the exponential fitting method. The concrete fitting formula is as follows.

$$\varepsilon_t = -32.31\mathrm{e}^{0.0005659H} + 33.33\mathrm{e}^{-0.0898H} \tag{2.252}$$

$$B = 7.229\mathrm{e}^{-0.0001943H} - 4.465\mathrm{e}^{-0.1129H} \tag{2.253}$$

When the magnetic induction intensity is 0.3325T, the corresponding magnetic field strength is $H_0 = 2.1$ kA/m. The magnetostriction of the thin nickel band is

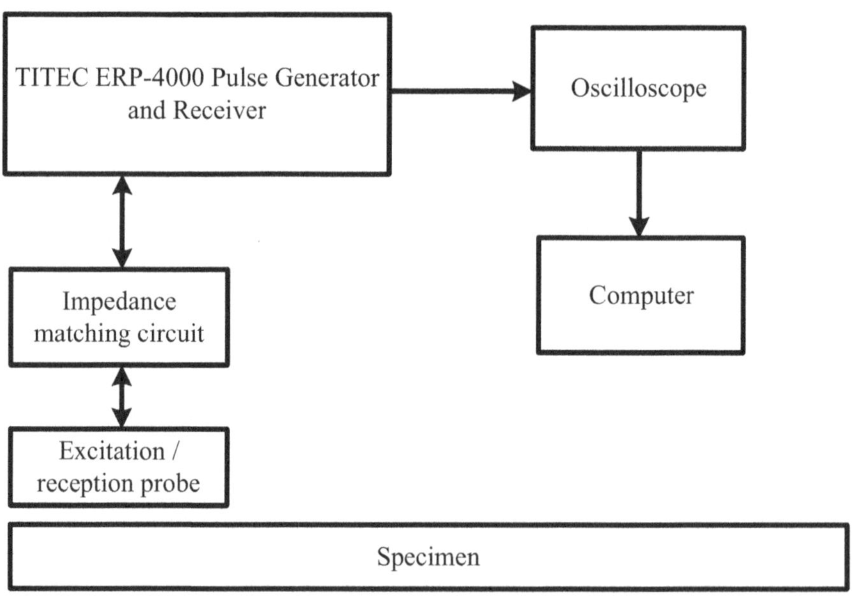

Fig. 2.59 Test connection diagram of spontaneous self-received Lamb wave EMAT

$\varepsilon_M = -4.7467$ ppm. The slope of the magnetostriction curve of thin nickel belt is $\lambda = -2.5171$ ppm. The dynamic permeability of the nickel band is 1.5833×10^{-4} H/m (relative permeability is 126). When the direction of the dynamic magnetic field is by the direction of the bias magnetic field, the total magnetic field only changes in the amplitude and the direction is constant. Therefore, the permeability of the nickel band along the bias magnetic field is the differential permeability. According to the magnetization curve in the figure, differential permeability of nickel band is 3.963×10^{-4} H/m (relative permeability is 315.4) when $H_0 = 2.1$ kA/m. The elements of the stiffness matrix of the nickel band are $c_{11} = 3.906 \times 10^{12}$ Pa, $c_{13} = 6.554 \times 10^{11}$ Pa, $c_{11} = 0$ Pa, $c_{11} = 1.972 \times 10^{11}$ Pa.

In this case, the nickel strip will be affected by the Lorenz force and the magnetostriction force. The formulas are shown, respectively.

$$f_L = -\sigma B_0 \frac{\partial A_z}{\partial t}$$
$$f_{Msx} = c_{44} \frac{1}{\mu_{xx}} \frac{3\varepsilon_M}{H_0} \frac{\partial^2 A_z}{\partial y^2} - \left[c_{13} - \frac{1}{2}(c_{11} + c_{12}) \right] \lambda \frac{1}{\mu_{yy}} \frac{\partial^2 A_z}{\partial x^2} \tag{2.254}$$

When recceiving ultrasonic signals, the Lorenz force cueiving ultrasonic signals, the Lorenz force current density and the hysteresis current density of the samples are calculated, respectively (Figs. 2.60 and 2.61).

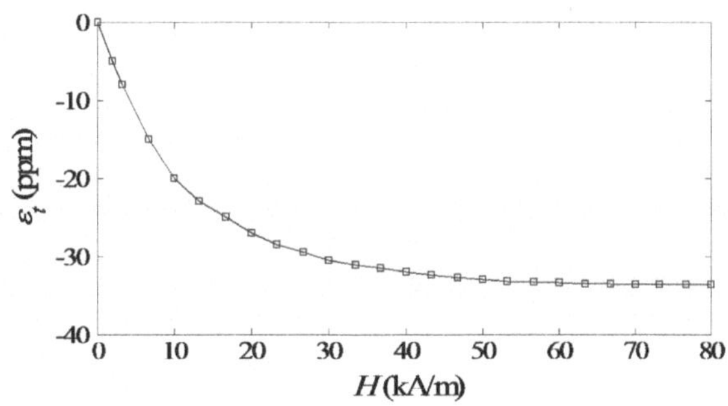

Fig. 2.60 Magnetostriction curve of pure nickel

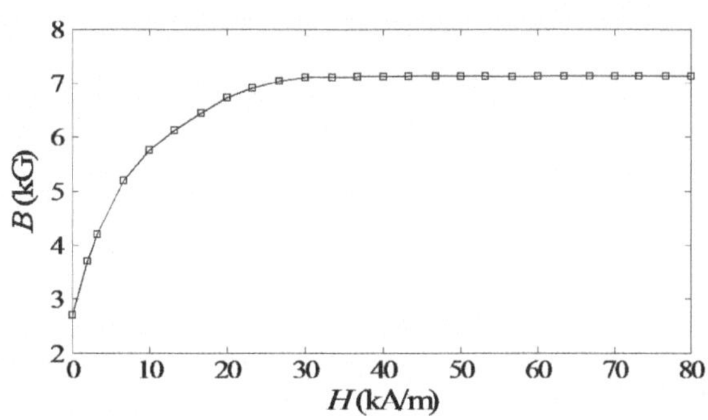

Fig. 2.61 Magnetization curve of pure nickel

$$J_{\mathrm{L}} = B_0 \sigma \frac{\partial u_x}{\partial t} \tag{2.255}$$

$$\begin{aligned} J_{Ms} = & \left[\frac{1}{\mu_{yy}} (c_{11} - c_{13}) \lambda - \frac{1}{\mu_{xx}} c_{44} \frac{3\varepsilon_{\mathrm{M}}}{H_0} \right] \frac{\partial^2 u_y}{\partial x \partial y} \\ & + \frac{1}{\mu_{yy}} \left(c_{13} - \frac{1}{2} c_{11} - \frac{1}{2} c_{12} \right) \lambda \frac{\partial^2 u_x}{\partial x^2} - \frac{1}{\mu_{xx}} c_{44} \frac{3\varepsilon_{\mathrm{M}}}{H_0} \frac{\partial^2 u_x}{\partial y^2} \end{aligned} \tag{2.256}$$

The frequency of the ultrasonic wave to be selected is 783.3 kHz, the number of excitation signals is 3, and the waveform of the excitation signal is (Fig. 2.62)

The excitation current is utilized by COMSOL Multiphysics software, and EMAT is modeled and numerically calculated. In the calculation, the number of cells in each ultrasonic length inside the sample is 8, and the time step is 2ns. In order to display the calculation results more conveniently, the observation point P is selected inside the sample, which is located at the left end 60 mm of the sample and 0.01 mm on the surface of the sample.

Figure 2.63 shows the distribution of the vector magnetic potential line generated by the excitation current within the coil of 10 μs. It is visible that the coil produces a dynamic magnetic field around it and the inside of the tested sample; and in the skin depth of the tested sample, the induced eddy produces Lorentz force under the remanence of the nickel band, and the dynamic magnetic field produced by the coil and the coil is produced. The remanence produces the magnetostriction force. Under the action of these two forces, the sample's internal point vibrates and produces ultrasonic waves. Figure 2.64 shows the distribution of the equivalent line distribution of the Lorenz force current density and the magnetostriction current density generated by the measured sample of the sample at the 25 μs. The pulsed magnetic field generated by the current density and the magnetostriction current density of the tested sample can induce the voltage in the coil and achieve the reception of the signal.

Figures 2.65 and 2.66 show the Lorenz force density and magnetostriction force density at the point of the sample, respectively. It is found that the Lorenz force used

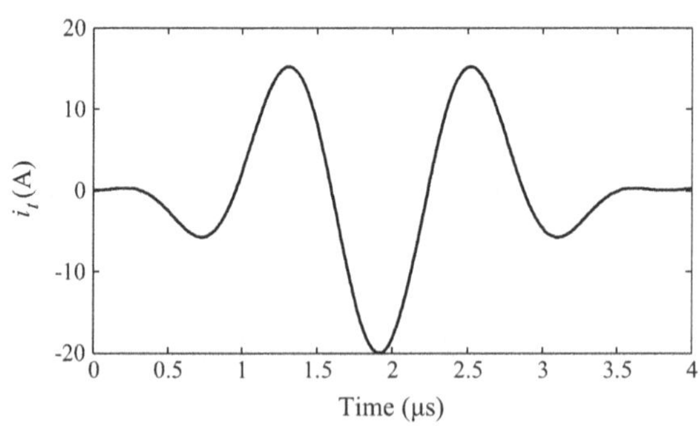

Fig. 2.62 Waveform of the exciting current

to excite the Lamb wave in the nickel strip is less than the magnetostriction force, which shows that the magnetostriction effect plays a decisive role in the excitation of the Lamb wave.

In Figs. 2.67 and 2.68, the x and y components of the particle shifts at the point are given respectively. From these two diagrams, it can be seen that the x component of the point displacement at the point of the point is greater than the y component; the group velocity of the two components is consistent, indicating that the ultrasonic wave of other modes is not excited.

Figure 2.69 shows the cloud diagram of the x component of the particle displacement at different times. It can be clearly seen from Fig. 2.69 that, with the change of time, the ultrasonic propagation in the sample, the reflection after the end face and the reflection after the reflection are propagated to the tested sample position again, and it can be found that the time from ultrasonic excitation to the realization of the first reflection wave is about 25 μs.

Figure 2.70 shows the waveform of the voltage signal received by the probe in the receiving mode. Figure 3.19 shows the signal received by the probe obtained by current excitation using the same number of cycles and frequencies. In order to compare the two waveforms, the amplitude of the first waveform is used as a benchmark to normalize the two waveforms. Comparing the two waveforms, it is found that the relationship between waveform and time is more consistent. In these two waveforms, the first wave packet is the receiving signal when the coil is excited by the ultrasonic wave and begins to propagate; the second, third, fourth wave packets are the received signals after the first, second, third reflection of the ultrasonic end of the ultrasonic wave.

According to the propagation distance and the propagation time, the group velocity of the ultrasonic wave is calculated. The distance between the second and the third wave packets corresponds to the time difference, and the distance of ultrasonic propagation is 120 mm. The corresponding numerical results of theoretical design wave velocity, simulated wave velocity, and experimental wave velocity are given in Table 2.12 (Fig. 2.71).

It can be seen from Table 2.12 that the wave velocity obtained by simulation is very close to the theoretical wave velocity, and the wave velocity obtained from the

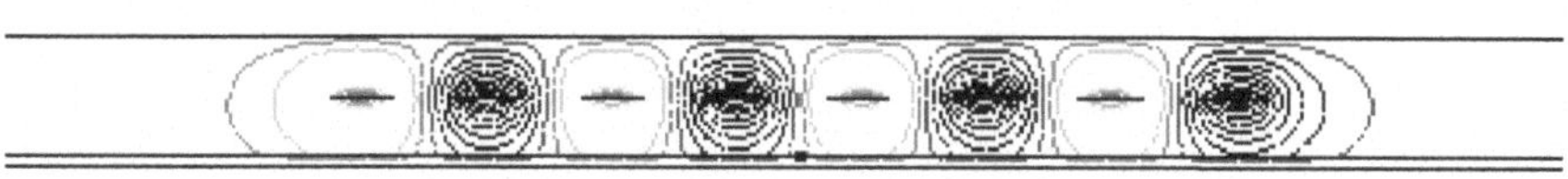

Fig. 2.63 Equipotential line of the magnetic vector potential generated by the current in the coil at 10 μs

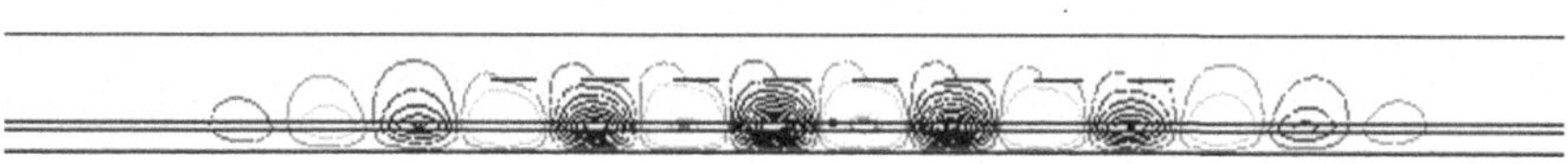

Fig. 2.64 Vector magnetic potential equipotential line produced by current density in the test sample at 25 μs

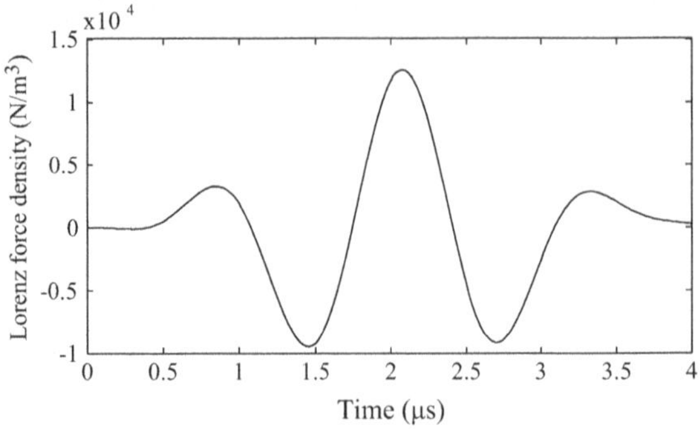

Fig. 2.65 Lorenz force density at the point of the sample

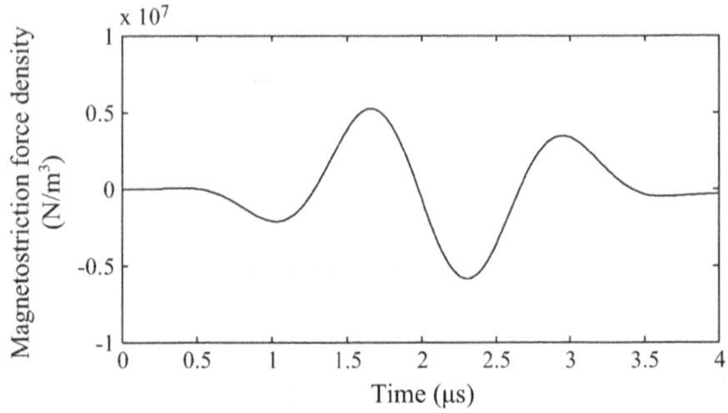

Fig. 2.66 *x* component of the magnetostriction force density at the point of the sample

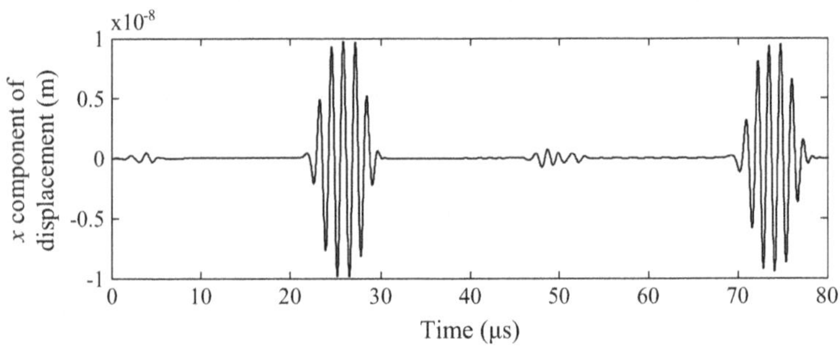

Fig. 2.67 Variation of the *x* component of particle displacement at point *P*

Table 2.12 Comparison of wave velocity values obtained by three methods

Theoretical wave velocity (m/s)	Simulation velocity (m/s)	Experimental velocity(m/s)
5067	5038	5217

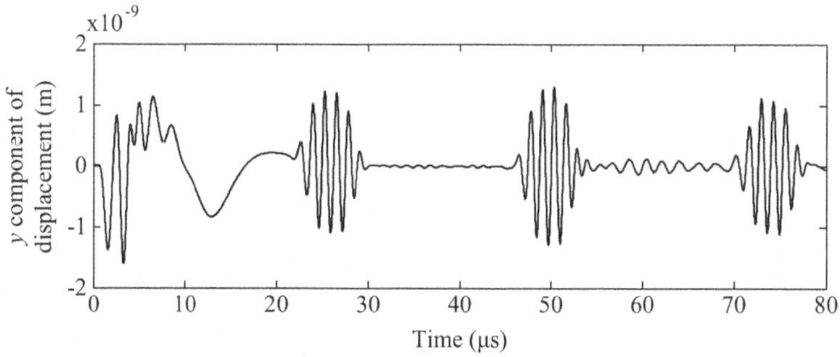

Fig. 2.68 Variation of y component of particle displacement at point P

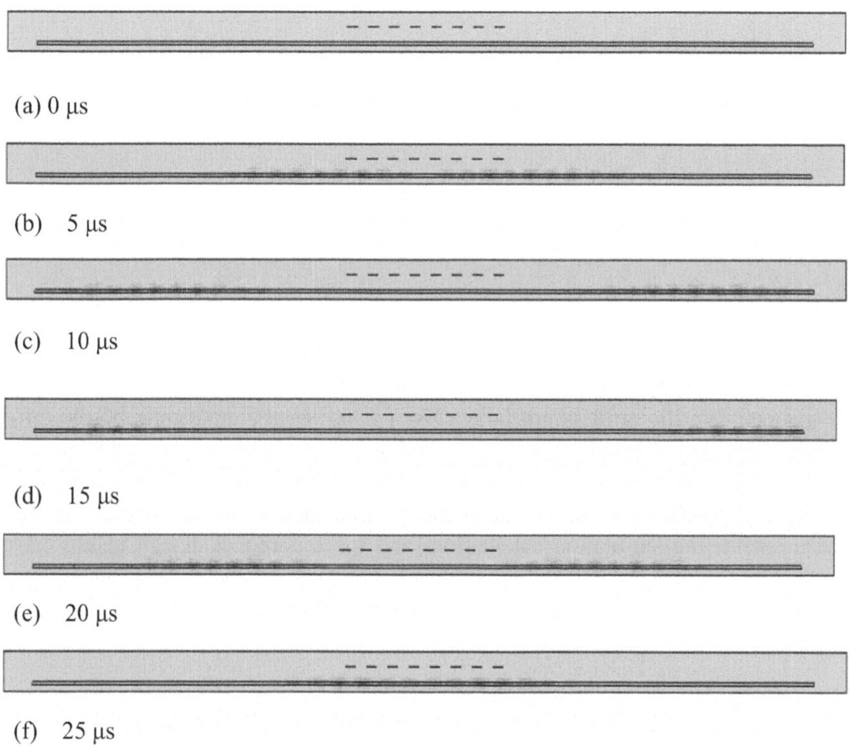

Fig. 2.69 x component of particle displacement at a different time

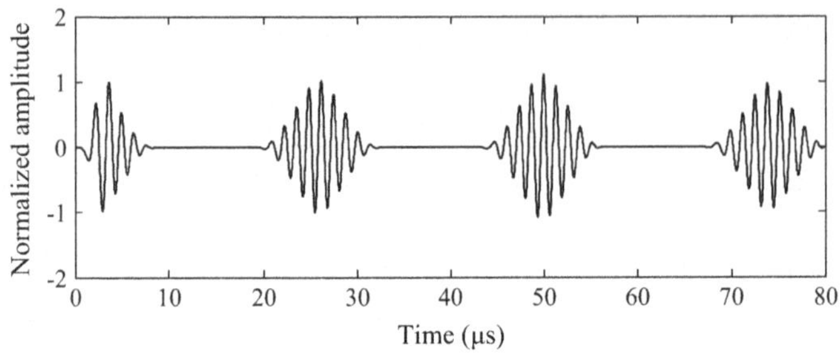

Fig. 2.70 Simulation value of receiving signal of the coil

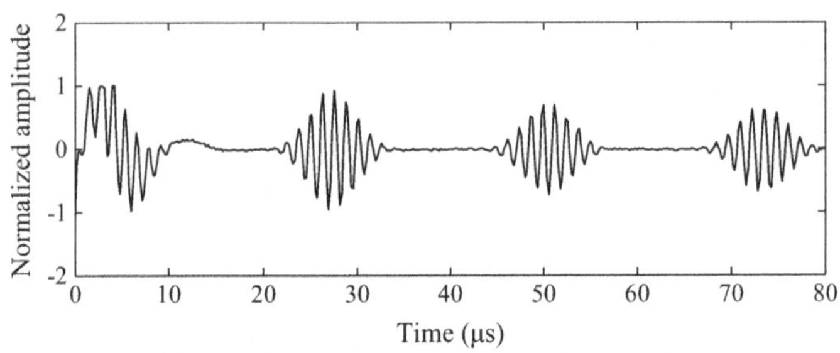

Fig. 2.71 Value of coil receiving signal test

experiment has some error with the theoretical wave velocity and the simulation wave speed, but the error is not large. The reason for the error may be the error between the material parameters of the tested nickel strip and the parameters used in the theoretical calculation. The Lamb wave of S0 mode of design frequency and wave velocity is obtained by simulation and experimental measurement. The experimental measurement, the simulation calculation, and the theoretical design results have a good consistency. The correctness and effectiveness of the simulation analysis method based on the magnetostriction mechanism of EMAT are verified.

2.2.3 Analytical Modeling and Calculation of SH Guided Wave EMAT

SH guided wave is an ideal choice for ferromagnetic materials by electromagnetic ultrasonic method. Because the reflection of SH wave produces only SH waves, no mode conversion will occur. It is very convenient for the reception and signal processing of the ultrasonic wave. The SH wave is evenly distributed at different depths of the ferromagnetic sheet; that is, it will propagate evenly in the plate, and the cracks will reflect at any position. Therefore, it can detect crack defects in any position. In the SH wave mode, the wave speed of the SH0 mode will not change

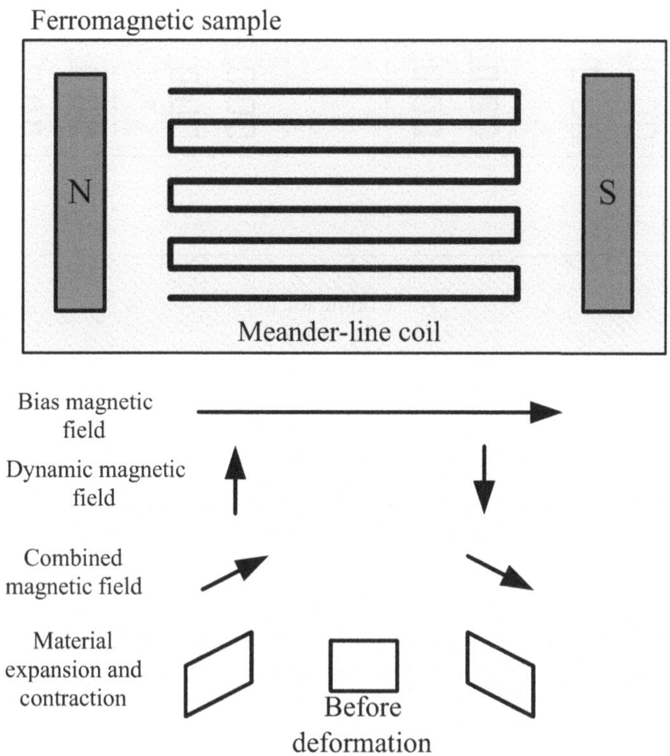

Fig. 2.72 Structure and working principle of SH guided wave EMAT

with the thickness of the plate, so it is suitable for the detection of different thickness of the plate and has strong generality.

The schematic diagram and working principle of SH guided wave EMAT are shown in Fig. 2.72. The direction of the bias magnetic field is parallel to the direction of the coil conductor, and the bias magnetic field is perpendicular to the direction of the dynamic magnetic field produced by the coil. The superposition of the two makes the ferromagnetic sample to produce periodic expansion and deformation, thus stimulating the ultrasonic wave. In the EMAT of the SH guided wave, the direction of the bias magnetic field is parallel to the direction of the eddy generated in the test sample, the test sample is not affected by the Lorenz force, and the electromagnetic ultrasound is produced only by the magnetostriction mechanism (Fig. 2.73).

Consider the case of multilayer folded multi splitting coil placed above the flat sample.

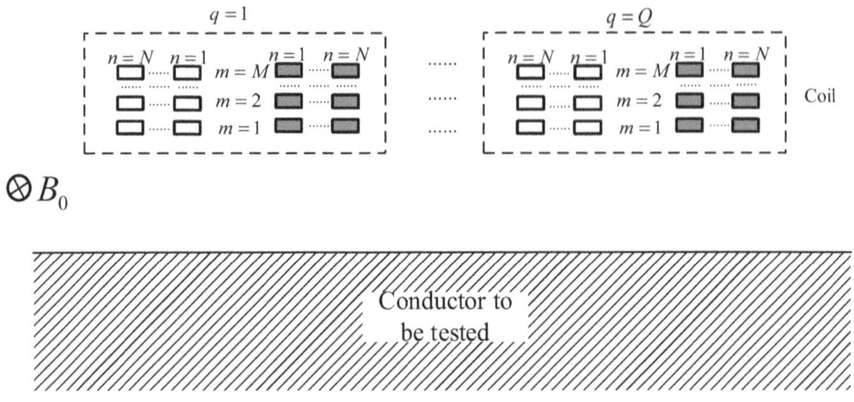

Fig. 2.73 Model of multilayer and multiple meander line coil

The magnetic induction intensity in the sample is

$$\begin{aligned}\mathbf{B}_{(c)} = {} & \frac{\mu_0 I}{2\pi^2 (h_2 - h_1) w} \int\limits_{-\infty}^{+\infty} \int\limits_{-\infty}^{+\infty} \left[\frac{\mathrm{e}^{\lambda z}}{-k^2} \times \frac{2k\mu_{\mathrm{r}}\lambda}{k\mu_{\mathrm{r}} + \lambda} \times (j\alpha\hat{\mathbf{x}} \right. \\ & \left. + j\beta\hat{\mathbf{y}} + \lambda\hat{\mathbf{z}}) \right] \sum_{m=1}^{M} \left(\mathrm{e}^{-kh_{2m}} - \mathrm{e}^{-kh_{1m}}\right) \\ & \times \frac{1}{\alpha\beta} \sum_{n=1}^{N} P_n \sum_{q=1}^{Q} \mathrm{e}^{-j\alpha x'_q} \mathrm{e}^{\mathrm{j}\alpha x} \mathrm{e}^{\mathrm{j}\beta y} \mathrm{d}\alpha \mathrm{d}\beta \end{aligned} \tag{2.257}$$

$\hat{\boldsymbol{x}}$ and $\hat{\boldsymbol{y}}$ are direction vectors.

The impedance of the coil is

$$Z = Z_d + Z_0 + \Delta Z \tag{2.258}$$

In the formula, Z_d is DC impedance; Z_0 is the impedance of free space; ΔZ represents impedance variation caused by the eddy current.

The analytical expressions for each impedance are shown, respectively.

$$
\begin{aligned}
Z_0 = {} & \frac{2j\omega\mu_0}{\pi^2(h_2-h_1)^2 w^2}\sum_{m'=1}^{M}\int_{-\infty}^{+\infty}\int_{-\infty}^{+\infty}\{2(h_{2m'}-h_{1m'}) \\
& +\frac{2}{k}\left[e^{-k(h_{2m'}-h_{1m'})}-1\right]+\sum_{m=1}^{m'-1}\frac{1}{k}\left(e^{kh_{2m}}-e^{kh_{1m}}\right)\left(e^{-kh_{1m'}}\right. \\
& \left.-e^{-kh_{2m'}}\right)+\sum_{m=m'+1}^{M}\frac{1}{k}\left(e^{-kh_{2m}}-e^{-kh_{1m}}\right)\left(e^{kh_{1m'}}-e^{kh_{2m'}}\right)\Bigg\} \\
& \times\frac{1}{(\alpha\beta)^2}\left(\sum_{n=1}^{N}P_n\right)^2\left(\sum_{q=1}^{Q}e^{-j\alpha x'_q}+\sum_{q=1}^{Q}e^{j\alpha x'_q}\right)\mathrm{d}\alpha\mathrm{d}\beta
\end{aligned}
\tag{2.259}
$$

$$
\begin{aligned}
\Delta Z = {} & \frac{j2\omega\mu_0}{\pi^2(h_2-h_1)^2 w^2}\int_{-\infty}^{+\infty}\int_{-\infty}^{+\infty}\frac{1}{k(\alpha\beta)^2}\left[\sum_{m=1}^{M}\left(e^{-kh_{2m}}\right.\right. \\
& \left.\left.-e^{-kh_{1m}}\right)\right]^2\left[\left(\sum_{n=1}^{N}P_n\right)^2\left(\sum_{q=1}^{Q}e^{-j\alpha x'_q}\right.\right. \\
& \left.\left.\times\sum_{q=1}^{Q}e^{j\alpha x'_q}\right)\right]\times\frac{k\mu_r-\lambda}{k\mu_r+\lambda}\mathrm{d}\alpha\mathrm{d}\beta
\end{aligned}
\tag{2.260}
$$

$$
Z_d = \frac{L}{\sigma_c(h_2-h_1)w}
\tag{2.261}
$$

L is the total length of the winding conductor, and σ_c is the conductivity of the coil conductor.

When the direction of the static bias field is parallel to the sample surface and parallel to the coil conductor, EMAT can excite and receive SH guided waves. At this time, the excitation of SH guided waves is determined by the shear strain S_6 produced by the dynamic magnetic field component H_x and the static magnetic field H_{0z}. That is, in the EMAT of SH guided wave, the tested sample satisfies the following magneto-elastic constitutive relation.

$$
S_6 = s_{66}\sigma_{66} + d_{16}H_x
\tag{2.262}
$$

$$
B_x = d_{16}\sigma_6 + \mu_{rev}H_x
\tag{2.263}
$$

The superposition of the dynamic magnetic field and the static bias magnetic field causes the ferromagnetic sample to produce dynamic strain, which is the source of the ultrasonic wave. The inverse magnetostriction effect reflects the dynamic stress generated by the dynamic stress in the test sample, which can be applied to the reception of electromagnetic ultrasonic signals.

According to the constitutive relation of magnetostriction and inverse magnetostriction, the magnetostriction strain S_{Ms6} in the sample is

$$S_{\text{Ms6}} = \frac{3\varepsilon_{\text{M}}}{H_0} H_x \tag{2.264}$$

Moreover, shear stress is

$$T_{\text{Ms6}} = -c_{44} \frac{3\varepsilon_{\text{M}}}{H_0} H_x \tag{2.265}$$

The specific value of the upper two formula determines the magnitude of the SH guided wave generated; that is, the magnitude of the SH guided wave is directly proportional to the piezomagnetic coefficient of the material. The magnitude of the SH guided wave can be quantified by the upper two formulae.

The frequency-domain expressions of magnetostriction strain and stress relaxation are obtained.

$$\begin{aligned} S_{\text{Ms6}} = & \frac{3\varepsilon_{\text{M}}}{H_0} \frac{I}{2\pi^2\mu_{\text{r}}(h_2 - h_1)w} \int_{-\infty}^{+\infty}\int_{-\infty}^{+\infty} \left[\frac{\text{e}^{\lambda z}}{-k^2} \times \frac{2k\mu_{\text{r}}\lambda}{k\mu_{\text{r}} + \lambda} \times j\beta \right] \\ & \sum_{m=1}^{M} \left(\text{e}^{-kh_{2m}} - \text{e}^{-kh_{1m}} \right) \times \frac{1}{\alpha\beta} \sum_{n=1}^{N} P_n \sum_{q=1}^{Q} \text{e}^{-j\alpha x'_q} \text{e}^{j\alpha x} \text{e}^{j\beta y} \text{d}\alpha \text{d}\beta \end{aligned} \tag{2.266}$$

$$\begin{aligned} T_{\text{Ms6}} = & -c_{44} \frac{3\varepsilon_t}{H_{0z}} \frac{I}{2\pi^2\mu_{\text{r}}(h_2 - h_1)w} \int_{-\infty}^{+\infty}\int_{-\infty}^{+\infty} \left[\frac{\text{e}^{\lambda z}}{-k^2} \times \frac{2k\mu_{\text{r}}\lambda}{k\mu_{\text{r}} + \lambda} \right. \\ & \left. \times j\alpha \right] \sum_{m=1}^{M} \left(\text{e}^{-kh_{2m}} - \text{e}^{-kh_{1m}} \right) \\ & \times \frac{1}{\alpha\beta} \sum_{n=1}^{N} P_n \sum_{q=1}^{Q} \text{e}^{-j\alpha x'_q} \text{e}^{j\alpha x} \text{e}^{j\beta y} \text{d}\alpha \text{d}\beta \end{aligned} \tag{2.267}$$

The single-layer unsplit folded coil and thin nickel plate were used as the coil of the SH guided wave EMAT, and the sample was tested. The direction of the static bias magnetic field was along the direction of the coil conductor.

The calculation model is a folded coil, which is placed above the thin nickel strip. The liftoff value between the coil and the thin nickel strip is 1 mm (Fig. 2.74).

In the analytical calculation, FFT-IFFT is used to calculate the *SMs*6. The excitation of the coil is tone-burst signal with a frequency of 500 kHz and a cycle number of 3, whose waveform is shown in Fig. 2.75.

The results S_{Ms6}/d_{16} of the left first conductor on the left side of the coil and the depth 0.0001 mm of the sample are calculated by analytical and time-step finite element methods, as shown in Fig. 2.76.

The analytical results are in good agreement with those calculated by the finite element method, which shows the correctness of the calculation formula. The values S_{Ms6}/d_{16} calculated here are constant for specific bias voltage and excitation frequency, and the value of shear strain S_{Ms6} is proportional to it.

The amplitude of the SH ultrasonic guided wave induced by the EMAT of the SH guided wave is related to two main factors: One is the piezomagnetic coefficient determined by the material stiffness matrix, the magnetostriction characteristic, and the static bias magnetic field, and the other is the amplitude of the dynamic magnetic field produced by the coil.

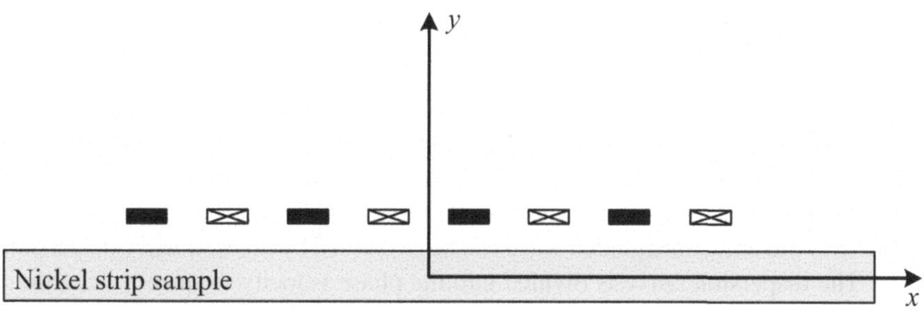

Fig. 2.74 Meander line coil is placed above the nickel strip

Fig. 2.75 Waveform of the exciting current

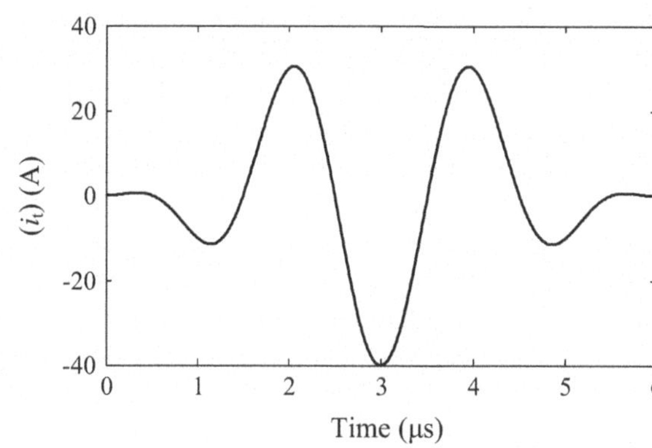

Fig. 2.76 Analysis and the result of finite element calculation

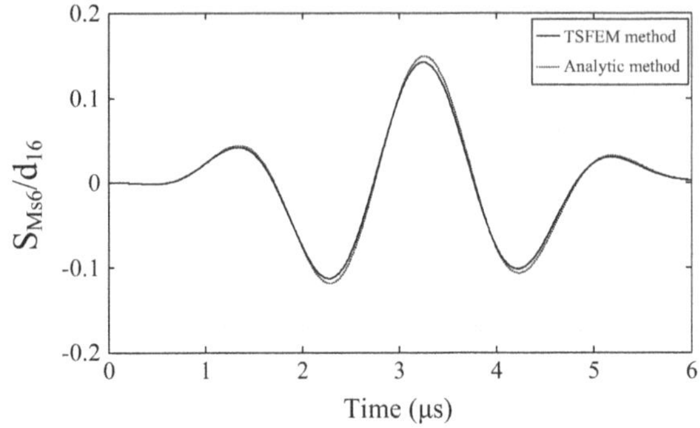

2.2.4 SH Guided Wave of Steel Plate and EMAT Theory Foundation Based on Magnetostriction

The frequency dispersion characteristic of ultrasonic guided wave means that the phase velocity and group velocity of the guided wave vary with the change of the working point of the guided wave. The working point of the guided wave is determined by the excitation frequency and the waveguide thickness. The dispersion characteristic of the ultrasonic guided wave can be expressed by the frequency dispersion curve. The dispersion curve is divided into the phase velocity frequency according to the phase velocity and the group velocity of the guided wave—dispersion curves and group velocity dispersion curves. In actual detection engineering, the guided wave detection signal received by the receiving transducer is actually a wave packet formed by the guided wave signal, so the wave velocity of the guided wave usually refers to the group velocity of the guided wave in the actual detection engineering, and the phase velocity dispersion curve of the guided wave is generally used in the identification of the special guided wave mode. The dispersion curve of the guided wave is an important basis for studying the propagation characteristics of guided waves. The relationship between the propagation velocity of the guided wave in a specific waveguide and the mode existence form and the working frequency of the guided wave in the waveguide is described. Also, the working point of the guided wave also indirectly determines the wavelength of the guided wave propagating in a specific waveguide. The wavelength of the guided wave has an important influence on the design and calculation of the distance between EMAT coils. Therefore, in order to stimulate the guided wave mode which satisfies the specific requirements in the steel plate, it is necessary to determine the suitable working point of the guided wave on the dispersion curve of the guided wave. The selection of the ultrasonic guided wave working point is the key link in the design of the excited guided wave EMAT, which is of great guiding significance for the design of the EMAT structure size parameters.

This section will explain the selection process of SH guided wave working point by analyzing the propagation characteristics of SH guided wave of steel plate, and by

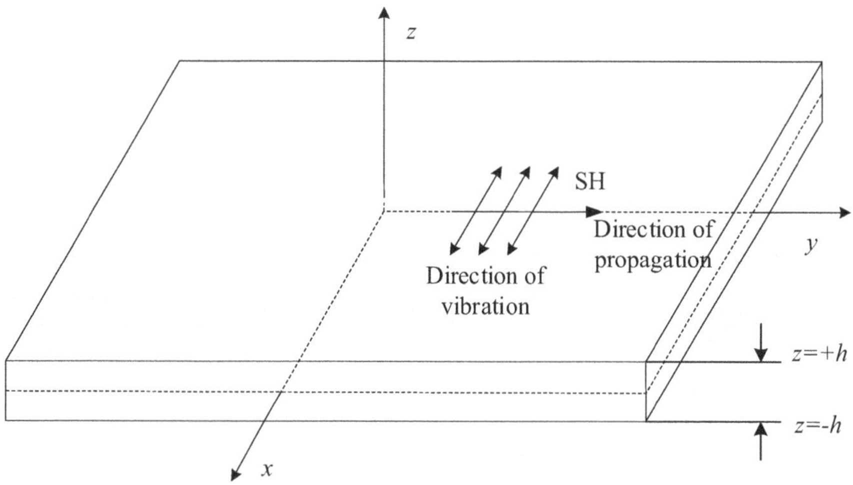

Fig. 2.77 Schematic diagram of the propagation of SH guided waves in free boundary steel plates

analyzing the working mechanism of EMAT based on magnetostriction, it provides a theoretical basis for the structure design of EMAT controlled wave direction in the direction of the steel plate SH.

In order to study the propagation characteristics of SH guided waves in steel plates, a propagation model of SH guided waves in free boundary steel plates was established (Fig. 2.77).

The thickness d of the free boundary steel plate is $2h$, and the free-form surface of the steel plate is $z = \pm h$ two planes. In the steel plate, the linear vibration source of SH guided wave is located on the plane of $y = 0$, its displacement is only in the direction of x, and the displacement in the direction of x is consistent and infinite extensibility. Therefore, the SH guided wave produced by the linear vibration source with the above characteristics is only displaced in the direction of x. This displacement is represented here by u_x. Because u_x has consistency in the direction of x, UX is independent of the variable x, and the two-order partial derivative of u_x to x is 0. Therefore, the displacement field equation of the SH guided wave generated in the free boundary steel plate is

$$\frac{\partial^2 u_x}{\partial^2 y} + \frac{\partial^2 u_x}{\partial^2 z} = \frac{1}{c_T^2} \frac{\partial^2 u_x}{\partial^2 t} \tag{2.268}$$

c_{T} is the shear wave velocity in the free boundary steel plate. The SH guided wave propagates along the y-direction and has a certain distribution in the z-direction. By solving the equation, the expression of u_x is obtained.

$$u_x^s(y, z, t) = B \cos(qz) \times e^{i(ky-\omega t)} \tag{2.269}$$

$$u_x^{\mathrm{a}}(y, z, t) = A \sin(qz) \times \mathrm{e}^{i(ky-\omega t)} \tag{2.270}$$

A and B are constants. The expression of the variable q is

$$q = \sqrt{\frac{\omega^2}{c_T^2} - k^2} \tag{2.271}$$

k is the wavenumber of SH wave, and ω is the angular frequency.

$$\omega = 2\pi f \tag{2.272}$$

$$k = \frac{2\pi}{\lambda} = \frac{\omega}{c_\mathrm{p}} \tag{2.273}$$

f is the frequency of the SH wave, the frequency of the SH wave in the dispersion curve. λ is the wavelength of the SH wave, and c_p is the phase velocity of the SH wave.

In the above two expressions on u_x, each expression is independent of the variable x, and all contain a traveling wave component along the y-direction and a standing wave component along the z-direction. These two expressions describe the symmetric modes and antisymmetric modes of SH waves, respectively.

According to the displacement field equation of SH wave symmetry mode and antisymmetric mode, and free surface boundary condition of free boundary plate, the dispersion equation of symmetric mode and antisymmetric mode SH wave can be obtained, respectively.

$$\sin(qh) = 0 \tag{2.274}$$

$$\cos(qh) = 0 \tag{2.275}$$

By solving the above dispersion equation, the explicit solution of the equation can be obtained.

$$\sin(qh) = 0 \; s.t. \; qh = n\pi, n = 0, 1, 2, \ldots \tag{2.276}$$

$$\cos(qh) = 0 \; s.t. \; qh = \frac{n\pi}{2}, n = 1, 3, 5, \ldots \tag{2.277}$$

And

$$qh = \frac{n\pi}{2}, \; n = 0, 1, 2, \ldots \tag{2.278}$$

When n is odd, it corresponds to antisymmetric mode. When n is even, it corresponds to symmetrical mode. The dispersion equation can be written as

$$\frac{\omega^2}{c_T^2} - \frac{\omega^2}{c_p^2} = \left(\frac{n\pi}{d}\right)^2 \tag{2.279}$$

d is the thickness of the steel plate and satisfies the $d = 2\ h$. fd is called the frequency–thickness product, and the phase velocity c_p can be expressed as a function of frequency–thickness product fd.

$$c_p(fd) = 2c_T \frac{fd}{\sqrt{4(fd)^2 - (nc_T)^2}} \tag{2.280}$$

It is known that when $n = 0$, the phase velocity of the $c_p = c_T$ mode wave is not changed with the frequency–thickness product, and the phase velocity of the SH0 mode wave is not dispersive. According to the expression of the phase velocity of the above SH wave, the phase velocity dispersion curve of the SH wave can be obtained (Fig. 2.78).

According to the SH wave velocity dispersion curve of the above picture, the phase velocity of the SH wave of the SH1, SH2 and higher-order mode changes with the frequency, and the phase velocity of the SH0 modal wave is 3200 m/s and is independent of the frequency. In the higher frequency region, there will be a variety of modes of SH wave, and the possible SH wave in the lower frequency region. The number of modalities is less.

The cutoff frequency thick product of the corresponding modal SH wave is

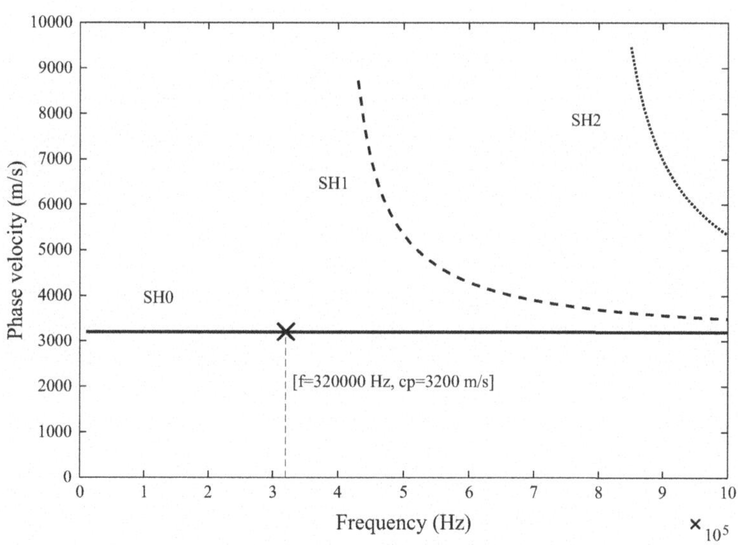

Fig. 2.78 Phase velocity dispersion curves of SH wave in 4-mm-thick steel plate

$$fd = \frac{nc_{\mathrm{T}}}{2} \tag{2.281}$$

The cutoff frequency of SH1 mode wave propagated in 4 mm steel plate can be calculated as 400 kHz according to the upper formula. Therefore, if only one SH0 mode is needed in the steel plate, the frequency of the working point of the SH0 mode wave should be lower than 400 kHz, and 320 kHz is selected as the working point frequency of the SH0 mode wave.

Dispersion equation can be expressed as

$$\frac{\omega^2}{c_T^2} - k^2 = \left(\frac{n\pi}{d}\right)^2 \tag{2.282}$$

The difference between the two sides is

$$\frac{2\omega \mathrm{d}\omega}{c_T^2} = 2k\mathrm{d}k \tag{2.283}$$

According to the definition of group velocity, group velocity can be obtained.

$$c_g = \frac{\mathrm{d}\omega}{\mathrm{d}k} = \frac{kc_T^2}{\omega} \tag{2.284}$$

The expression of the group velocity c_g on the frequency–thickness product *fd*

$$c_g(fd) = c_T\sqrt{1 - \left(\frac{nc_T}{2fd}\right)^2} \tag{2.285}$$

It can be seen from the formula that when $n = 0$ is obtained, the group velocity of $c_p = c_T$, namely SH0 mode wave, does not vary with the frequency–thickness product, and the group velocity of SH0 mode wave does not disperse. According to the expression of SH wave velocity above, the group velocity dispersion curve of the SH wave can be obtained (Fig. 2.79).

According to the velocity dispersion curve of the SH wave group velocity above, the group velocity of the SH wave of SH1, SH2, and higher-order mode changes with the frequency, and the group velocity of the SH0 modal wave is 3200 m/s and is independent of the frequency. In the higher frequency region, there will be a variety of modes of SH wave and the possible SH wave in the lower frequency region. The number of modalities is less.

Similar to the analysis of the phase velocity of the SH wave, the working frequency of SH0 mode wave is 320 kHz, and its group velocity is 3200 m/s.

To sum up, for the 4-mm-thick steel plate, the frequency of selecting SH0 mode guide wave is 320 kHz, the phase velocity and the group velocity are 3200 m/s, and the calculation can get the wavelength of 10 mm.

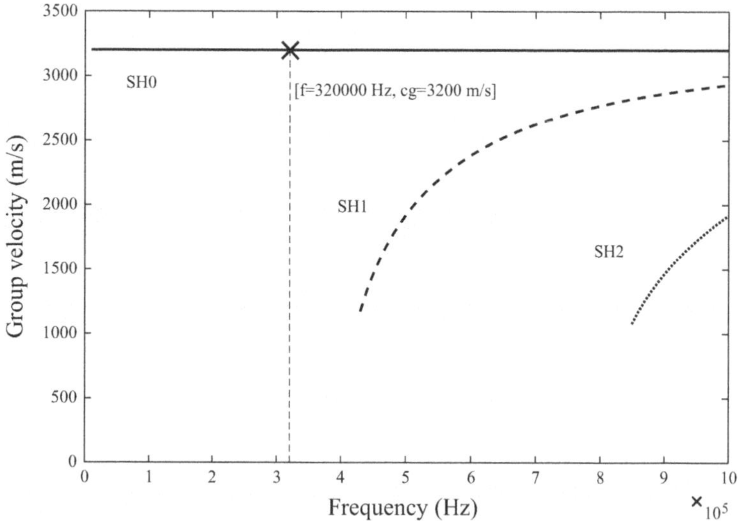

Fig. 2.79 Group velocity dispersion curves of SH wave in 4-mm-thick steel plate

In this section, through theoretical analysis of the propagation characteristics of SH guided waves, the dispersion curves of SH guided waves are obtained, and the choice of SH guided wave operating points are explained in detail and systematically. Next, based on the propagation characteristics of SH guided wave and working point parameters, the theoretical basis of SH guided wave EMAT based on magnetostriction steel plate is analyzed.

As ferromagnetic steel plates have magnetostriction properties, the mechanism of EMAT in the production of ultrasonic guided waves in the steel plate generally includes Lorenz force and magnetostriction, but for ferromagnetic materials, the role of magnetostriction mechanism plays a dominant role. The changes in the size or shape of ferromagnetic materials due to the change of magnetization state are called magnetostriction benefits, and the change in magnetization state caused by changes in the size or shape of ferromagnetic materials is called magnetostriction effect and also called piezomagnetic effect. The magnetostriction effect and its inverse effect indicate the relationship between the mechanical properties and magnetic properties of ferromagnetic materials. Magnetostriction effects also have strong nonlinear characteristics due to the nonlinear characteristics of ferromagnetic materials. But because the magnetostriction EMAT generally uses a strong bias magnetic field and relatively weak dynamic magnetic field, the fluctuation of the synthetic magnetic field near the bias magnetic field is relatively small, so the magnetostriction of the magnetic field is relatively small. The effect can be approximately expressed as a linear relationship. It can be expressed in the following formula.

$$\varepsilon = s^{\mathrm{H}}\sigma + \mathrm{d}H \tag{2.286}$$

$$B = d^{\mathrm{T}}\sigma + \mu^{\sigma} H \tag{2.287}$$

ε is magnetostriction quantity, σ represents stress, B represents magnetic flux density, H means magnetic field intensity. s^{H} shows the elastic flexibility matrix measured when the magnetic field strength H is constant. d (6×3) is a magnetostriction matrix, which indicates the magneto-mechanical coupling. d^{T} is a transposition of the magnetostriction matrix d. Formula (2.286) represents the direct magnetostriction process; that is, the application of the magnetic field causes the change in the shape or size of the material, which is used to stimulate the SH wave. Equation (2.287) describes the magnetostriction inverse effect, which is used to receive the SH wave and produce the guided wave detection voltage signal, and the change of the material stress causes the magnetization. The change causes the time-varying flux density and induces the electric field or electromotive force in the coil by Faraday's law of electromagnetic induction. Under the assumption of strong bias magnetic field (relative to dynamic magnetic field), there are only two independent variables in the magnetostriction matrix d: One variable d_{22} is proportional to the derivative of the working point of the magnetostriction curve, and the other d_{61} is proportional to the ratio of the magnetostriction strain to the bias magnetic field, based on the magnetostriction in the actual testing project. The SH wave EMAT relies mainly on variable d_{61} to excite ultrasonic waves.

In the ferromagnetic material region below the back coil, the ferromagnetic material is magnetized to produce a static bias magnetic field; the alternating current is connected in the coil, and the alternating current induces an alternating magnetic field in the ferromagnetic material region of its attachment, that is, the dynamic magnetic field shown in Fig. 2.4; the local area of the ferromagnetic material is in static bias magnetic field and dynamics. Under the common effect of a magnetic field, the shape of the magnetic field is based on the magnetostriction effect periodically, which causes the vibration of the local region. Finally, the SH wave is formed in the ferromagnetic material and propagates vertically in the direction of the bias magnetic field.

Based on the established coordinate system and the principle of magnetostriction generation of SH wave, Fig. 2.80 shows the structure and working principle of magnetostriction SH wave EMAT.

The EMAT excited by the SH wave in the steel plate is composed of a permanent magnet that provides a static bias magnetic field, a back coil consisting of a conductor with alternating current and a partial region of the steel plate under the bias magnetic field. Among them, permanent magnets can be replaced by electromagnets or premagnetized materials with magnetostriction properties. The static bias magnetic field H_x is in the x-direction, straight wire is placed along the static bias magnetic field and carries alternating current. The dynamic magnetic field H_y is generated along the y-direction. Because the static bias magnetic field H_x direction is perpendicular to the H_y direction of the dynamic magnetic field, for the synthetic magnetic field, during the periodic change of the alternating current, only the direction of the synthetic

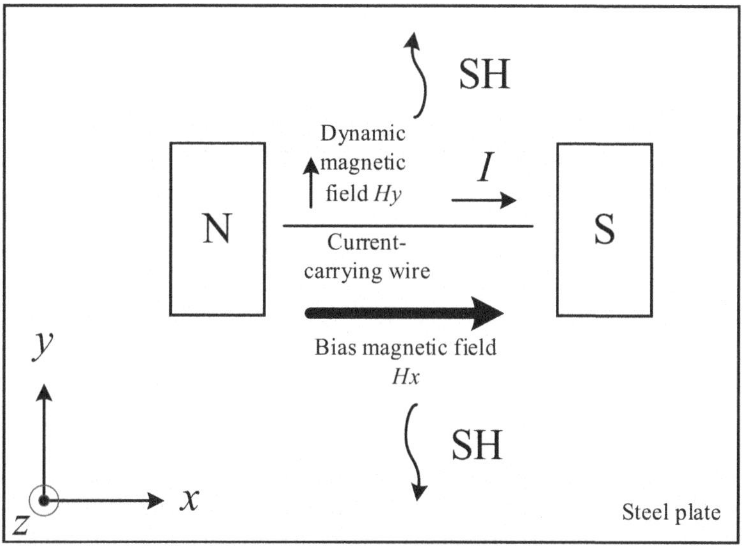

Fig. 2.80 Schematic diagram of EMAT structure based on magnetostriction generating the SH wave

magnetic field changes, the shear magnetostriction of ε_6 will be produced in the plate plane, and the SH wave propagating along the y-direction is finally produced.

$$\varepsilon_6 = s_{66}\sigma_6 + d_{61}H_y \tag{2.288}$$

The above formula shows that the generation and size of the magnetostriction volume are directly proportional to the parameters d_{61} and the H_y of the dynamic bias magnetic field. The above single current carrying wire is just to explain the working principle of the steel plate SH wave EMAT. In actual inspection engineering, the folding coil shown in Fig. 2.4 is generally used. The EMAT structure is only applicable to ferromagnetic materials, and its performance depends largely on the magnetostriction of the materials to be tested. In order to solve this problem, the scope of application of this structure EMAT can be expanded, and the materials with very strong magnetostriction can be bonded to the material to be detected. The EMAT can be used to detect the SH wave for the non-ferromagnetic material structure parts. The commonly used strong magnetostriction material includes nickel and cobalt alloy.

Based on the above principle, the permanentized nickel band is used to provide the bias magnetic field, and the dynamic alternating magnetic field is provided by the carrier flow back coil. The SH wave EMAT structure is designed as shown in Fig. 2.81.

The pre-magnetized nickel strip along the x-direction is bonded to the steel plate with an epoxy resin adhesive, and the residual magnetism in the nickel strip is the static bias magnetic field H_x. The current carrying coil is placed on the surface of the

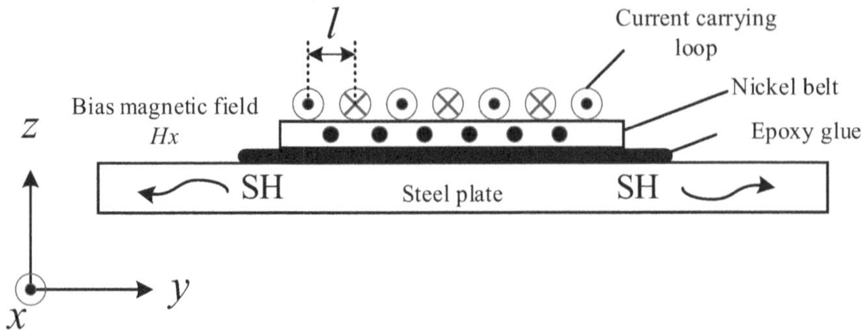

Fig. 2.81 Schematic diagram of EMAT structure based on nickel strip and meander line coil

nickel strip. The current direction of the single turn wire is parallel to that of the nickel strip, and the distance between adjacent conductors is l. When the alternating current of frequency f is applied to the coil, the dynamic alternating magnetic field y in the direction of the H_y will be induced in the nickel strip. In this way, a periodically varying magnetostriction force $f_{\mathrm{M}x}$ will be generated along the x-direction in the nickel strip. The magnetostriction force $f_{\mathrm{M}x}$ in the nickel strip under the single turn wire can be expressed as

$$f_{Mx} = -2(1+\upsilon)c_{66}\frac{\varepsilon_6}{H_x}\frac{\partial H_y}{\partial y} \tag{2.289}$$

v is the Poisson's ratio of the nickel strip, and c_{66} is the element in the elastic stiffness matrix. The magnetostriction force produces ultrasonic vibration in the x-direction of the nickel band, and the ultrasonic vibration is propagated through the epoxy resin adhesive and propagates into the steel plate. The vibration source of the SH wave propagating along the y-direction is formed in the steel plate. The vibration source is only vibrated in the direction of x, the vibration in the direction of x is consistent, and the vibration and displacement of the surface are not produced. The characteristic also determines that the SH wave generated by the EMAT will not be affected by the bearing medium of the steel plate during propagation.

According to Fig. 2.1, the origin of coordinates is placed at the center of the steel plate under the geometric center of the folding coil. The vibration source of the EMAT in the steel plate only vibrates in the direction of x, and the vibration in the direction of x is consistent. Therefore, the requirements of the propagation model of the SH guided wave in the free boundary steel plate in Fig. 2.1 for the vibration source are satisfied. The size of the reset coil in the direction of y is y_0

$$u_x^s(y,z,t) = B_0(y)\cos(qz)\times \mathrm{e}^{i[k(y-y_0)-\omega t]} \tag{2.290}$$

$$u_x^{\mathrm{a}}(y,z,t) = A_0(y)\sin(qz)\times \mathrm{e}^{i[k(y-y_0)-\omega t]} \tag{2.291}$$

The amplitude $B_0(y)$ and $A_0(y)$ will attenuate with the increase of distance y, and the law of attenuation is satisfied.

$$B_0(y) \propto \frac{1}{\sqrt{y}}, A_0(y) \propto \frac{1}{\sqrt{y}} \tag{2.292}$$

In the meander line coil of the above SH guided wave EMAT, it is assumed that the two adjacent wires W_1 and W_2 are located at y_1 and y_2 in the y-direction and satisfy $Y1 < Y2$. The two conductors will generate SH waves in the steel plate, respectively, then in any y-direction y_a ($y_a > y_2$), and the vibration phases induced by the two SH waves are, respectively

$$\varphi_1 = \frac{2\pi(y_a - y_1)}{\lambda} - \omega t + \varphi_{10} \tag{2.293}$$

$$\varphi_2 = \frac{2\pi(y_a - y_2)}{\lambda} - \omega t + \varphi_{20} \tag{2.294}$$

λ is the wavelength of the SH wave; φ_{10} and φ_{20} are the initial phase of the vibration caused by two SH waves; and the relationship is satisfied.

$$\varphi_{10} - \varphi_{20} = \pm\pi \tag{2.295}$$

The difference $\pm\pi$ between the two is due to the special structure of the meander line coil, which leads to the opposite direction of the instantaneous current of the two adjacent conductors W_1 and W_2 at any time, which leads to the opposite direction of the induced dynamic alternating magnetic field H_y; that is, its phase difference is $\pm\pi$. Therefore, the phase difference $\Delta\varphi$ of the vibration caused by these two SH waves is

$$\begin{aligned} \Delta\varphi &= \varphi_1 - \varphi_2 \\ &= \frac{2\pi(y_2 - y_1)}{\lambda} + (\varphi_{10} - \varphi_{20}) \\ &= \frac{2\pi l}{\lambda} \pm \pi \end{aligned} \tag{2.296}$$

l is the distance between the adjacent two turns of the EMAT coil.

In order to make the vibration caused by these two SH waves superimposed to enhance the amplitude of the SH wave, the phase difference $\Delta\varphi$ between them should be satisfied.

$$\Delta\varphi = 2n\pi(n = 1, 2, 3, \ldots) \tag{2.297}$$

The relationship between the distance between adjacent two turns of the EMAT coil and the l should be satisfied.

$$l = (2n \pm 1)\frac{\lambda}{2}(n = 1, 2, 3, \ldots) \tag{2.298}$$

According to the above, the distance between the two adjacent two turns of the return coil should be an odd number of times of the half wavelength to be excited, which can make the SH waves excite each turn of the EMAT and enhance the SH wave energy transmitted in the steel plate, which improves the signal-to-noise ratio and the characteristic information of the signal of the SH guide wave detection at the receiver. Accuracy plays an important role. In addition, although the SH wave detection working point selected in this paper does not exist the problem of multimodal SH wave, the design of the back coil designed to meet the above relationship can reduce the other modal SH waves propagating in the steel plate to a certain extent, because of the dynamic bias magnetic field induced by the other two adjacent winding wires for other modal SH waves and the return coil. The phase difference can no longer satisfy the relation that can make SH waves superimposition each other, and it is easier to weaken each other, so the other mode SH waves are equivalent to a kind of suppression behavior. Therefore, the design method mentioned above has an important guiding function for EMAT to excite single-mode guided waves and mode control.

This section analyses the theoretical basis of the SH guided wave EMAT design based on magnetostriction steel plate. The model of SH wave excitation in the steel plate based on magnetostriction mechanism is studied, and the SH wave magnetostriction EMAT of steel plate based on nickel and refolding coils is given. The magnetostriction EMAT is applied to the magnetostriction EMAT based on the SH guiding working point of the steel plate determined in the previous section. The key parameter design method of the loop is deduced theoretically.

Next, based on the SH wave propagation characteristics of steel plate and the working mechanism of SH wave magnetostriction EMAT of steel plate, the structure design of SH wave direction controllable magnetostriction EMAT for steel plate is elaborated in detail.

2.2.5 Structure Design of SH Guided Wave Direction Controlled EMAT for Steel Plate

The SH guided wave direction controlled EMAT based on magnetostriction steel plate is expected to be realized mainly in the following aspects: (1) The excitation and reception of SH guided waves can be carried out in the steel plate in a certain direction according to the strong direction; (2) the direction of the excitation and receiving SH guided waves can be accurately controlled and adjusted; (3) in the angle range of the excitation and reception of SH guided waves, a single point arrangement and full angle range detection can be realized as far as possible to excite and receive all angles of the SH guide wave in the angle range of excitation and reception of the SH guided wave; (4) the same excitation in the same direction when the steel plate SH guided wave is excited in all directions. The amplitudes or signal intensities of the excited SH guided waves should be in good agreement with the excitation parameters such as excitation voltage and excitation frequency. The above requirement or expectation is also the innovation of EMAT steering wave steering controlled by SH in this paper. Next, based on the above functions, we design the structure of guided wave controllable EMAT of steel plate SH designed in this paper.

In this paper, the basic principle of directional controllable EMAT is designed based on the structure and principle of EMAT of nickel strip and winding coil (Fig. 2.82).

The overall shell of the controllable EMAT is cylindrical, and the upper section is cross section along the diameter of the upper surface of the shell. The direction of the center of the surface of the housing is equipped with a directional adjusting knob for precisely adjusting the direction of the SH guided wave excited or received by EMAT. Therefore, the outer shape of this controllable EMAT is a cylindrical shell and a direction adjusting knob on the upper surface of the shell. The direction adjusting knob is directly connected with the rotating shaft and directly drives the rotating shaft to rotate circumferentially, and the rotation of the direction adjusting knob is consistent with the rotation time of the rotating shaft. The upper part of the shaft is mounted on the upper part of the outer shell by a sliding bearing. The central part is installed in the support frame through a sliding bearing, and the end of the support frame is installed on the side wall of the shell, which plays a role of fixed position and structural support for the rotating shaft. A rotating slider is installed below the rotating shaft, and its shape is a rectangular body, which follows the rotating shaft and the direction knob together, and the rotation of the rotating slider is consistent with the rotation moment of the rotating axis and the direction adjustment knob. There is an arc groove along the circumference of the lower surface of the rotating slider. Its function is close to the open metal ring in the lower part and is the same as the orbit

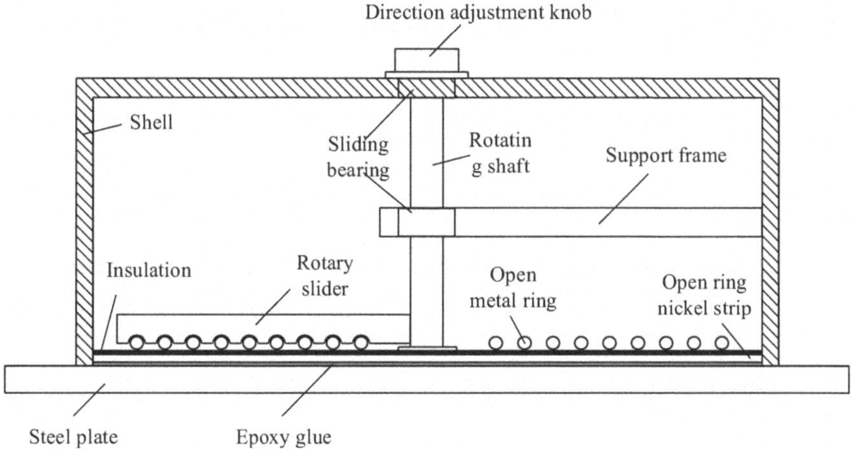

Fig. 2.82 Direction controlled EMAT section

of the open metal ring, to ensure a good coupling effect with the open metal ring. Figure 2.83 shows the more detailed structural arrangement of the rotating slider section.

The lower surface of the rotating slider is arranged along the radial direction with a mosaic conductor, and the inlaid conductor is alternately distributed along the radial direction. Its function is to selectively connect the open metal ring to make the open metal ring and the inlaid conductor form the back coil on the surface of the rotating slider. The open metal ring is consistent with the opening size and position of the open ring nickel belt, and the open metal ring and the open circle ring nickel belt coincide, and the insulating layer is arranged between the two. The pre-opening magnetization along the circumferential direction has been carried out before the installation of the nickel ribbon of the open ring; that is, the direction of remanence is circumferential. The open ring nickel strip is bonded to the surface of the steel sheet by epoxy resin. Figure 2.84 is the overlook detail diagram of the direction controlled EMAT.

The alternating voltage is applied to the opposite ends of the inner ring of the open metal ring and the outer ring. If there are no rotating sliders and inlaid conductors, the open metal ring cannot form a current path, and the circuit is in an open circuit. Because of the existence of the mosaic conductor in the rotating slider, the inlaid conductor interconnects the adjacent open metal ring, making the open metal ring part of the lower part of the rotating slider and the current path formed with the inlaid conductor, and the current path satisfies the form of the back coil, so the dotted line part in the rotating slider is produced in Fig. 2.9. The alternating current, the direction of the alternating current at this time, is shown in Fig. 2.85.

The circumferential direction of the opening ring nickel strip is magnetized at the front of the installation, resulting in a circumferential bias magnetic field H0. The alternating current induces an alternating magnetic field in the open ring nickel belt in the lower part of the rotating slider. Based on the magnetostriction effect, the SH0 mode guided wave is produced in the opening circular nickel belt in the lower part of the rotating slider. The guided wave is coupled to the steel plate through epoxy resin glue, thus producing SH0 modal guided waves in the radial direction of the rotating slider.

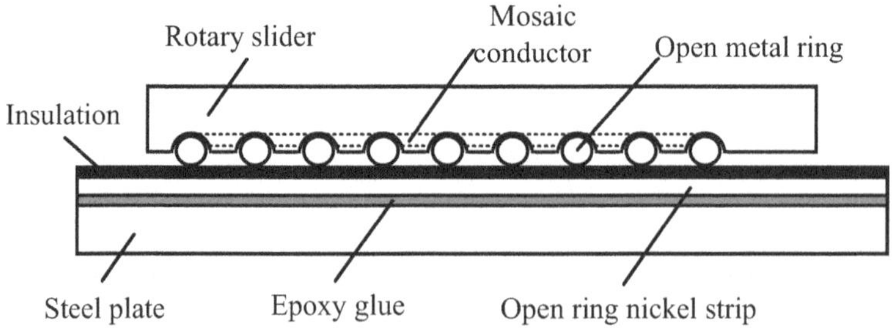

Fig. 2.83 Part detail structure diagram of directionally controllable EMAT rotary slider

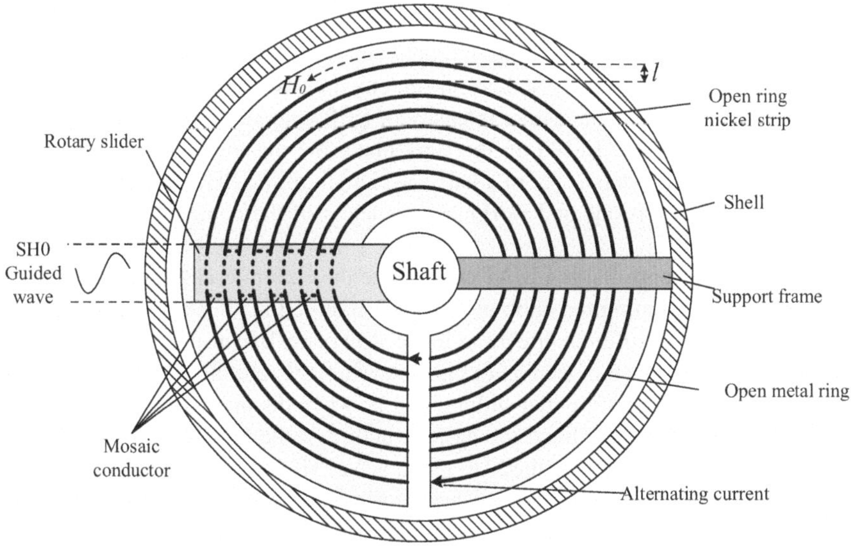

Fig. 2.84 Top view of direction controlled EMAT

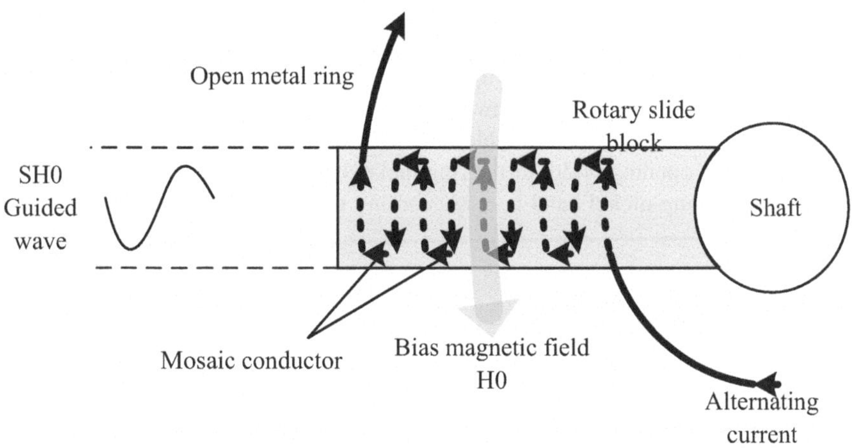

Fig. 2.85 Schematic diagram of current controlled path controlled by EMAT

The rotating slider is synchronized with the axis of the rotating shaft and the direction adjustment knob of the top. After installing the controllable EMAT in the steel plate, the circumferential position of the rotating slider can be changed by adjusting the knob in the direction of rotation; thus, the SH0 guide wave is excited in the different directions in the steel plate to realize the controllable direction excitation of the guided wave of the steel plate SH0. The following diagram shows the three-dimensional structural schematic diagram of the controllable EMAT in this direction (Fig. 2.86).

According to the working point of the selected SH0 guided wave of steel plate, for 4-mm-thick steel plate, the working frequency of selecting SH0 mode guide wave is 320 kHz, the phase velocity and group velocity are 3200 m/s, and the wavelength is 10 mm. In order to enhance the intensity of SH0 guided wave signal excited by the direction controlled EMAT in the steel plate, the distance l between the two adjacent open metal rings is set to a half wavelength of the SH0 guided wave. The key geometric dimensions and working parameters of the controllable EMAT are shown in Table 2.13.

The direction controllable EMAT can change the circumferential position of the rotary slider by adjusting the knob of the rotation direction, thereby stimulating the SH0 guided wave in different directions in the steel plate. Because of the existence of the ring nickel belt and the opening of the metal ring, the rotating slider has a certain blind zone in the circumferential sliding; that is, when the opening is located at the lower part of the rotating slider, the basic structure of the magnetostriction EMAT cannot be formed to stimulate the SH0 guide wave. Therefore, the SH0 guide wave cannot be excited in the steel plate. These positions constitute the circumferential detection blind zone of the direction-controllable EMAT(Fig. 2.87).

The position of the direction controlled EMAT circumference detection blind area should be the position of the edge of the rotating slider and the opening edge of the ring nickel belt. At this time, the rotating slider cannot cross the opening edge of the ring nickel belt and enter the opening area. Therefore, the opening edge of the two-ring nickel belt corresponds to the two edges of the circumferential rotation range of the rotating slider. The angle range between the two boundaries is the circumferential detection blind area with controllable EMAT direction. The width of the rotating slider is w_s, the opening width of the ring nickel belt is w_n, and the inner diameter of the opening ring nickel band is r_n. According to the definition of the geometric

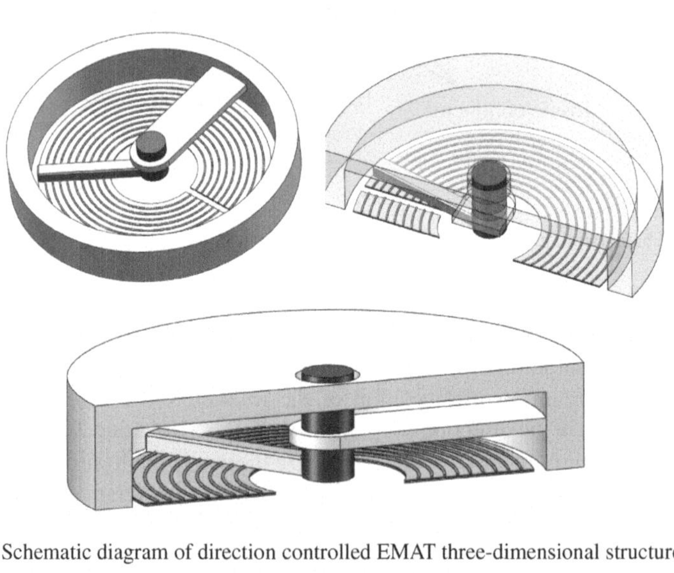

Fig. 2.86 Schematic diagram of direction controlled EMAT three-dimensional structure

Table 2.13 Key parameters of magnetostriction direction controllable EMAT

Parameter	Value
Excitation of SH wave mode	SH0
Working frequency f (kHz)	320
Open metal ring spacing L (mm)	5
Circumferential slider circumferential width WS (mm)	10
Circular nickel band opening width wn (mm)	5
Inner radius RN (mm) of an open ring nickel band	30
Open round ring radius of nickel (mm)	80
Radial width (mm) of an open ring nickel band	50
Thickness of open ring nickel band (mm)	0.5
Insulating layer thickness (mm)	0.5
Number of open metal rings	9
Open metal ring diameter (mm)	1
Liftoff value of open metal ring and nickel band (mm)	1.25
Radius of inner ring open metal ring (mm)	35
Radius of outer ring open metal ring (mm)	75
The inner boundary distance between the metal ring and the open ring nickel strip in the inner ring (mm)	5
The distance between the outer boundary of the metal ring and the opening ring of the outer ring (mm)	5
Rotating slider height (mm)	7
Inlaid conductor height in rotating slider (mm)	0.5
The width of the inlaid conductor (mm) in the rotating slider	1
Radius of rotating shaft (mm)	10
Direction adjusting knob radius (mm)	60

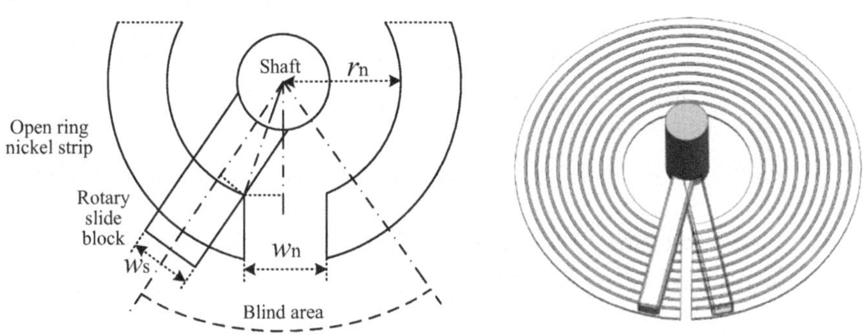

Fig. 2.87 A schematic diagram of direction controlled EMAT circumference detection blind area

relation and the circumferential detection blind area, the angle range of the angle detection blind area can be expressed by the angle range of θ_d, so the blind area angle θ_d in the direction controlled EMAT can be calculated in the next type.

$$\theta_d = 2\left(\arcsin\frac{\frac{w_s}{2}}{r_n} + \arcsin\frac{\frac{w_n}{2}}{r_n}\right) \tag{2.299}$$

According to the above parameters of the controllable EMAT in the direction, the angle of detection blind area θ_d of the direction controllable EMAT can be calculated to be 28.74°. For the 360° full angle range detection, the full angle detection percentage D_a is defined, indicating that the direction controlled EMAT can effectively stimulate the ratio of the angle range of the SH0 guide wave to the full angle range.

$$D_\mathrm{a} = \frac{360° - \theta_d}{360°} \times 100\% \tag{2.300}$$

According to the above formula, the total angle detection percentage of the direction controlled EMAT is 92%, and the SH0 guided wave can be excited at most angles in the full angle range.

In this section, the structure design of the guided wave direction controlled EMAT for steel plate SH is described in detail. In particular, the function and the relationship of each component of the direction controlled EMAT are detailed, and the key design parameters and the values of the controlled EMAT in this direction are given, and the blind area of the square controlled EMAT is studied. In order to describe the proportion of direction controlled EMAT in full angle detection more intuitively, the percentage of all angle detection of controlled EMAT in this direction is defined and calculated. In this section, the functions and properties that are expected to be realized in the direction of EMAT are expected, including the direction polyfocal of the excitation and reception of the SH guided waves in the steel plate in a certain direction, the amplitude of the excitation of the SH guided wave or the consistency of the signal intensity in all directions, and the direction of the guided wave of the steel plate is carried out. And the steel plate SH guided wave controllable EMAT key performance test platform is designed, and related performance verification test and result analysis are carried out.

2.2.6 Experimental Verification of Directional SH Guided Wave EMAT in Steel Plate

In this section, the experimental method is used to study the key properties and parameters of the guided wave direction controlled EMAT in the steel plate SH, including the direction of the excitation and reception of the SH guided waves in the steel plate in a certain direction, the consistency of the amplitude or signal intensity of the plate SH guided wave excitation in all directions, and the full angle range detection. The blind area of measuring time is discussed in detail, such as the construction of the test platform, the configuration of test parameters, and the analysis and discussion of the results.

In order to study and verify the performance and parameters of the direction controlled EMAT, a direction controlled EMAT performance verification and testing platform was built on the 4-mm-thick steel plate, and a series of related testing tests were carried out by the test platform. The schematic diagram of the direction controllable EMAT performance verification and test platform is shown in Fig. 2.88.

A simple polar coordinate system containing only the circumferential angle is established on the 4-mm-thick steel plate, and the direction controlled EMAT is installed in the origin position of the coordinate system. The opening center of the circular nickel belt in the direction controlled EMAT is corresponding to the 270° angle position. Twenty-four magnetostriction omnidirectional receiving EMAT are evenly arranged around the direction controlled EMAT in a circular array, and the angle interval between the two adjacent omnidirectional receiving EMAT is 15°, and the distance between the omnidirectional receiving EMAT and the direction controlled EMAT is 0.45 M. The alternating voltage is applied to the output of a power amplifier at both ends of the open metal ring of the direction controlled EMAT inner and outer ring. The alternating voltage is applied to the back coil composed of an open metal ring and a mosaic of the rotating slider, and the alternating current is produced in it. The excitation parameters of the power amplifier are configured by a computer, including excitation voltage amplitude, excitation frequency, excitation cycle num-

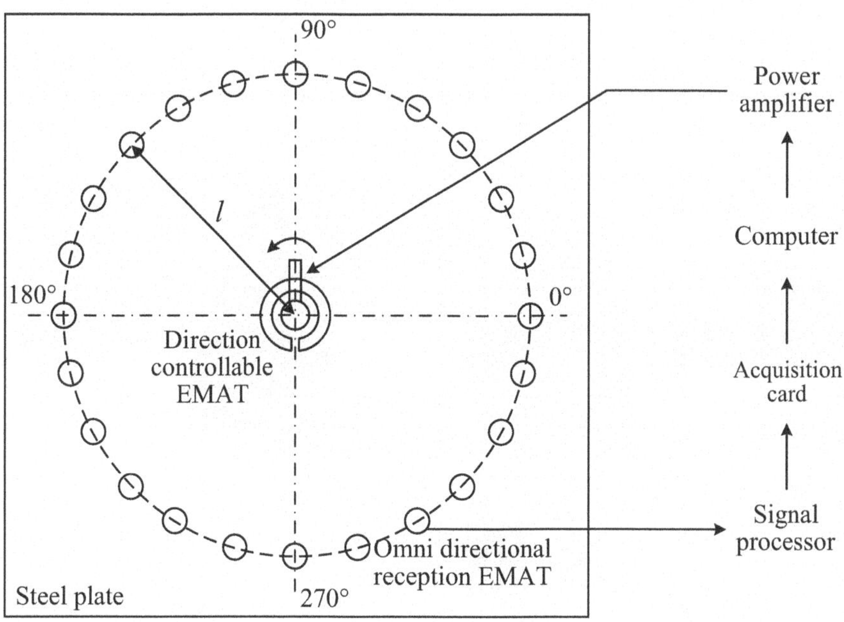

Fig. 2.88 Schematic diagram of the performance test platform for directional controllable EMAT

ber, and so on. After the power amplifier stimulates the direction controlled EMAT, based on the magnetostriction principle, the direction controlled EMAT excites the SH guided wave in the steel plate in a certain direction. The omnidirectional receiving EMAT receives the SH guided wave propagating in the steel plate and converts it into a voltage signal into the signal processor. After the signal processor amplifies narrowband filtering, the signal processor is sent into the acquisition card for acquisition and digital processing. Finally, the guided wave detection signal and data are sent to the computer for storage, analysis, and processing.

Based on the magnetostriction, SH guided wave omnidirectional reception EMAT of steel plate is made up of circular wound coils, open ring nickel strips, and steel plates. The opening circular ring nickel is bonded to the surface of the steel plate with an epoxy resin adhesive. The circular coiled coil is made of thin plate PCB, the circular coiled coil is placed above the open ring nickel belt, and the center of the two coils is reclosing. Using this omnidirectional receiving EMAT, the SH guided wave from various angles can be received, to obtain as much information as possible for the detection of the guided wave of the steel plate. It is beneficial to the performance verification test of the direction controlled EMAT and the more abundant information of the subsequent defect detection and imaging test. Figure 2.89 shows the physical map of the direction controllable EMAT performance verification and testing platform.

For the focusing test of the direction controlled EMAT to stimulate the SH guided wave at a certain angle, the experimental method is to adjust the direction knob of the direction controlled EMAT, so that it can launch the SH0 guided wave at a certain angle, collect all the guided wave detection signal received by the omnidirectional receiving EMAT, and compare the relation of the signal amplitude. For the consistency test of SH guided waves excited by direction controlled EMAT at various angles, the test method is to adjust the direction knob of direction controlled

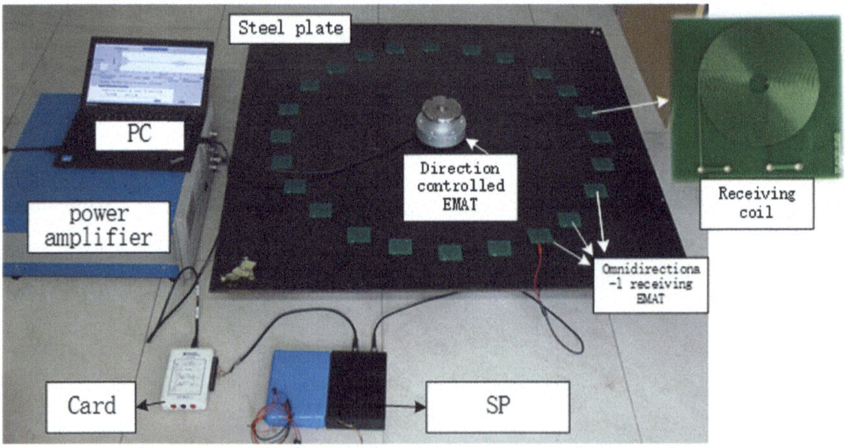

Fig. 2.89 Object-oriented EMAT performance verification and testing platform

EMAT, so that it is aimed at each omnidirectional receiving EMAT, respectively, the SH0 guided wave is excited, respectively, only the guided wave detection signal of the aligned omnidirectional receiving EMAT is collected, and each of them will be collected. A guided wave detection signal received by EMAT omnidirectional receiver is compared and analyzed. The test parameters set during the test are shown in Table 2.14.

According to the setting of the above test parameters, we can carry out the direction controlled EMAT performance verification and testing test, and we will analyze and discuss the test results below.

In the focusing test of the direction controlled EMAT in a certain angle to stimulate the SH guided wave, the directional control knob of the direction controlled EMAT is directed to the 90° direction for the excitation of the guided wave of the SH0, and all the guided wave detection signals received by the omnidirectional receiving EMAT are collected. Figure 2.15 shows the detection waveform received by the omnidirectional receiving EMAT at the 75° position, the 90° position, and the 105° position. Since the other position omnidirectional receiving EMAT has hardly received the detection waveform, it is not displayed here (Fig. 2.90).

It is known from the above figure that the amplitude of the detection signal in the 90° position is obviously larger than that of the adjacent 75° and 105°, and the energy of the guided wave signal is mainly concentrated on the direction path of the 90-degree radiation. In order to quantitatively express and study the distribution of guided wave signal energy at all angles, the signal intensity P of the guided wave detection signal is defined. For a guided wave detection signal, the voltage amplitude of each point is expressed as X (n), $n = 1,2,3,\ldots N$, where N is the total number of guided wave detection signals, and the signal strength P of the guided wave detection signal is defined as

Table 2.14 Performance verification and parameter settings for directional controllable EMAT

Parameter	Value
Peak of excitation voltage (V)	240
Excitation frequency f (kHz)	320
Number of excitation cycles (a)	12
Length of steel plate (mm)	1000
Width of steel plate (mm)	1000
Thickness of steel plate (mm)	4
Receiving coil diameter (mm)	35
The number of turns (turns) of the receiving coil	30
Direction controlled EMAT and omnidirectional receiving EMAT distance (mm)	450
Signal processor magnification	5000
Narrowband filter center frequency (kHz) of signal processor	320
Narrowband filter bandwidth (kHz) of signal processor	20

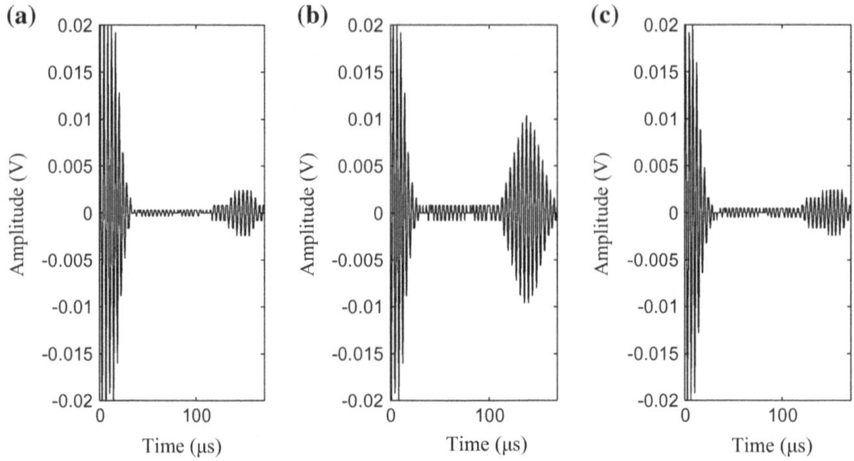

Fig. 2.90 Direction controlled EMAT excitation SH guided wave focusing detection waveform: **a** 75° position omnidirectional receiving EMAT; **b** 90° position omnidirectional receiving EMAT; **c** 105° position omnidirectional receiving EMAT

Fig. 2.91 Normalized waveguide detection signal intensity distribution on a circumferential angle

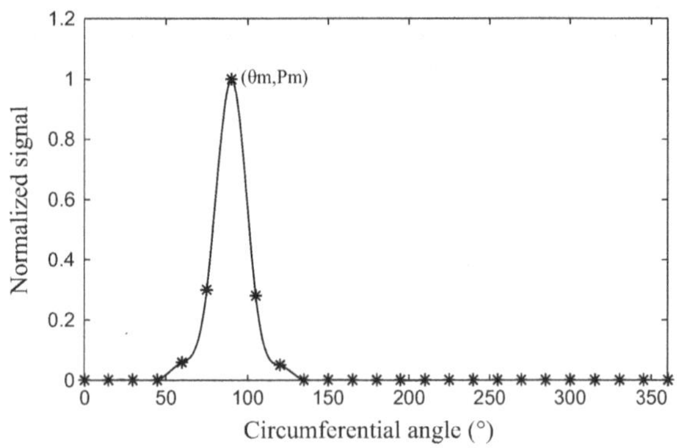

$$P = \sqrt{\frac{1}{N}\sum_{n=1}^{N}|x(n)|^2} \tag{2.301}$$

The signal intensity of the SH0 guided wave detection signal received by each omnidirectional receiving EMAT is calculated by the upper formula, and after the normalization of its value, the normalized distribution diagram of the circumferential position of the signal intensity of the guided wave detection signal is made, as shown in Fig. 2.91.

In the distribution of the circumferential angle of the guided wave intensity, the circumferential angle is expressed with θ, and the maximum intensity of the detected signal in the upper figure is (θ_m,P_m). In order to quantitatively describe the focusing angle of SH guided wave excited by controllable EMAT in this direction, the focusing signal intensity threshold P_t is defined.

$$P_t = a^* P_m \tag{2.302}$$

A is a focusing intensity threshold factor, and its range of value is $0 < a < 1$. It is a parameter of user value. It is indicated that the guided wave detection signal is an effective signal when the ratio of the signal intensity of the guided wave detection signal is a, and the corresponding angle is included in the excited guided wave of the controlled EMAT. Focus the angle range. Based on the focusing signal intensity threshold, the two boundary values of the focusing angle of the guided wave can be calculated by the following two formulas.

$$\theta_{t1} = \arg\min|P(\theta) - P_t| \cap (\theta < \theta_m) \tag{2.303}$$

$$\theta_{t2} = \arg\min|P(\theta) - P_t| \cap (\theta > \theta_m) \tag{2.304}$$

According to the above formula, the steering angle of the guided wave can be expressed as EMAT with controllable direction.

$$\theta_f = \theta_{t2} - \theta_{t1} \tag{2.305}$$

In this experiment, the emission angle of the direction controlled EMAT is 90°, the focusing intensity threshold factor is 0.8, the calculated focus angle boundary value of the excited guided wave is 87.8° and 92°, respectively. Therefore, the direction controlled EMAT excited guided wave focusing angle of this test is calculated to be 4.2°.

In order to study the consistency of the direction controlled EMAT in all angles, the direction of the direction controlled EMAT is adjusted to adjust the direction knob in the direction controlled EMAT, so that it is aligned to every omnidirectional receiving EMAT, and the SH0 guided wave is excited under the same excitation parameters, and only the guided wave detection signal of the aligned omnidirectional receiving EMAT is collected and calculated, and their signal strength is calculated separately. According to the angle and installation of the blind area of the direction controlled EMAT, the direction adjusting knob can rotate the circumferential angle range from 284.37° to 255.63°; that is, the omnidirectional receiving EMAT in the 270° position is located in the detection blind area of the direction controlled EMAT, so a total of 23 omnidirectional receiving EMAT receives the directed wave signal and sends its letter. The intensity is normalized and is drawn according to the circumferential position (Fig. 2.92).

From the above diagram, we can see that the intensity of SH guided wave excited by EMAT cannot be very different from each other and has a good consistency. In order to quantify the consistency of the SH guided wave signal intensity at all

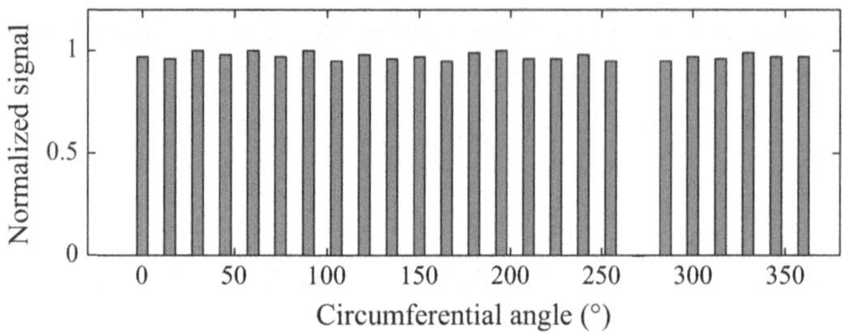

Fig. 2.92 Intensity distribution of directional EMAT guided wave coherence signal

angles, the direction controlled EMAT is used to define the consistency error E_c of the direction controlled EMAT guided wave in different directions.

$$E_{\mathrm{c}} = \frac{1}{M}\sum_{m=1}^{M}|P(m) - P_E| \quad (2.306)$$

$$P_E = \frac{1}{M}\sum_{m=1}^{M}P(m) \quad (2.307)$$

M is an effective number of EMAT for omnidirectional reception. The consistency error of the direction controlled EMAT induced by the direction controlled SH guided wave signal intensity at different angles is 1.4%, indicating that the direction controlled EMAT in the direction of the direction is in good agreement with the excitation signal intensity of the SH guided wave in the detection area.

In this section, a key performance index for evaluating the direction controlled EMAT of the guided wave in a certain angle is proposed to excite the SH guided wave at a certain angle. Moreover, direction controlled EMAT is excited at all angles. The focusing angle of the guided wave in the direction controlled EMAT is defined, and the consistency error of the guided wave in different directions is excited in different directions. In the same performance parameters, a direction controlled EMAT performance verification and the testing platform are built, and the experimental method is used to calculate the focusing angle and consistency error of the direction controlled EMAT excited guided wave. The results show that the direction controlled EMAT can excite the SH guided wave in a certain direction with a good focus, and the SH guide wave is excited in all directions. The consistency error is very small, so it has good consistency. The direction controlled EMAT has good performance indexes on the focusing and consistency of SH guided wave, which can provide an important theoretical basis and method support for the directional emission link in the directed emitter array guided wave scattering imaging method of omnidirectional receiving.

References

1. P.D. Wilcox, J.S.M. Lowe, P. Cawley, The excitation and detection of Lamb waves with planar coil electromagnetic acoustic transducers. IEEE Trans. Ultrason. Ferroelectr. Freq. Control **52**(12), 2370–2383 (2005)
2. R. Ludwig, X.W. Dai, Numerical and analytical modeling of pulsed eddy currents in a conducting half-space. IEEE Trans. Magn. **26**(1), 299–307 (1990)
3. J.L. Rose, *Ultrasonic waves in solid media* (Cambridge University Press, New York, 1999)
4. F.W.J. Olver, D.W. Lozier, R.F. Boisvert et al., *Handbook of mathematical functions* (Cambridge University Press, New York, 2010)
5. S. Wang, S.L. Huang, Z. Wei, Simulation of Lamb wave's interactions with transverse internal defects in an elastic plate. Ultrasonics **51**(4), 432–440 (2011)
6. S. Wang, S.L. Huang, Y. Zhang et al., Modeling of an Omni-directional electromagnetic acoustic transducer driven by the Lorentz force mechanism. Smart Mater. Struct. **25**(12), 125029 (2016)
7. S. Wang, S.L. Huang, W. Zhao et al., Approach to Lamb wave lateral crack quantification in elastic plate based on reflection and transmission coefficients surfaces. Res. Nondestr. Eval. **21**(4), 213–223 (2010)
8. Y. Zhang, S. Wang, S.L. Huang et al., Mode recognition of Lamb wave detecting signals in metal plate using the Hilbert-Huang transform method. J. Sens. Technol. **5**, 1–14 (2015)
9. S.L. Huang, Y. Zhang, S. Wang et al.: Multi-mode electromagnetic ultrasonic Lamb wave tomography imaging for variable-depth defects in metal plates. Sensors. **16**(5), 628, 1–10 (2016)
10. Y. Zhang, S.L. Huang, S. Wang et al., Recognition of overlapped Lamb wave detecting signals in aluminum plate by EMD-Based STFT flight time extraction method. Int. J. Appl. Electromagnet. Mech **52**(3–4), 991–998 (2016)
11. Y. Zhang, S.L. Huang, S. Wang et al., Direction-controllable electromagnetic acoustic transducer for SH waves in steel plate based on magnetostriction. Prog. Electromagnet. Res. **50**, 151–160 (2016)

Chapter 3
Time-of-Flight Extraction Method for the Electromagnetic Ultrasonic Guided Wave Detection Signal

Given the multimode and dispersion characteristics of an electromagnetic guided wave, guided wave modal research and control are mainly conducted from two aspects of work mode or work point selection and guided wave detection signal processing. When the ultrasonic guided wave is in the multimodal working mode, various modal guided waves are propagated in the steel plate according to the pre-set working point and are converted into voltage signals by receiving EMAT through the role of defects, and are amplified, filtered, and the guided wave detection data are formed after the acquisition and processing. The guided wave detection data exist in the form of signal wave packets in the time domain. The guided wave detection signals of various modes are aliased in the same wave packet in the time domain and cannot be directly analyzed and processed in the time domain. It is necessary to separate the modal guided wave detection signals in the time domain wave packet to identify the corresponding modal guided wave in the guided wave detection signal, to make a clear and reasonable explanation for the guided wave detection signal.

On the other hand, when the ultrasonic guided wave is in the single modal excitation mode, the excitation end theoretically produces only one mode of the guided wave, but after the guided wave and the defect act, a modal transition may occur to enable the reception of the EMAT. Received guided waves of different modalities; for the SH0 modal guided wave of the steel plate, if the operating point is selected in the low-frequency region, that is, a single SH0 modal guided wave region, since the working region can theoretically only exist a mode of SH0 modality. In the guided wave, after the SH0 modal guided wave and the defect are generated, other mode SH waves will not be generated in the SH wave category, but other types of guided waves such as Lamb waves may be generated. However, since the receiving EMAT is specifically targeted A special type of transducer designed for SH0 mode guided wave, which cannot receive Lamb wave, cannot generate the signal component of Lamb wave mode in the guided wave detection signal; therefore, it is used to detect the single SH0 modal guided wave of the steel plate. The advantage is that the mode conversion process after the guided wave and the defect is not considered, thereby simplifying the analysis and processing of the guided wave detection signal. From versatility and uniformity of the method of processing, the guided wave detection

S. Huang et al., *Theory and Methodology of Electromagnetic Ultrasonic Guided Wave Imaging*, https://doi.org/10.1007/978-981-13-8602-2_3

signal is reasonable and necessary to perform pattern recognition of the guided wave detection signal before the TOF or amplitude extraction, and it is possible to perform each guided wave mode for a multimodal working mode. Separation of the state is a prerequisite for the subsequent guided wave detection signal TOF or amplitude extraction; for a single modal excitation work mode, the guided wave of other modes should not be decomposed theoretically, and its TOF or amplitude extraction before performing modal identification can verify the theoretical conclusions. If other modal guided waves are not resolved, the next step can be performed. Therefore, the modal identification of the electromagnetic guided wave detection signal plays an important role in both the multimode operation mode and the single-mode excitation operation mode, which is conducive to making a clear and reasonable explanation for the guided wave detection signal, and it provides an important preprocessing work for the follow-up guided wave detection signal TOF or amplitude extraction.

TOF is important information and feature of the ultrasonic guided wave detection signal, and it is the direct input of the guided wave scattering imaging method of the subsequent magnetic sound array. In the guided wave scattering imaging model of the magneto-acoustic array, the small variation of the TOF may bring the model solution result. The larger error, that is, the model solution results in higher sensitivity to TOF, and the TOF extraction accuracy directly determines the defect contour imaging accuracy of the guided wave scattering imaging method of the magneto-acoustic array. Therefore, the detection signal of the electromagnetic ultrasonic guided wave is high. Accuracy extraction is important for defect detection and imaging. However, the distribution of the electromagnetic guided wave detection signal in the time domain usually occupies a large time width, and the arrival time of the ultrasonic guided wave detection signal cannot be accurately obtained in the time domain, and thus the ultrasonic guided wave detection signal cannot be determined; In addition, the excitation side of the ultrasonic guided wave uses burst excitation signals of multiple cycles, the frequency band of which is narrow, and the receiving side of the guided wave uses a narrowband filter circuit to filter the amplified guided wave detection signal, and thus enters the computer. The guided wave detection data to be analyzed are a non-stationary signal with significant narrowband characteristics. Therefore, the electromagnetic traveling wave detection signal with narrowband non-stationary characteristics is accurately extracted to provide accurate TOF input quantity for the guided wave scattering imaging model of the magneto-acoustic array and to reduce the calculation error of the guided wave scattering imaging model of the magneto-acoustic array. It is of great significance to improve the imaging accuracy of defect contours.

This chapter first analyzes the principle of EMD modal identification method and the feasibility of modal separation of time-domain aliasing guided wave detection signals and separately analyzes the EMD mode in the SH guided multimode working mode and single-mode working mode. The state recognition method is verified by experiments, and it is pointed out that the EMD modal identification method plays an important role in the guided wave scattering imaging method of magneto-acoustic arrays. Secondly, the theoretic basis, principles, and steps of time–frequency energy density extraction TOF extraction method are described in detail. The time–frequency energy density precipitation TOF extraction method combined with narrowband electromagnetic ultrasonic wave detection signal characteristics and advantages of discrete short-time Fourier transform in-depth analysis conducted and the time-to-hour extraction accuracy of the time–frequency energy density extraction TOF extraction method was verified by experiments. Finally, this chapter will build a steel plate defect guided wave scattering detection test platform. The modal identification ability of the EMD modal identification method for the defect guided wave scattering detection signal and the time–frequency energy density extraction TOF extraction method for the defect guided wave detection are separately tested. The accuracy of the signal's TOF extraction is verified by experiments, and it is pointed out that the EMD modal identification method proposed in this chapter and the time–frequency energy density precipitation TOF extraction method plays an important role in the guided wave scattering imaging method of magneto-acoustic arrays.

3.1 Time-Domain Aliasing Guided Wave Detection Signal EMD Modal Identification Method

The guided waves of different modalities have different particle vibration modes, which is the basic starting point for the separation and identification methods of guided wave detection signals. For the SH guided wave, the entire displacement field of the SH guided wave is divided into a symmetrical component and an antisymmetric component. That is, the particle vibration is divided into two parts: symmetrical vibration and antisymmetric vibration, which represent the symmetrical mode and antisymmetric mode, respectively. The SH guided wave, which makes the SH guided wave detection signal in the time domain aliased, can start from different modes of the SH guided wave with different modes of vibration, and it can use the different modes of the SH guided wave that is aliased in the same signal wave packet. Specifically, different particle vibration modes cause different guided wave propagation modes, namely guided wave modes, and then cause different EMF distributions on the receiving EMAT, thereby forming voltage signals with different vibration modes. Therefore, the problem of modal separation and identification of time-domain aliasing guided wave detection signals is essentially the problem of modal separation and identification when non-stationary signals with different vibration modes are aliased in the time domain.

In 1998, Huang of NASA proposed an empirical mode decomposition method [1] for analyzing nonlinear and non-stationary signals. First, the non-stationary signal was subjected to empirical mode decomposition to obtain its intrinsic modal function. Intrinsic mode functions (IMF), then perform a Hilbert transform (HT) on each IMF and obtain the time–frequency distribution of the original signal. Because this method does not rely on any priori basis function, the signal decomposition is completely based on the signal's vibration mode and characteristics, and it can theoretically be applied to the signal decomposition of any type and vibration mode, so it has better self-adaptiveness. This method is currently widely used in different engineering fields and is often referred to as the Hilbert-Huang transform (HHT).

The electromagnetic guided wave detection signal belongs to a typical non-stationary signal. The time-domain aliased guided wave detection signal includes the guided wave of different vibration modes. The EMD method can analyze the non-stationary signal and obtain its natural modal function. It can separate the guided wave detection signals of different vibration modes that are aliased in the time domain. Therefore, the EMD method is used to perform modal separation and identification of time-domain aliasing detection signals of different modes of the electromagnetic ultrasonic guided wave. Combined with the signal characteristics of the time-domain aliased electromagnetic ultrasonic guided wave detection signal and the advantages of the EMD method, the principle of the EMD algorithm is feasible. The following is a detailed description of the theoretical method for the modal separation and identification of time-domain aliasing guided wave detection signals using the EMD method.

3.1.1 Principle of EMD Modal Identification Method

The main part to the modal separation of the time-domain aliasing guided wave detection signal by the EMD method is to solve the intrinsic modal function IMF of the original guided wave detection signal. The intrinsic modal function IMF solved must meet two conditions: (1) For all its data points, the difference between the number of local extremum points and the number of zero crossings does not exceed 1; (2) The local mean value of the data is 0, which is expressed as the average value of the local maximum envelope (upper envelope) and the local minimum envelope (lower envelope) is 0. For the original guided wave detection signal $x(n)$, $n = 1, 2, 3 \ldots N$, where N is the total number of points of the guided wave detection signal, and its upper and lower envelopes can be expressed as $x_{\max}(n)$ and $x_{\min}(n)$, the mean $x_{m11}(n)$ of the upper and lower envelopes can be expressed as

$$x_{m11}(n) = \frac{x_{\max}(n) + x_{\min}(n)}{2} \tag{3.1}$$

Subtract this upper and lower envelope mean $x_{m11}(n)$ from the original guided wave detection signal $x(n)$ to obtain a first estimate of the first intrinsic modal function IMF1 $h_{11}(n)$

$$h_{11}(n) = x(n) - x_{m11}(n) \tag{3.2}$$

$h_{11}(n)$ is regarded as the original guided wave detection signal $x(n)$, and it is iterated according to the processing method of the original guided wave detection signal $x(n)$ described above, and the second one of the first intrinsic modal function IMF1 is obtained. The estimated value $h_{12}(n)$ can be expressed as

$$h_{12}(n) = h_{11}(n) - x_{m12}(n) \tag{3.3}$$

where $x_{m12}(n)$ is calculated as

$$x_{m12}(n) = \frac{h_{11}\max(n) + h_{11}\min(n)}{2} \tag{3.4}$$

Similar to the above, $h_{12}(n)$ is regarded as the original guided wave detection signal $x(n)$ and iterated in the same manner. When the kth iteration is performed, the kth estimated value $h_{1k}(n)$ of the first intrinsic modal function IMF1 is obtained, and the iteration termination condition is

$$\sum_{n=1}^{N} \frac{\left|h_{1(k-1)}(n) - h_{1k}(n)\right|^2}{h_{1(k-1)}^2(n)} \le \varepsilon \tag{3.5}$$

Among them, ε is usually in the range of 0.2–0.3. The above formula shows that the first intrinsic mode can be terminated when the first two intrinsic modal functions IMF1 have a small enough difference between the two adjacent estimation values $h_{1k}(n)$ and $h_{1(k-1)}(n)$. The iteration of the function IMF1 is solved. The first intrinsic modal function IMF1 is $c_1(n)$

$$c_1(n) = h_{1k}(n) \tag{3.6}$$

The first intrinsic modal function IMF1 $c_1(n)$ is solved. The first intrinsic modal function IMF1 $c_1(n)$ is subtracted from the original guided wave detection signal $x(n)$ to obtain the residual signal $r_1(n)$ of the first intrinsic modal function.

$$r_1(n) = x(n) - c_1(n) \tag{3.7}$$

Next, the residual signal $r_1(n)$ of the first intrinsic modal function is regarded as the original guided wave detection signal, and the second intrinsic modal function IMF2 $c_2(n)$ is solved according to the above similar iterative process, and its residual signal $r_2(n)$. By analogy, finally, a total of M intrinsic modal functions and a residual signal $r_M(n)$ are obtained, where M is a positive integer, then the original guided wave detection signal $x(n)$ can be expressed as

$$x(n) = \sum_{i=1}^{M} c_i(n) + r_M(n) \tag{3.8}$$

Here, $i = 1, 2, 3, \ldots M, M$ is a positive integer and represents the number of intrinsic modal functions obtained by decomposing the original guided wave detection signal $x(n)$.

The original guided wave detection signal $x(n)$ is decomposed into a superposition of an M intrinsic modal function IMF and a residual signal $r_M(n)$. Each intrinsic modal function IMF represents a difference in the original guided wave detection signal. Vibration mode component. For the guided wave detection signals with different modal time-domain aliasing, the EMD modal decomposition method can separate the multimodal aliasing guided wave detection signals in principle and use the modal guided wave detection signals as their natural modes. The form of the function is characterized to realize the separation and identification of the detection signals of the aliasing guided wave in different modes.

The proposed EMD modal identification method will be experimentally verified under the SH guided multimode and the single working mode. In the SH guided multimode working mode, the EMATs used for emission and reception are all omnidirectional EMATs based on the magnetostriction mechanism used in Chap. 2 and consist of notched ring nickel tape and densely wound coil PCB. This type of EMAT is basically not limited by the working point and working mode of the SH guided wave, that is, this type of EMAT is not designed for a certain modal SH guided wave operating point or operating frequency, so it is applied in the SH guided wave multimodal working mode. In the guided wave single-mode operation mode, the EMAT used for the transmission is the direction-controllable EMAT proposed in Chap. 2, and the working point is the design working point of the EMAT with controllable direction. The receiving EMAT is based on the magnetic induction used in Chap. 2. The stretching mechanism, the omnidirectional EMAT formed by the notched toroidal nickel tape and the densely wound coil PCB, is consistent with the emission transducer, receiving transducer, and SH guided wave working modes and operating points in Chap. 2. It is mainly consistent with the launching/receiving EMAT working mode and working point of subsequently guided wave scattering imaging work, laying the foundation for the construction of the following guided wave scattering imaging test platform At the same time, the key performances of the magnetostrictive controllable EMAT and the omnidirectional EMAT proposed in this chapter can be verified by the verification of the performance of the direction-controllable EMAT and the omnidirectional EMAT receiving single mode SH0 guided waves. Also, the change of the test conditions of the two working modes, especially the change of the emission EMAT, will not negatively affect the experimental verification of the EMD modal identification method. The main reason is that the focus of the EMD modal identification method is the time-domain aliasing multimode guided wave detection signal or the single-mode guided wave detection signal. No matter which emission EMAT is used and in which operating mode and working point, the EMD modal identification method can separate the modal guided wave signals from the principle of time-domain aliasing of the multimode guided wave detection signals. For a single-mode guided wave detection signal, the EMD modal identification method should be able to embody this modal guided wave in its natural modal function. Therefore, the EMD modal identification method is immune to the omnidirectional or directional type of the transmitted transducer and has good adaptability to the guided wave detection signal.

3.1.2 EMD Modal Identification Method Test Verification

The experimental verification of the EMD modal identification method will be carried out in the SH guided multimode working mode, and the single modal working mode, respectively, and the test verification under these two working modes will be discussed in detail below.

According to the group velocity dispersion curve of the SH guided wave in the 4-mm-thick steel plate shown in Fig. 3.1, the operating frequency of the selected SH guided wave is 700 kHz. At this operating frequency, two modes of the SH guided wave are theoretically generated. They are SH0 and SH1, respectively, and their theoretical group speeds are $c_{g0} = 3200$ and $c_{g1} = 2626$ m/s, respectively.

On the 4 mm steel plate, the experimental verification platform for the EMD modal identification method under the multimode working mode of the SH guided wave was constructed, as shown in Fig. 3.2. Both the transmitting and receiving EMATs are based on magnetostrictive omnidirectional EMATs, and the distance between them is 0.2 m.

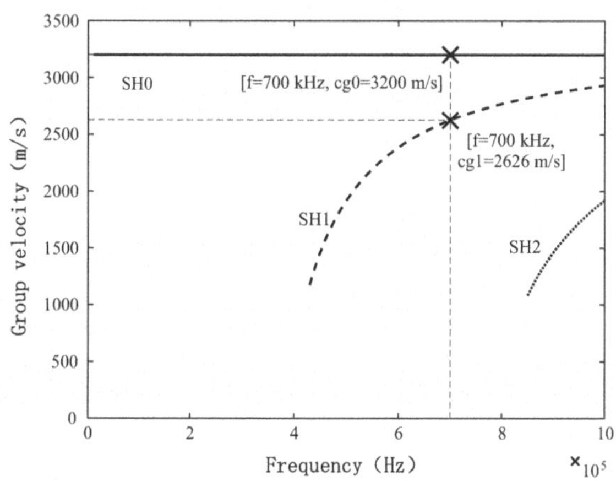

Fig. 3.1 Multimode working point of SH guided wave for 4 mm thick steel plate used for EMD modal identification test verification

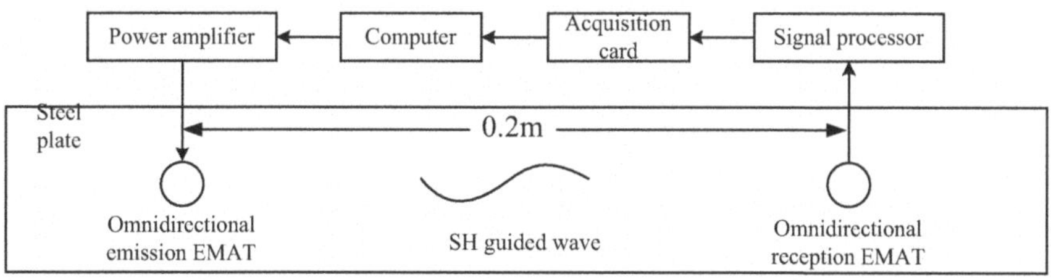

Fig. 3.2 Block diagram of the experimental verification platform for EMD modal identification in the mode of multimode operation of SH guided wave

The working principle of the test mentioned above and verification platform is similar to that of the direction-controllable EMAT performance verification and detection test platform in Chap. 2, which will not be repeated here. The test parameter settings during the test are shown in Table 3.1.

According to the setting of the above test parameters, the EMD modal identification method verification test in the mode of SH guided multimode operation is performed. The test results are analyzed and discussed below.

According to the operating point of the SH guided wave and the theoretical propagation velocity of the SH0 and SH1 modal guided waves, the theoretical propagation time is 62.5 and 76.16 μs, respectively, under the above experimental parameters. Under the above operating mode and test parameter settings, the waveforms of the guided wave detection data collected are as shown in Fig. 3.3.

From Fig. 3.3, it can be seen that the SH0 modal guided wave with a theoretical propagation time of 62.5 μs and the SH1 guided wave with 76.16 μs are aliased

Table 3.1 Test parameter setting of EMD modal identification method in SH guided multimode operation mode

Parameter	Value
Excitation voltage peak (V)	100
Excitation frequency f (kHz)	700
Number of incentive cycles	12
Steel plate length (mm)	1000
Steel plate width (mm)	1000
Plate thickness (mm)	4
EMAT and EMAT distances (mm)	200
Signal processor magnification	5000
Signal processor narrowband filter center frequency (kHz)	700
Signal processor narrowband filter bandwidth (kHz)	20

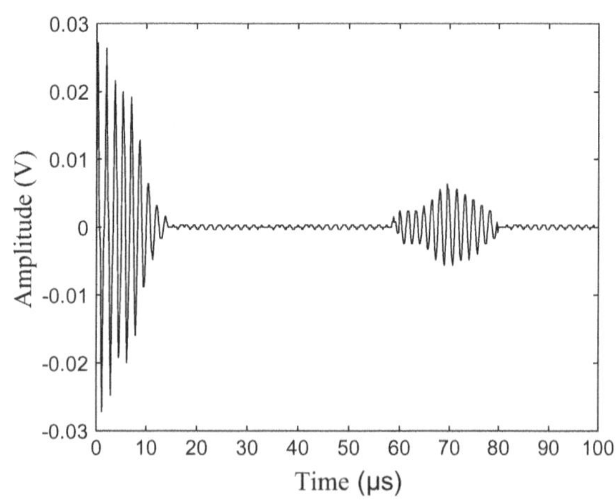

Fig. 3.3 Guided wave detection data waveforms in the SH guided wave multimode operating mode

in a waveguide detection signal wave packet in the time domain and cannot be distinguished in the time domain. If the test parameters and the size of the transducer position are unknown, the guided wave detection signal will be misjudged to be a direct wave of some modal guided wave to a large extent.

The EMD modal identification method is used to analyze the above-mentioned time-domain aliased guided wave detection signals, and its intrinsic modal function IMF is iteratively obtained. In each screening of the IMF, the screening iteration termination condition is $\varepsilon \leq 0.25$. The termination condition of the whole process of modal decomposition is that no new IMF will be generated. The solution to the intrinsic mode function of the time-domain aliased guided wave detection signal is shown in Fig. 3.4.

In the solution result of the above-mentioned time-domain aliasing guided wave detection signal, IMF, IMF1, and IMF2 are separated as main intrinsic modal functions in the time domain, and they have different vibration modes of the particles. In the time domain, it also occupies different time ranges, respectively; IMF3 mainly contributes to the larger middle part of the time-domain aliasing wave detection signal wave packet; the IMF4 to IMF7 have smaller amplitudes. The impact and contribution to the original guided wave detection signal is also relatively small.

To describe the differences between IMF3 and IMF1 and IMF2, the power spectral densities of IMF1 to IMF3 were calculated. The results are shown in Fig. 3.5. The power spectral density distributions of IMF1 and IMF2 are relatively close and their numerical ranges are mostly between −50 and −100 dB. However, the power spectral density of IMF3 has appeared to be below −100 dB in value. Also, the power spectral density distributions of IMF1 and IMF2 in the low angular frequency region are more concentrated, while the power spectral density distribution of IMF3 is relatively uniform, and there is an obvious phenomenon of low-angle frequency region distribution and concentration. Therefore, the power spectral density distribution of IMF3 has a significant difference concerning the power spectral density of IMF1 and IMF2, which is the result of the superposition of the distortion of the tail of IMF1 and the IMF2. It should not be used as the guided wave mode contained in the original guided wave detection signal.

In order to accurately obtain the time information of IMF1 and IMF2 and identify the two intrinsic modal functions, STFT short-time Fourier transforms (STFT) time–frequency analysis method is used to extract IMF1 and IMF2. The result of the extraction is shown in Fig. 3.6.

Among them, the TOF extraction result of IMF1 is 65.49 μs, which is close to the theoretical propagation time of SH0 mode guided wave of 62.5 μs. The IMF2 TOF extraction result is 73.79 μs, which is close to the theoretical propagation time of an SH1 mode guided wave of 76.16 μs. In order to quantitatively compare the TOF extraction result t_{ex} of the guided wave of SH and the theoretical propagation time t_{the}, the relative error e_{r} of the guided wave TOF extraction result and the theoretical propagation time is defined.

$$e_{\text{r}} = \frac{|t_{\text{ex}} - t_{\text{the}}|}{t_{\text{the}}} \times 100\% \tag{3.9}$$

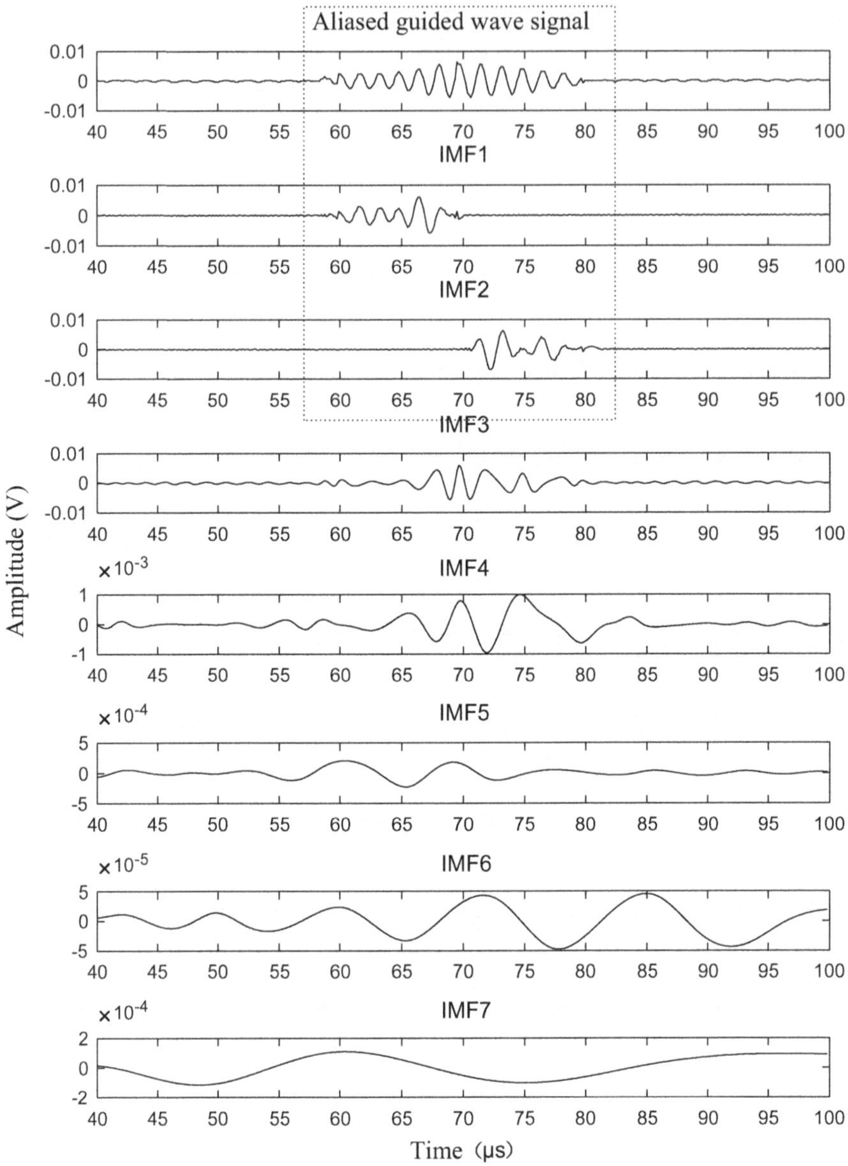

Fig. 3.4 Intrinsic mode function of time-domain aliasing guided wave detection signal

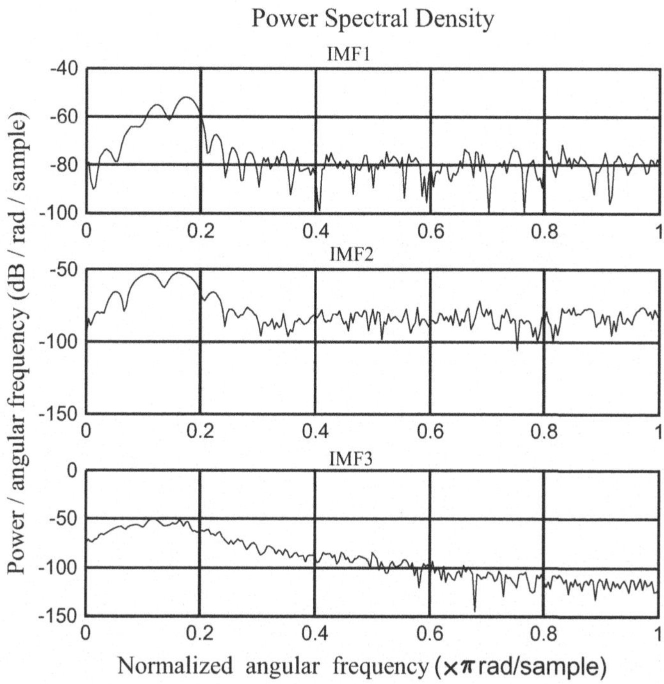

Fig. 3.5 Power spectrum density calculation results of time-domain aliasing guided wave detection signals IMF1 to IMF3

The results of the TOF extraction of the above SH guided wave and the theoretical propagation time are compared, and the results are as follows in Table 3.2.

According to Table 3.2, the relative error between the IMF1 TOF extraction result and the theoretical propagation time of the SH0 modal guided wave is small, the relative error between the IMF2 TOF extraction result and the theoretical propagation time of the SH1 modal guided wave is small, and the relative error between the two is less than 5% (Fig. 3.4). In the IMF solution of the time-domain aliasing guided wave detection signal, the IMF1 after EMD decomposition is the SH0 mode guided wave mixed in the wave detection signal packet. IMF2 is the SH1 mode guided wave that is aliased in the wave detection signal packet. Therefore, EMD modal identification method is used to separate and identify guided wave detection signals with different time-domain aliasing modes. The different mode SH guided waves aliased in the same guided wave detection signal packet are separated by different intrinsic mode functions IMF. The modal separation and identification of time-domain aliasing guided wave detection signal packets in the multimode operation mode of SH guided wave are realized, and the EMD modal identification method is verified in the multimodal working mode of SH guided wave.

In the following, the proposed EMD modal identification method will be experimentally verified under the single-mode working mode of SH guided wave.

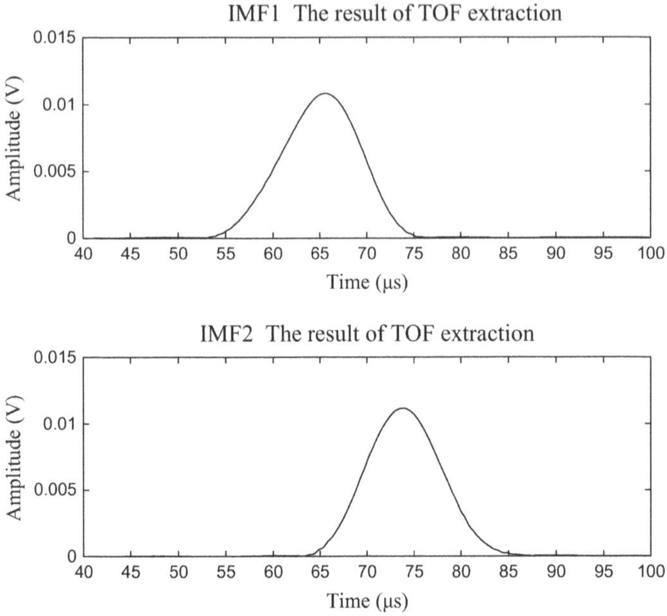

Fig. 3.6 Timing aliased guided wave detection signals IMF1 and IMF2 timing results

Table 3.2 Time comparison of TOF extraction and theoretical propagation time in multimode operation of SH guided wave

SH guided mode	Theoretical propagation speed (m/s)	Theoretical propagation time (μs)	TOF extraction result (μs)	Corresponding IMF	Relative error (%)
SH0	3200	62.5	65.49	IMF1	4.78
SH1	2626	76.16	73.79	IMF2	3.11

In the EMD modal identification verification test of the SH guided single-mode operation mode, the emission EMAT is the direction-controllable EMAT based on magnetostriction presented in Chap. 2. The receiving EMAT is an omnidirectional EMAT based on nickel tape and densely wound coil PCB. The verification test platform is constructed using the direction-controllable EMAT performance verification and detection test platform proposed in Sect. 2.2.6 . The test parameters are configured the same as the platform. The excitation frequency and the narrowband filter frequency of the signal processing module are both 3200 kHz. It can be seen from the SH-wave curve shown in Fig. 2.79 in the group velocity dispersion curve of a 4-mm-thick steel plate. Theoretically, there is only one mode of the guided wave at the working point. Under the test platform and test parameters set up in Sect. 2.2.6 of Chap. 2, the SH0 guided wave excited by the 90° direction of EMAT with direction controllable in Sect. 2.2.6 is analyzed.

The EMD modal identification method is used to analyze the SH guided wave detection signal collected by those mentioned above 90° position omnidirectional reception EMAT, and its intrinsic modal function IMF is iteratively obtained, in each selection of the IMF screening process. The screening iteration termination condition is $\varepsilon \leq 0.25$. The termination condition of the modal decomposition process is that no new IMF is generated. The solution of the natural modal function of this SH guided wave detection signal is shown in Fig. 3.7.

In the above SH0 single-mode guided wave detection signal IMF solution results, only the amplitude magnitude of the IMF1 signal is equivalent to that of the original SH0 single-mode guided wave detection signal. The amplitudes of IMF2 to IMF5 are small relative to the original guided wave detection signal and have little effect

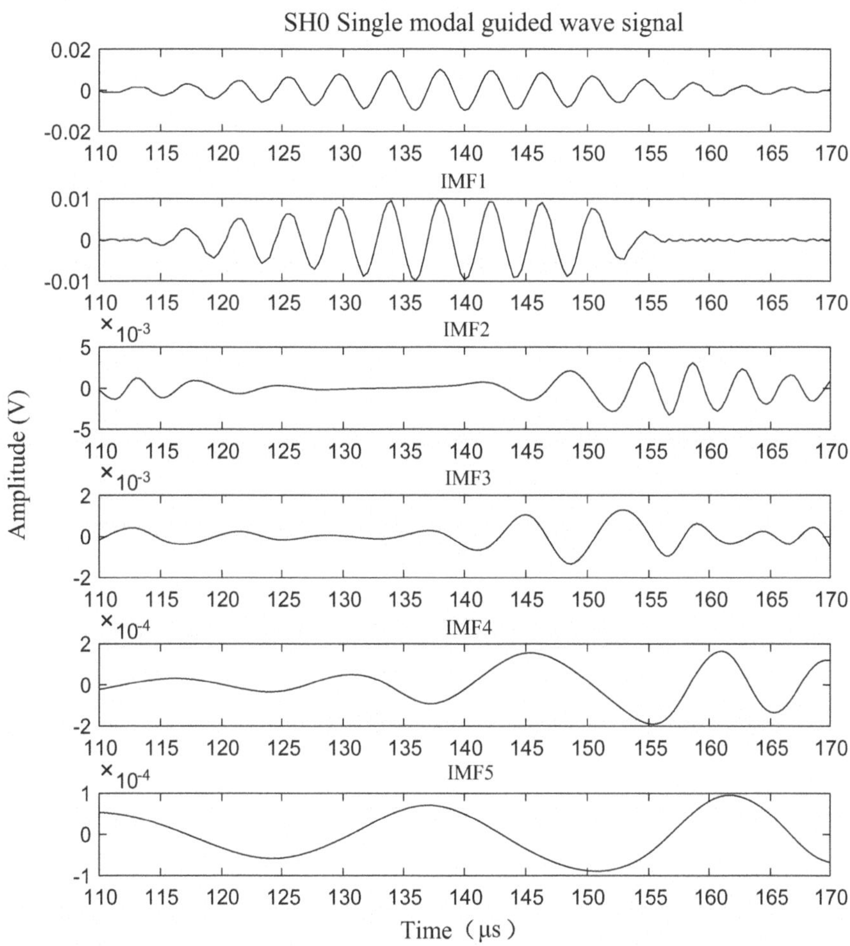

Fig. 3.7 Intrinsic modal function of SH0 single-mode guided wave detection signal

on the original guided wave detection signal. Therefore, the SH mentioned above guided wave detection signal is subjected to EMD modal decomposition, only one dominant natural mode function IMF1 is decomposed, and other intrinsic modal functions that are similarly dominant are not present, indicating that the original guided wave detection signal has only one modal guided wave.

To further illustrate the representative role of the dominant intrinsic modal function IMF1 on the original SH0 guided wave detection signal, it is similar to that of the previous section. The original SH0 guided wave detection signal and the decomposed intrinsic modal function IMF1 are extracted by the STFT TOF, and the extraction result is shown in Fig. 3.8.

In the above TOF extraction results, the original TOF of the SH0 guided wave detection signal is 138.3 μs, and the TOF extraction result of the intrinsic mode function IMF1 is also 138.3 μs. The intrinsic modal function IMF1 represents the representative effect of the original SH0 guided wave detection signal, and also proves that the original guided wave detection signal only contains the SH0 modal guided wave.

According to Eq. (3.9), calculate the relative error between the extracted TOF and the theoretical propagation time of the guided wave SH0. The result is as follows in Table 3.3.

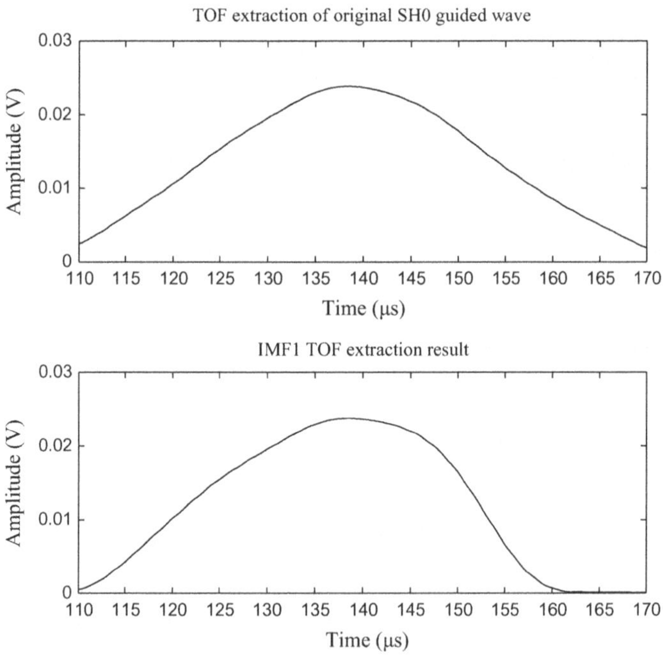

Fig. 3.8 Original SH0 guided wave detection signal and intrinsic modal function IMF1 TOF extraction result

Table 3.3 Time comparison of TOF extraction and theoretical propagation time in single guide mode of SH guided wave

SH guided mode	Theoretical propagation speed (m/s)	Theoretical propagation time (μs)	TOF extraction result (μs)	Corresponding IMF	Relative error (%)
SH0	3200	140.625	138.3	IMF1	1.65

According to Table 3.3, the relative error between the IMF1 TOF extraction result and the theoretical propagation time of the SH0 single-mode guided wave is small, and the IMF1 TOF extraction result is the same as the original SH0 guided wave detection signal TOF extraction result, indicating that the original guided wave detection signal only contains a mode of guided wave of SH0. Therefore, using the EMD modal identification method to analyze single-mode guided wave detection signals will only decompose the guided wave detection signals of the included modes. It is embodied in the form of an intrinsic mode function IMF of the same order of magnitude as the original guided wave detection signal, and the TOF extraction result of this intrinsic modal function IMF is the same as that of the original single-mode guided wave detection signal. This shows that the intrinsic modal function IMF has a good representative effect on the original single-mode guided wave detection signal. The above-mentioned experimental verification results for the single-mode operation mode EMD modal identification method of the SH guided wave can be used as an important basis for determining whether a guided wave detection signal is a single mode and identifying a specific mode type, that is, EMD decomposition of a guided wave detection signal. If only an intrinsic modal function IMF of the order of magnitude of the same magnitude as the original guided wave detection signal is present, and the result of the TOF extraction of the intrinsic modal function IMF is the same as the result of direct TOF of the guided wave detection signal. The guided wave detection signal is a single-mode guided wave signal, and its specific modal type can be determined by calculating its actual propagation speed and comparing it with the relevant operating point of the dispersion curve.

3.2 Time–Frequency Energy Density Precipitation TOF Extraction Method for Narrowband Guided Wave Detection Signal

The electromagnetic guided wave detection signal is a typical non-stationary signal, and its frequency components will change with time [2–8]. The traditional Fourier transform can only obtain what frequency components it contains. Time–frequency characterization of signals is one of the effective processing methods for non-stationary signals. It can describe the characteristics of the frequency components of non-stationary signals over time, namely their time-varying spectrum characteristics.

Short-term Fourier transform STFT is a relatively efficient class of time–frequency characterization methods for non-stationary signals. It has the advantages of fast calculation speed, clear physical meaning, and wide controllability of window functions. It is widely used in the analysis of non-stationary signals.

The basic principle of the short-time Fourier transform is to decompose the signal $x(t)$ in the time domain t into signal components with multiple widths and small enough signals. After the windowing process is performed on each signal component, a Fourier transform is applied:

$$\mathrm{STFT}_x(t,\omega) = \int x(\tau)g^*(\tau - t)\mathrm{e}^{-j\omega\tau}\mathrm{d}\tau \tag{3.10}$$

where $g(t)$ is the window function applied to the signal during the short-time Fourier transform. It can be seen from the above equation that the time-domain signal $x(t)$ is characterized in the time-domain t and frequency domain ω after short-time Fourier transform. The time-domain one-dimensional representation t of the non-stationary signal $x(t)$ is converted into a two-dimensional representation of the time–frequency domain (t, ω). The variation of the intensity of different frequency components of a non-stationary signal with time can be described in the time domain t, and the frequency component distribution of the non-stationary signal at different times can also be described in the frequency domain ω. In order to describe the energy distribution of the non-stationary signal $x(t)$ in the two-dimensional time–frequency domain (t, ω), the energy density of the non-stationary signal $x(t)$ in the two-dimensional time–frequency domain (t, ω) is defined as $E(t, \omega)$

$$E(t,\omega) = |\mathrm{STFT}_x(t,\omega)|^2 \tag{3.11}$$

The short-time Fourier transform is a linear transformation of the non-stationary signal $x(t)$, and the product of the time width and the bandwidth in the transform process is a constant. If you expect a higher time resolution, you must take a smaller time width. Since the product of time width and bandwidth is constant, a smaller time width corresponds to a larger bandwidth, while a larger bandwidth corresponds to a lower frequency resolution. This is an essential feature determined by the definition of the short-time Fourier transform, which basically cannot meet both high time resolution requirements and high frequency resolution requirements.

The narrowband electromagnetic guided wave detection signal used in this chapter is a non-stationary signal with a very narrow frequency band, and the excitation end of the guided wave uses a multicycle narrowband excitation signal voltage. A narrowband filter circuit is used in the signal processor to narrowband filter the received guided wave detection signal. According to the operating point of the controllable EMAT in Chap. 2, the signal processor narrow-band filter center frequency is 320 kHz, narrow-band filter bandwidth is 20 kHz, and bandwidth is only 6.25% of the filter center frequency. Therefore, most of the guided wave detection signal energy is concentrated in the vicinity of the center frequency of 320 kHz. Therefore, the band of narrowband electromagnetic guided wave detection signals used in this chapter is very narrow, and the information in the frequency domain is relatively scarce. The main information is reflected in the time domain—for example, the TOF of each guided wave detection signal. The time–frequency characteristics of those mentioned above narrowband electromagnetic ultrasonic guided wave detection signal exactly match the characteristics of the short-time Fourier transform signal processing: Due to the lack of frequency domain information in the narrowband electromagnetic ultrasonic wave detection signal, a wider bandwidth can be used in the short-time Fourier transform. Due to the more attention paid to the time-domain information of the narrowband electromagnetic guided wave detection signal and its more time-domain information, a narrower time width can be used in the short-time Fourier transform. The requirement of narrow bandwidth and narrow time width in the short-time Fourier transform of the narrowband electromagnetic ultrasonic guided wave detection signal meets the limitation of the time width and the bandwidth of the short-time Fourier transform. In the short-time Fourier transform of the narrowband electromagnetic ultrasonic guided wave detection signal, sufficient and detailed time-domain information can be obtained with the shortest possible period, without having to worry about the loss of frequency domain information. Therefore, the short-time Fourier transform is very suitable for the time–frequency analysis of the non-stationary narrowband electromagnetic ultrasonic guided wave detection signal on its principle and obtains its relatively accurate time–frequency domain, especially the time domain information.

This chapter proposes a time–frequency energy density extraction TOF extraction method for the narrowband electromagnetic ultrasonic guided wave detection signal and embeds the short-time Fourier transform into it. The time–frequency characteristics of narrowband electromagnetic ultrasonic wave detection signals in the time domain and lack of information in the frequency domain are fully taken into account—the advantage of short-time Fourier transform, such as fast calculation speed, clear physical meaning, and wide controllability of window function, etc. In principle, it can accurately characterize the time–frequency characteristics of narrowband electromagnetic guided ultrasonic wave detection signals. In the following, the principle and steps of the proposed time–frequency energy density precipitation TOF extraction method will be described in detail.

3.2.1 Principle and Steps of Time–Frequency Energy Density Precipitation Extraction Method

The time–frequency energy density extraction TOF extraction method proposed in this chapter addresses the problem that the distribution of the narrowband electromagnetic ultrasonic wave detection signal in the time domain usually occupies a large time width and cannot accurately obtain the arrival time of the ultrasonic guided wave detection signal in the time domain. The time–frequency energy density distribution of the original guided wave detection signal is obtained by using the discrete short-time Fourier transform. The TOF of the guided wave detection signal is obtained by extracting the energy density curve at the center frequency. The specific principles and steps are as follows:

(1) The narrowband electromagnetic guided wave detection signal collected is expressed as $x(m)$, where $m = 1, 2, \ldots, M$, and M are positive integers. A discrete short-time Fourier transform is performed on the narrowband electromagnetic ultrasonic wave detection signal $x(m)$. The time coordinate obtained by the transformation is t_i ($i = 1, 2, \ldots, M$) and the frequency coordinate is f_k ($k = 1, 2, \ldots, N$), where N is a positive integer. The energy density at a certain set of determined time points and frequency points (t_i, f_k) is $E(t_i, f_k)$. The discrete short-time Fourier transform DSTFT(i, k) of the narrowband electromagnetic ultrasonic wave detection signal $x(m)$ is

$$\mathrm{DSTFT}(i,k) = \sum_{m=1}^{M} x(m)g(i-m)\mathrm{e}^{-\mathrm{j}\frac{2\pi km}{M}} \tag{3.12}$$

where DSTFT(i, k) is the result of the discrete short-time Fourier transform of the electromagnetic guided wave detection signal $x(m)$; $g(i)$ is the window function of the discrete short-time Fourier transform.

Then, in the two-dimensional time–frequency distribution (t_i, f_k) of the discrete short-time Fourier transform, the time–frequency energy density $E(t_i, f_k)$ is defined as

$$E(t_i, f_k) = |\mathrm{DSTFT}(i,k)|^2 \tag{3.13}$$

(2) In the time–frequency energy density distribution obtained in step (1), for a discrete time point t_i, the corresponding frequency coordinate is f_k ($k = 1, 2, \ldots, N$). Determine whether there is a value of the frequency point f_k equal to the value of the center frequency f_c, if yes, extract the time–frequency energy density $E(t_i, f_c)$ corresponding to the center frequency f_c, and perform step (5); if not, perform step (3).
(3) For the frequency coordinate f_k ($k = 1, 2, \ldots, N$) corresponding to the discrete time point t_i in step (2), the frequency points f_l and f_h closest to it are extracted on both sides of the center frequency f_c. Where $f_l < f_c < f_h$

$$f_l = \arg\min|f_k - f_c|, \;\; f_k < f_c \tag{3.14}$$

$$f_h = \arg\min|f_k - f_c|, \;\; f_k > f_c \tag{3.15}$$

The time–frequency energy density $E(t_i, f_l)$ at (t_i, f_l) and the time–frequency energy density $E(t_i, f_h)$ at (t_i, f_h) are extracted, respectively.

(4) $E(t_i,f_l)$ and $E(t_i,f_h)$ obtained according to step (3). Using the linear interpolation method, find the time–frequency energy density $E(t_i,f_c)$ corresponding to the discrete time point t_i and the center frequency f_c.

$$E(t_i, f_c) = (f_c - f_l)\frac{E(t_i, f_h) - E(t_i, f_l)}{f_h - f_l} \tag{3.16}$$

(5) For all the discrete time points t_i ($i = 1, 2, \ldots, M$) obtained in step (1), determine whether the time–frequency energy density $E(t_i,f_c)$ corresponding to the center frequency f_c is all obtained. If yes, proceed to step (6); if not, consider discrete time point t_{i+1} and return to step (2).

(6) Using the time–frequency energy density $E(t_i,f_c)$ at all discrete time points t_i ($i = 1, 2, \ldots, M$) and the determined center frequency. By the least-squares fitting of the time–frequency energy density $E(t_i,f_c)$ at discrete time points t_i ($i = 1, 2, \ldots, M$). Set the energy density curve $E(t,f_c)$ in the time domain based on the center frequency f_c, set the time domain fitting function to $\varphi(t)$, and the energy density curve in the time domain can be expressed as

$$E(t, f_c) = \arg\min \sum_{i=1}^{M} |\varphi(t_i) - E(t_i, f_c)|^2 \tag{3.17}$$

(7) Extract the time t_p ($p = 1, 2, \ldots, M_1$) corresponding to each peak of the energy density curve $E(t,f_c)$ in the time domain.

$$t_p = \arg\max E(t, f_c) \tag{3.18}$$

where M_1 is a positive integer and represents the peak number of the energy density curve $E(t,f_c)$ in the time domain, satisfying $M_1 < M$.

(8) Taking the peak time t_0 of the energy density curve $E(t,f_c)$ in the time domain corresponding to the initial pulse wave of the original guided wave detection signal as the timing starting point of the guided wave detection signal TOF, the TOF t_n of each guided wave detection signal can be calculated as

$$t_n = t_p - t_0 \tag{3.19}$$

where $n = 1, 2, \ldots, M_1 - 1$, $M_1 - 1$ represents the total number of guided wave detection signal wave packets to be extracted.

The above is the theoretical step of the time–frequency energy density extraction TOF extraction method for the narrowband electromagnetic ultrasonic guided wave detection signal proposed in this chapter. The TOF of each guided wave detection signal is the difference between the peak value of the energy density curve in the time domain of the guided wave detection signal and the peak value of the energy density curve in the time domain of the initial pulse wave. The algorithm flow of the time–frequency energy density extraction TOF extraction method proposed in this chapter is shown in Fig. 3.9.

This chapter proposes a time–frequency energy density extraction TOF extraction method to accurately extract the TOF of the narrowband electromagnetic ultrasonic guided wave detection signal. First, discrete short-time Fourier transform is used to process the narrowband electromagnetic ultrasonic wave detection signal data,

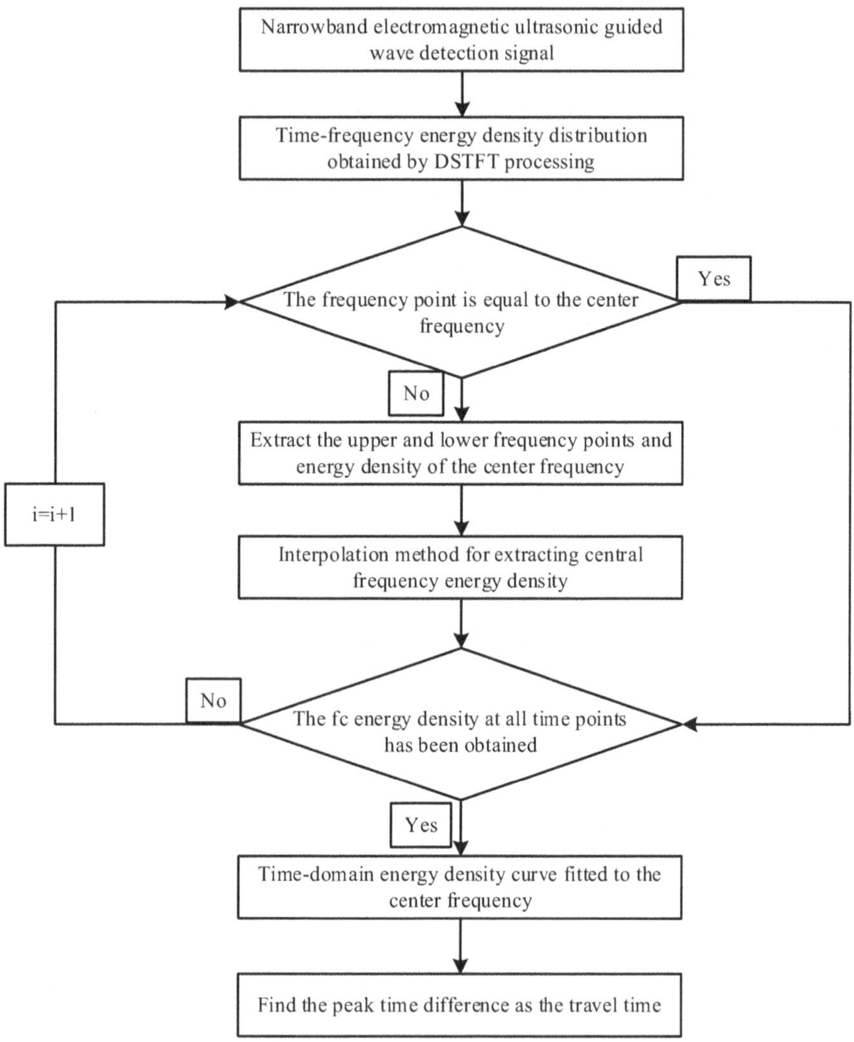

Fig. 3.9 Block diagram of an algorithm for extracting time–frequency energy density TOF

and the time–frequency energy density distribution of the guided wave detection signal data is obtained; and the time–frequency energy density distribution of the detection wave signal data is guided. A time–frequency energy density precipitation method based on data center frequency is established to extract the time–frequency dispersion curve of the data center frequency, and the energy density curve in the time domain is established by fitting the discrete curve. The difference between the time corresponding to each peak of the time domain energy density curve and the peak time of the initial pulse wave time domain energy density curve is taken as the TOF of each electromagnetic ultrasonic guided wave detection signal.

The processing method of the time–frequency energy density precipitation TOF extraction proposed in this chapter is the detection signal of the electromagnetic guided wave with narrowband characteristics. That is, the signal energy distribution range in the frequency domain is very narrow, and most of the signal energy is concentrated near the center frequency. At this time, the frequency energy density precipitation time extraction method is effective in that the object to be effectively treated is not limited by the specific guided wave modes. The main reason is that it uses the distribution of signal time–frequency energy density curves in the time domain. The TOF comes from time domain information corresponding to the local energy of the signal energy concentration because the premise is narrowband electromagnetic guided wave detection signals. Therefore, it has little relationship with the vibration mode of the signal itself, that is, it is not affected by the specific guided wave mode in principle. Therefore, the time–frequency energy density extraction TOF extraction method proposed in this chapter does not focus on the specific guided wave modes. It can be used as a general TOF extraction method for narrowband electromagnetic ultrasonic guided wave detection signals.

In the following, taking the TOF extraction of the narrowband SH guided wave detection signal in the steel plate as an example, the time and frequency energy density precipitation TOF extraction method proposed in this chapter are used to verify the accuracy and versatility of the narrowband electromagnetic guided wave detection signal extraction time.

3.2.2 *Time–Frequency Energy Density Precipitation Time Extraction Method Test Verification*

In order to verify the accuracy and versatility of the time–frequency energy density extraction TOF extraction method proposed in this chapter for extracting the TOF of the narrowband electromagnetic ultrasonic guided wave detection signal, the influence of the guided wave working point, propagation distance and dispersion characteristics on the TOF extraction accuracy is studied to carry out time–frequency energy density precipitation time extraction test validation. The test and verification platform are also carried out on a 4-mm-thick steel plate. Its components and working principles are similar to the EMD modal identification method test verification platform shown in Fig. 3.2 and will not be repeated here. On the 4 mm thick steel plate SH guided wave group velocity dispersion curve, a total of 4 working frequencies are set, which are 320, 460, 560, and 700 kHz, respectively, as shown in P1, P2, P3, and P4 in Fig. 3.10.

The operating point of the SH0 guided wave at the operating frequency of P1 is P10; similarly, at the operating frequency of P2, the working point of the SH0 guided wave is P20, and the working point of the SH1 guided wave is P21, and so on. According to the dispersion curve and operating frequency settings, the working point setting of the SH guided wave is shown in Table 3.4.

In combination with Fig. 3.10 and Table 3.4, except for the working point P10 corresponding to the SH0 guided wave at the working frequency of P1, the working points of the SH0 guided wave and the SH1 guided wave are, respectively, corresponded to the working frequencies of P2 to P4, so a total of 7 are obtained. The

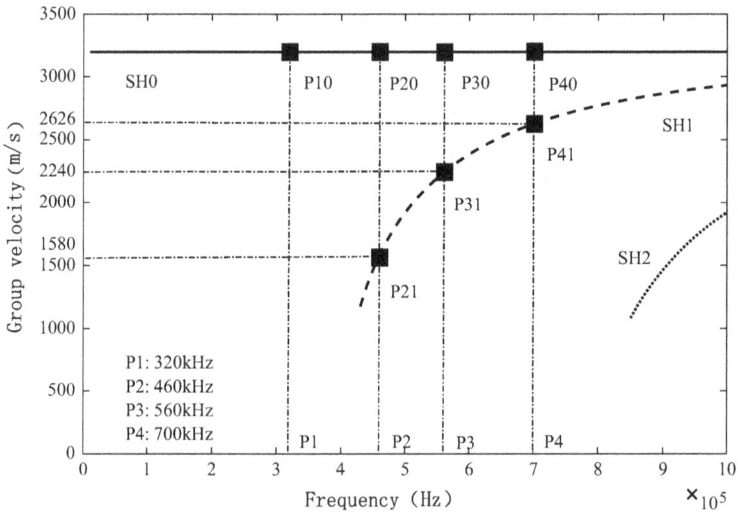

Fig. 3.10 Verification of the operating point of the verification platform for time–frequency energy density precipitation TOF extraction method

Table 3.4 Extraction method of time–frequency energy density TOF verification platform work point setting

Operating frequency number	Operating frequency value (kHz)	SH0 guided point number—group velocity (m/s)	SH1 guided wave working point number—group velocity (m/s)
P1	320	P10-3200	–
P2	460	P20-3200	P21-1580
P3	560	P30-3200	P31-2240
P4	700	P40-3200	P41-2626

operating points, namely P10, P20, P21, P30, P31, P40, and P41, corresponding to the corresponding modal guided wave group velocities for each operating point given in Table 3.4.

For the P10 operating point, the emitted EMAT is the direction-controllable EMAT developed in Chap. 2, and the EMAT used in the rest of the operating points is a magnetostrictive omnidirectional EMAT based on a nickel band and a closely wound coil. The reception EMAT for all operating points is those mentioned above magnetostrictive omnidirectional EMAT. In the verification test of each working point, the distance between the launching EMAT and the receiving EMAT was increased from 20 to 10 cm in steps of 10–110 cm, that is, ten proofing tests of the different propagation distances of the SH guided waves were required for each working point. The test parameter settings during the test are shown in Table 3.5.

For each of the ten detection distances for each of the above operating points, SH guided waves were excited, received, and collected according to the test parameters in Table 3.5. The time–frequency energy density extraction TOF extraction method proposed in this chapter was used to extract the guided time of SH guided waves. In order to compare the accuracy of the TOF extraction, under the same conditions, the widely used HHT method is also used to extract the acquired SH guided wave detection signal. The TOF extracted by the time–frequency energy density precipitation method is compared with the TOF extracted by the HHT method.

In the following, taking P20 and P21 operating points at a working frequency of 460 kHz at P2, and emitting EMAT and receiving EMAT spacing as 100 cm as examples, the application steps of the time–frequency energy density extraction TOF extraction method and the key intermediate link results are illustrated. Figure 3.11 shows the waveform of the guided wave detection signal collected during this test.

As shown in Fig. 3.11, the first waveform is the initial pulse wave received by the EMAT when the excitation voltage is applied to the transmitted EMAT. It can be seen from Fig. 3.10 that at the P2 operating frequency, the theoretical group velocity of the SH1 guided wave is smaller than the theoretical group velocity of the SH0 guided wave, so the theoretical arrival of the received EMAT should be the SH0

Table 3.5 Time–frequency energy density precipitation time extraction method test parameter setting

Parameter	Value
Excitation voltage peak-to-peak (V)	220
Excitation frequency f (kHz)	320, 460, 560, 700
Number of incentive cycles (1)	12
Plate thickness (mm)	4
EMAT and EMAT distance (cm)	20:10:110
Signal processor magnification	5000
Signal processor narrowband filter center frequency (kHz)	320, 460, 560, 700
Signal processor narrowband filter bandwidth (kHz)	20

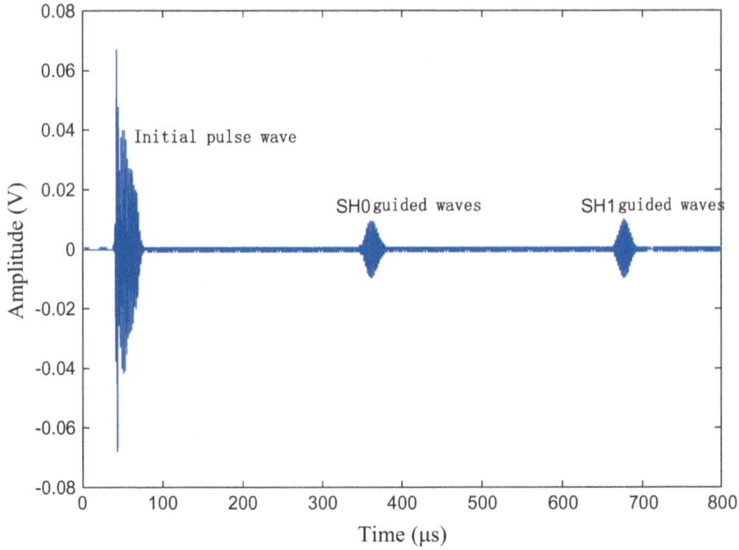

Fig. 3.11 Waveforms of guided wave detection signals for EMAT interval 100 cm at the P2 operating frequency

guided wave. That is, the second waveform is temporarily determined as the SH0 guided wave. In the same way, the third waveform is temporarily determined as the SH1 guided wave.

According to the theoretical steps of the time–frequency energy density extraction TOF extraction method, the original short guided wave detection signal is first subjected to discrete short-time Fourier transform, and its time–frequency two-dimensional distribution is obtained. The guided wave detection data contain a total of 5072 data points. In the discrete short-time Fourier transform, a Hamming window function is used to intercept the original guided wave detection signal, and the window function has a width of 317. That is, one-sixteenth of the data points of the original guided wave detection signal. According to step (1) of the time–frequency energy density precipitation TOF extraction method, the time–frequency energy density value of the original guided wave detection signal in the two-dimensional time–frequency distribution is calculated and plotted in a two-dimensional time–frequency distribution, as shown in Fig. 3.12.

From Fig. 3.12, it can be seen that the energy of the guided wave detection signal is mainly concentrated near the center frequency of 460 kHz. By performing step (2) to step (6) of the proposed time–frequency energy density precipitation TOF extraction method, a time–frequency energy density distribution curve in the time domain based on the center frequency is obtained, as shown in Fig. 3.13.

In Fig. 3.13, according to step (7) to step (8) of the time–frequency energy density precipitation TOF extraction method, the peak time of the initial pulse wave, the SH0 guided wave, and the SH1 guided wave on the time-domain energy density curve is

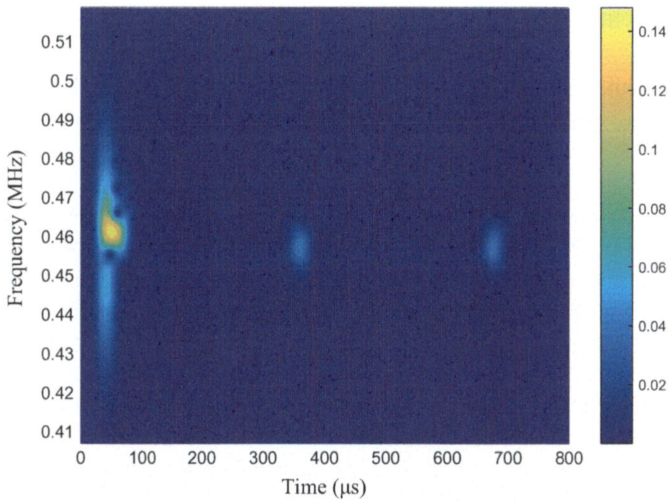

Fig. 3.12 Time–frequency energy density two-dimensional distribution of guided wave detection signal with EMAT spacing 100 cm at the P2 operating frequency

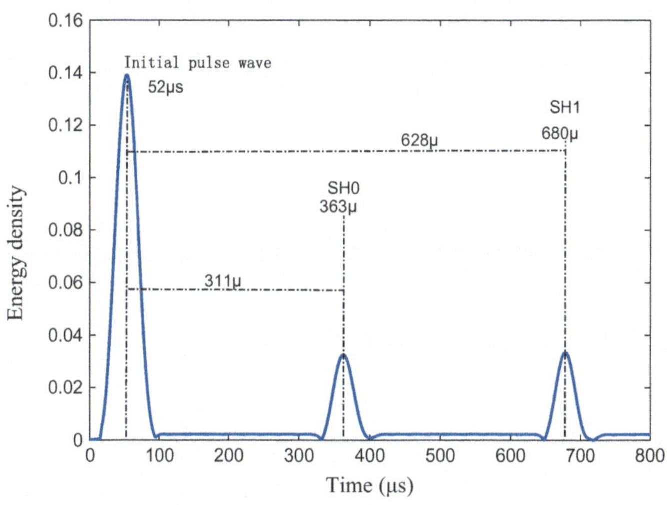

Fig. 3.13 Time-domain energy density distribution curve of the guided wave detection signal with EMAT spacing 100 cm at the P2 operating frequency

first extracted. They are 52, 363, and 680 μs, respectively. The time for re-calculating the SH0 guided wave concerning the initial pulse wave was 311 μs, and the time for the SH1 guided wave concerning the initial pulse wave was 628 μs. Therefore, using the time–frequency energy density precipitation TOF extraction method, the TOF extraction values of SH0 guided wave and SH1 guided wave in the guided wave detection signal is 311 and 628 μs, respectively.

Figure 3.14 shows the extracted result of the original guided wave detection signal obtained by the HHT method. The waveform is the Hilbert transform result of the first intrinsic modal function IMF in the HHT transform process. Compared with the resulting map obtained by the proposed time–frequency energy density precipitation TOF extraction method, many waveforms that are difficult to interpret physical meaning are generated in the result. At the top of the waveform, there is a very serious jitter phenomenon in the key area of the TOF, and this area produces many local peaks, which brings great difficulties to the extraction work. Only the time corresponding to the maximum peak value can be taken as the TOF of the waveform, which makes the analysis and processing of the waveform particularly cumbersome. Because there is a great degree of uncertainty in the TOF extraction process of such waveforms, there is a great negative impact on the TOF extraction work, and it is difficult to ensure the accuracy of TOF extraction.

From Fig. 3.14, the initial pulse wave HHT method extraction is 41 μs, the SH0 guided wave HHT method is extracted as 372 μs, the SH1 guided wave HHT method is extracted as 720 μs, the SH0 guided wave TOF calculation is 331 μs, and the SH1 guided wave TOF calculation is 679 μs.

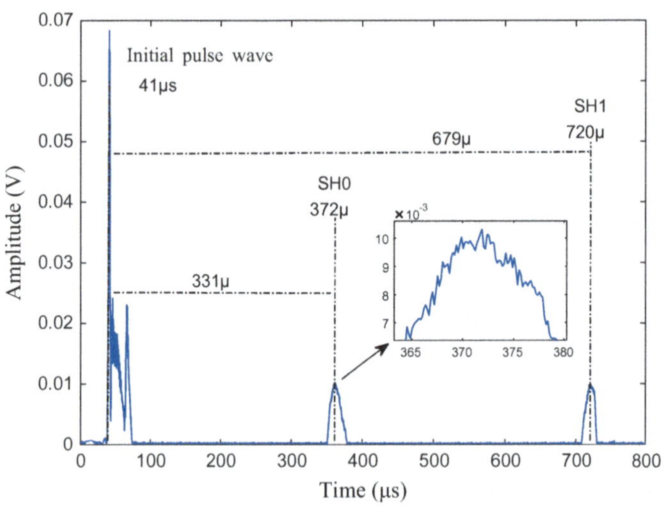

Fig. 3.14 Waveguide detection signal HHT method for transmitting and receiving EMAT spacing at the P2 operating frequency

Table 3.6 Comparison of TOF and TOF extraction results of SH guided wave theory (100 cm at P2 operating frequency)

SH guided mode	Theoretical propagation speed (m/s)	Transmission distance (cm)	Theoretical propagation time (μs)	TEDP TOF extraction result (μs)	HHT TOF extraction result (μs)	Relative error of TEDP TOF (%)	HHT TOF relative error (%)
SH0	3200	100	312.5	311	331	0.48	5.92
SH1	1580	100	632.91	628	679	0.78	7.28

Based on the theoretical group velocity of the SH0 and SH1 guided waves at the operating frequency of P2 in the dispersion curve, the theoretical propagation time of the SH0 guided wave and the SH1 guided wave at the EMAT and the received EMAT spacing of 100 cm can be calculated, respectively. The theoretical propagation time of the SH0 guided wave and the SH1 guided wave is compared with the time–frequency energy density precipitation time extraction method and the HHT TOF extraction method. The relative error of the theoretical TOF and the extraction TOF is calculated using the relative error defined in Eq. (3.9). The calculation results are shown in Table 3.6. Time–frequency energy density precipitation is represented by the abbreviation TEDP in subsequent tables and figures.

According to Table 3.6, the relative errors of the SH0 guided wave and the SH1 guided wave extracted by the time–frequency energy density precipitation TOF extraction method are relatively small, all within 1%, exhibiting high TOF extraction accuracy. The error of the extraction TOF obtained by the HHT method relative to the theoretical TOF is relatively large, mainly due to the problem of signal interpretation caused by the limitation of its algorithm and the problem of serious jitter in the extraction time zone. The time–frequency energy density precipitation TOF extraction method proposed in this chapter fully considers the characteristics of the narrowband electromagnetic guided wave detection signal, analyzes the time–frequency spectrum characteristics in depth, and incorporates the initial pulse wave into the analysis range of the algorithm. By calculating the time difference between the guided wave detection signal and the initial pulse wave, the TOF of each guided wave detection signal is obtained, which greatly improves the TOF extraction accuracy of the narrowband electromagnetic guided wave detection signal.

3.2.3 *Sensitivity Analysis of Time–Frequency Energy Density Extraction TOF Extraction Method*

In the following, this chapter further validates the proposed time–frequency energy density precipitation TOF extraction method for TOF extraction efficiency and accuracy of narrowband electromagnetic ultrasonic guided wave detection signals. At the same time, in-depth study of the characteristics of guided wave dispersion, operating point frequency and guided wave propagation distance and other factors affect the accuracy of time–frequency energy density extraction TOF extraction method to extract TOF.

According to the theoretical group velocity of the SH guided wave at each working point of the test platform in Table 3.4, the theoretical propagation time of the SH guided wave in each experiment at a propagation distance of 20–110 cm can be calculated. At the same time, the time–frequency energy density extraction TOF extraction method and the HHT method were used to extract the guided wave detection signals of the above experiments. The SH guided wave propagation time and TOF extraction results of each experiment are plotted in Fig. 3.15 according to the operating point frequency and guided wave propagation distance.

For the sake of illustration, in Fig. 3.15, the theoretical propagation times of the SH0 guided wave and SH1 guided wave are denoted by SH0T and SH1T, respectively. The time–frequency energy density extraction TOF extraction method is used to represent SH0 guided wave and SH1 guided wave extraction TOF using SH0E-TEDP and SH1E-TEDP, respectively. The TOF of the SH0 guided wave and the

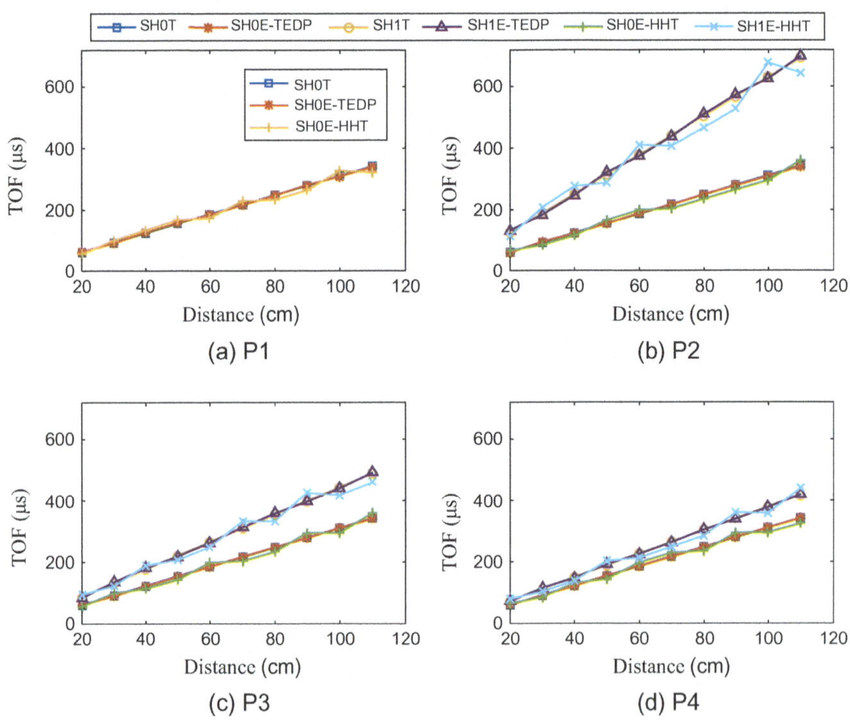

Fig. 3.15 Comparing the experimental results of the guided wave propagation time and the TOF extraction of the SH. **a** P1 operating frequency: 320 kHz, **b** P2 operating frequency: 460 kHz, **c** P3 operating frequency: 560 kHz, and **d** P4 operating frequency: 700 kHz

SH1 guided wave extraction using the HHT method is represented by SH0E-HHT and SH1E-HHT, respectively. Since there are only SH0 guided waves at the P1 operating point, their legends are marked separately in Fig. 3.15a, and b–d share a legend.

According to Fig. 3.15, the TOF extracted by the time–frequency energy density precipitation TOF extraction method is very close to the corresponding theoretical TOF. The TOF taken by the HHT method and the theoretical TOF greatly deviate. The discrete curve with the change of the guided wave distance and the discrete curve of the guided wave time travel with the distance cannot be closely matched. On the other hand, Fig. 3.15 can also reflect the relationship between the SH0 guided wave speed and the SH1 guided wave speed at the four working frequencies P1 to P4, that is, the SH0 guided wave speed does not change. In Fig. 3.15, the slope of the distance versus TOF is constant. The difference between the SH1 guided wave speed and the SH0 guided wave speed is the largest at the P2 operating frequency, followed by the P3 operating frequency, and the P4 working frequency is the closest to the SH1 guided wave speed, and the SH0 guided wave speed.

To quantitatively study the error relationship between time–frequency energy density precipitation time extraction method, HHT extraction time and SH guided wave in a theoretical time of each experiment. Equation (3.9) is used to calculate the relative error between the time–frequency energy density extraction TOF extraction method and the TOF of SH guided wave theory, TOF of HHT method and TOF of SH guided wave. Moreover, it is plotted in Fig. 3.16 according to guided wave mode and propagation distance.

In Fig. 3.16, the relative TOF error extracted by the HHT method is significantly greater than the relative TOF error extracted by the time–frequency energy density precipitation TOF extraction method. The HHT method TOF extraction relative error curve is located in the time–frequency energy density precipitation TOF extraction error curve. Above, the relative error of the HHT method is more than 5%, and the minimum relative error of the time–frequency energy density is 0.21%. The corresponding working point is P40, that is, the SH0 guided wave has a propagation distance of 110 cm at a working frequency of 700 kHz.

In the EMD mode identification method of this chapter, the operating point of the guided wave multimodal mode of operation verification is the P4 frequency operating point in this section, and the guided wave propagation distance is 20 cm. After EMD mode decomposition of the original guided wave detection signal, the original pulse wave was not considered in the extraction time, and only the peak value of the initial pulse wave was used as the timing starting point. The extracted relative errors of extracted SH0 guided wave TOF and SH1 guided wave TOF are 4.78 and 3.11%, respectively; while the time–frequency energy density extraction TOF extraction method takes into account the effect of the initial pulse wave under the same experimental conditions. The time difference between the arrival time of the guided wave detection signal and the initial pulse wave TOF is taken as the TOF of the guided wave detection signal. The relative errors of the SH0 guided wave TOF and the SH1 guided wave TOF are 1.5 and 2.95%, respectively. Relative to the original pulse wave signal is not considered, the relative error of TOF is significantly

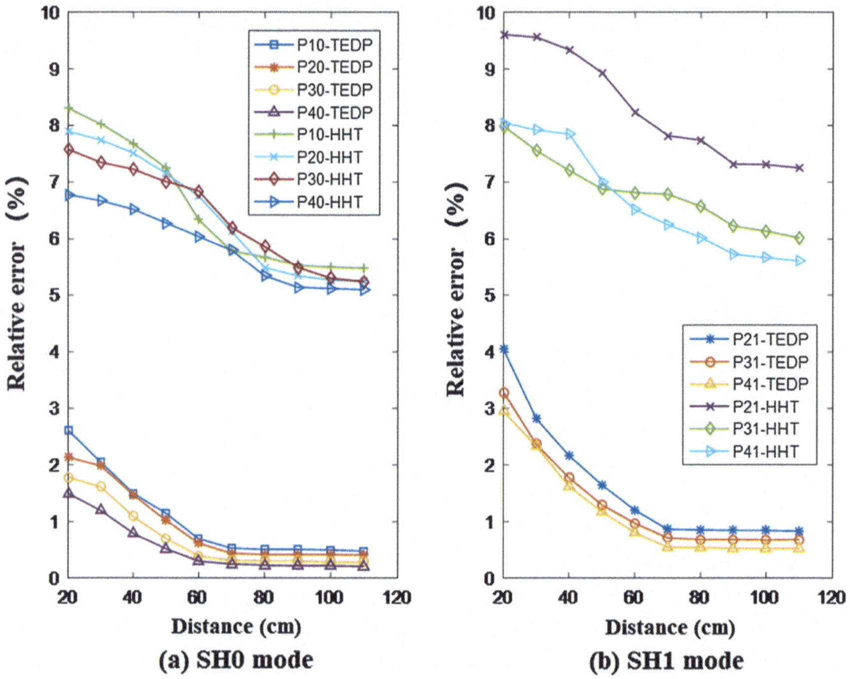

Fig. 3.16 Relative error of the propagation time and TOF extraction results of the SH guided wave theory for each test. **a** SH0 mode and **b** SH1 mode

reduced. It is demonstrated that the time–frequency energy density precipitation TOF extraction method proposed in this chapter plays an important role in the analysis and interpretation of the initial pulse wave signal in improving the extraction accuracy of the SH guided wave TOF.

With the increase of SH guided wave propagation distance, the relative error of time–frequency energy density extraction TOF decreases gradually, and when the propagation distance reaches 70 cm, the relative error changes gradually and basically remains within 1%. The relative error of the HHT method TOF showed a similar trend with the increase of the propagation distance, but the relative error did not show a clear trend, and there was a certain degree of fluctuation with the increase of the propagation distance. Therefore, the time–frequency energy density precipitation time extraction method is superior to the HHT TOF extraction method in the accuracy of TOF extraction and the stability of the algorithm. The relationship between the characteristics of the guided wave dispersion, the operating point frequency, and the guided wave propagation distance and the TOF extraction accuracy are all directed to the time and frequency energy density extraction TOF extraction method proposed in this chapter.

In Fig. 3.16, as the guided wave propagation distance increases, the relative error of TOF extraction gradually decreases, and the relative error presents a saturation trend after the propagation distance reaches a certain value. Two reasons include: (1) When the distance between the EMAT and the receiving EMAT is short, the high-voltage excitation signal applied to the EMAT to be transmitted will cause more serious interference to the normal guided wave detection signal received by the EMAT. The interference voltage is induced at both ends of the receiving EMAT coil through the spatial coupling so that the interference signal is introduced into the collected guided wave detection data, which has a negative impact on the extraction of the TOF of the desired modal guided wave detection signal. When the EMAT is far enough from the receiving EMAT, the high-voltage excitation signal applied on the EMAT does not substantially influence the reception of the EMAT and the ultrasonic guided wave is received, thereby causing less interference to the collected guided wave detection signal. (2) The initial pulse wave is formed by the induction of the receiving EMAT coil by the high-voltage excitation signal applied to emit EMAT, which occupies a large time range in the time domain. When the distance between the transmitting EMAT and the receiving EMAT is short, the required modal guided wave detection signal and the initial pulse wave signal are relatively close in the time domain. The initial pulse wave amplitude and signal intensity are generally significantly larger than the amplitude and signal strength of the guided wave detection signal, which has an adverse effect on the TOF extraction, resulting in relatively large TOF extraction error.

In order to comprehensively describe the relationship between TOF extraction error and distance, the relative errors of TOF taken by each working point under a certain distance are accumulated to obtain the relative error and distance area map of each work point, as shown in Fig. 3.17.

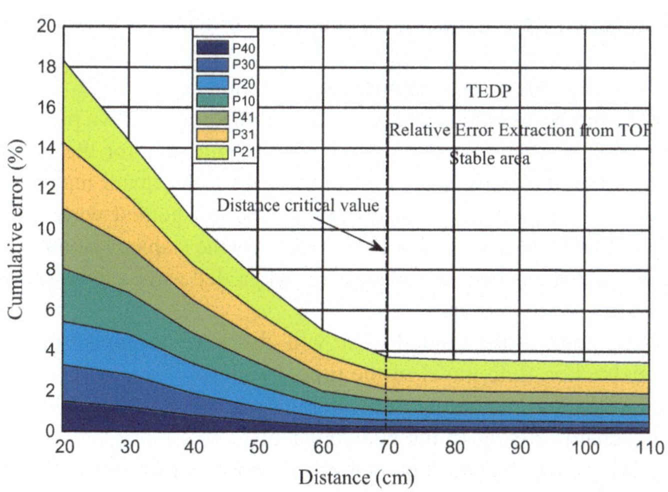

Fig. 3.17 Extracting relative error and guide wave propagation distance area for travel points at each operating point

Figure 3.17 shows that when the SH guided wave propagation distance exceeds 70 cm, the relative error of the TOF obtained by the time–frequency energy density precipitation TOF extraction method is gradually stable and tends to be saturated. This article is called the time–frequency energy density precipitation TOF extraction method to extract the stable region D_s relative error TOF. Moreover, define time–frequency energy density precipitation time extraction method relative error stable area distance critical value d_c

$$d_c = \arg\ \min d\ \text{ s.t. } \frac{\partial e_r}{\partial d} \le \varepsilon \tag{3.20}$$

where d is the guided wave propagation distance, and e_r is the relative error of the guided wave TOF extraction result and the theoretical propagation time defined by Formula (3.9). ε is a small positive number defined by the user. It is used to characterize the partial derivative of the relative error of the TOF to the guided wave propagation distance. When the numerical value is lower than ε, the relative error tends to saturate and no longer changes significantly. According to the above formula, the time–frequency energy density precipitation TOF extraction method is used to extract stable regions D_s with relative TOF error.

$$D_s = \{d | d \ge d_c\} \tag{3.21}$$

The real-time energy density precipitation TOF extraction method extracts the stable region D_s with relative TOF error including the guided wave propagation distance that exceeds the stable region distance d_c. The definition and calculation of relative error stable region and its critical distance value for time–frequency energy density precipitation TOF extraction method can provide an important reference for the rational design of EMAT and receiving EMAT spacing and optimization of topological size structure of the magnetic acoustic array.

The relationship between the relative error and the guided wave propagation distance during TOF extraction can provide valuable guidance for the EMAT layout and array size in guided wave scattering imaging for subsequent magneto-acoustic arrays. That is to say, the propagation distance of SH guided wave from EMAT transmitting to EMAT receiving should be located in the stable region Ds of extracting TOF error, so that the TOF information of SH guided wave can be extracted with higher accuracy.

The working point of the guided wave will also have an impact on the relative error of the extraction time travel of the time–frequency energy density extraction TOF extraction method, including the operating frequency of the guided wave and the dispersion characteristics of the guided wave.

When studying the influence of guided wave's working frequency on relative error of time–frequency energy density precipitation time extraction method to extract TOF, we should ensure that the dispersion characteristic of the guided wave does not work in it, that is, follow the principle of single variable or control variable. The four work points P0, P20, P30, and P40 on the SH0 can be used for research. The main reason is that the SH0 guided wave has no dispersion phenomenon. According to Fig. 3.16a, the P10-TEDP to P40-TEDP curves can be ranked as follows: P40, P30, P20, and P10. That is, the TOF extraction error of the P40 is relatively minimal, and the accuracy is relatively highest, while the P10 TOF extraction error is relatively largest and the accuracy is relatively lowest. The main reason is that when other conditions are the same, especially when the number of excitation signals is the same, the higher the guided wave operating frequency, the narrower the time width occupied by the guided wave waveform in the time domain. That is, the higher its aggregation at a certain moment, the better it is to extract its timing with higher accuracy. However, the increase in frequency will bring another problem: that is, the waveguided mode will increase in the high-frequency area. As shown in Fig. 3.10, there are two modes of SH-waves at P2 to P4 operating frequencies, which is contradictory to the expectation of detection using a single-mode guided wave. Therefore, in the second chapter, when designing SH guided wave direction control EMAT, it is considered that the controllable EMAT produces a single SH0 modal guided wave, and the excitation frequency is increased as much as possible to obtain a more accurate extraction TOF. Therefore, the P10 operating point of the SH0 guided wave was selected.

When studying the dispersion characteristics of the guided wave on the time–frequency energy density precipitation time extraction method to extract the relative error caused by the TOF, the same reason, we should ensure that the guided wave working frequency does not work. Therefore, the P20 and P21 operating points at the P2 operating frequency, the P30 and P31 operating points at the P3 operating frequency, and the P40 and P41 operating points at the P4 operating frequency can be compared and analyzed.

In Fig. 3.16, the P20-TEDP curve in (a) is located below the P21-TEDP curve in (b), which shows that the relative error of the TOF extraction at the P20 working point is smaller than that at P21, and the TOF extraction accuracy is higher. According to the same method, analyze the other two groups, that is, the relative error of P30 working point extraction is smaller than that of P31, and the relative error of P40 working point is smaller than that of P41. The main reason is the dispersion characteristics of the SH1 guided wave, that is, its group velocity will change with the guided wave operating frequency, although the center frequency-based excitation signal is used at the excitation end. However, it will inevitably introduce other frequency components near the center frequency, which will cause the actual propagation velocity of the SH1 guided wave to deviate from the propagation speed of the theoretical working point. Therefore, the actual extraction time will be biased against the theoretical TOF under the theoretical operating point. For the three working points P41, P31, and P21 of the SH1 guided wave, the partial derivative of the group velocity to the working frequency satisfies

$$\left.\frac{\partial v_{\mathrm{g}}}{\partial f}\right|_{\mathrm{P41}} < \left.\frac{\partial v_{\mathrm{g}}}{\partial f}\right|_{\mathrm{P31}} < \left.\frac{\partial v_{\mathrm{g}}}{\partial f}\right|_{\mathrm{P21}} \tag{3.22}$$

That is the group velocity at the working point of P21 changes significantly with the frequency, which is also the reason that the time–frequency energy density at the working point of P21 is relatively large.

Therefore, the non-dispersive SH0 guided wave is selected for the subsequent research of guided wave scattering imaging of magneto-acoustic arrays, which lays the foundation for accurately extracting guided wave TOF information.

Comprehensive analysis of the operating point and distance on the time–frequency energy density precipitation TOF extraction method to extract the relative error of TOF, according to the working point and guided wave propagation distance relative error calculated value is plotted in Fig. 3.18. The relative error is formed on the distance-working point in the two-dimensional plane.

The above analysis of the influence of guided wave propagation distance guided wave operating frequency and guided wave dispersion characteristics on the relative error of TOF extraction is more apparent in Fig. 3.18. For a certain working point, the relative error of the TOF extraction decreases with distance; the relative error of the TOF extraction from the P10 to P40 working point gradually decreases; the relative error of the P20 work point TOF is smaller than the P21. In Fig. 3.17, the vertical height occupied by different working point area maps at a certain distance means that the corresponding TOF extraction relative error and its analysis can also get similar conclusions, which will not be repeated here.

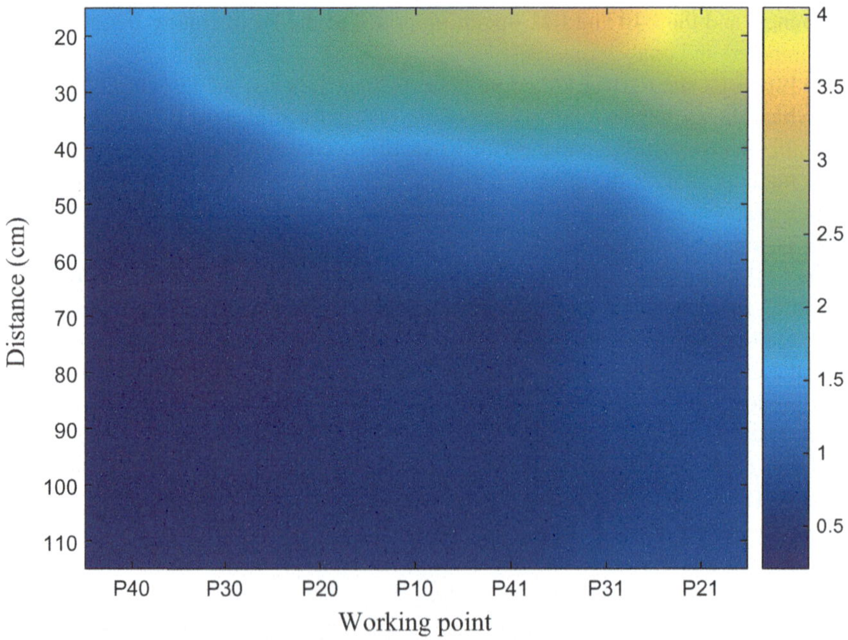

Fig. 3.18 Time–frequency, energy density precipitation TOF extraction method, extracts TOF relative error distribution in distance-working point plane

3.3 Modal Identification and TOF Extraction Test Verification of Guided Wave Scattering Detection Signal of Steel Plate Defect

In order to verify the feasibility and accuracy of the proposed EMD modal identification method and time–frequency energy density precipitation TOF extraction method for defect guided wave scattering detection signals, a relevant test verification platform was built on a 4-mm-thick steel plate. Its components and working principle are shown in Fig. 3.19.

In the above test platform, EMAT is emitted using the direction-controllable EMAT proposed in Chap. 2, and the received EMAT is a magnetostrictive omnidirectional EMAT based on a nickel tape and a closely wound coil. Moreover, the direction controllable emission EMAT direction adjustment knob is aligned with the edge of the square defect on the steel plate. The omnidirectional reception EMAT is arranged at the rear of the controllable EMAT and is used to receive only the SH guided wave scattered by the square defect.

The working point where the direction-controllable emission EMAT excites the SH guided wave is the P10 operating point in Fig. 3.10, and only the oscillating guided wave of SH0 is excited. The excited SH0 guided wave propagates in the direction of the square defect in the direction of the direction adjustment knob. It scatters after encountering a square defect, propagates in the opposite direction, and is received by the omnidirectional receiver EMAT and sent to the signal processor for processing. The guided wave propagation path has been indicated in Fig. 3.19; the working principle of the acquisition card, computer, and power amplifier and its role in this test platform are similar to those of the test above verification platform and will not be repeated here. The setting of test parameters is also basically the same as the previous relevant tests in this chapter, as shown in Table 3.7.

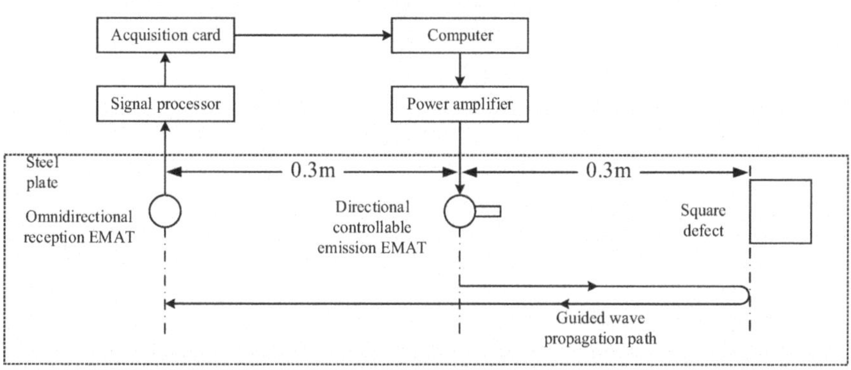

Fig. 3.19 Block diagram of test verification platform for EMD modal identification and TEDP TOF extraction guided wave scattering detection signal

Table 3.7 EMD mode identification and TEDP TOF extraction guided wave scattering detection signal test parameter setting

Parameter	Value
Excitation voltage peak-to-peak (V)	320
Excitation frequency *f* (kHz)	320
Number of incentive cycles (1)	12
Plate thickness (mm)	4
Square defect side length (mm)	100
Defect depth of square (mm)	1.5
Directional controllable launch EMAT and square defect distance (cm)	30
Controllable emission EMAT and omnidirectional reception EMAT distance (cm)	30
Guided wave working point theoretical group velocity (m/s)	3200
Signal processor magnification	5000
Signal processor narrowband filter center frequency (kHz)	320
Signal processor narrowband filter bandwidth (kHz)	20

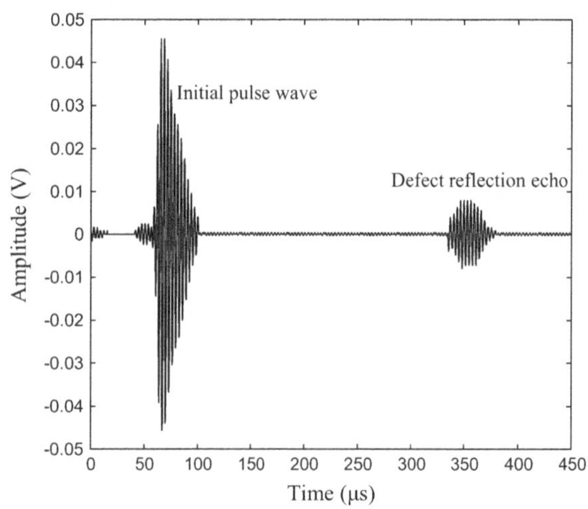

Fig. 3.20 Scattered wave detection signal waveform diagram for square defect of steel plate

Relying on the above test platform, the EMD modal identification and the time–frequency energy density extraction TOF extraction method according to the test mentioned above parameters were tested and verified for the modal identification and TOF extraction of guided wave scattering detection signals. The test results will be analyzed below.

After processing by the signal processor, the data waveform of the guided wave scatters detection of the defect of the steel plate is shown in Fig. 3.20.

First, the EMD modal identification method was used to analyze the data of the square wave guided scatter detection signal of the above steel plate, and its intrinsic modal function IMF was calculated iteratively. In each IMF selection process, the screening iteration termination condition is $\varepsilon \leq 0.25$. The termination condition of the modal decomposition process is that no new IMF is generated. The solution of the intrinsic modal function of the guided wave scattering detection signal data for the square defect of the steel plate is shown in Fig. 3.21.

In the solution of the intrinsic modal function IMF of the square wave defect detected wave scatter detection signal, only the amplitude magnitude of the IMF1 signal is the same as the amplitude magnitude of the original defect reflection echo signal. The amplitudes of IMF2 to IMF4 are small relative to the original defect reflection echo signal and have no effect on the original defect reflection echo signal. Therefore, the EMD mode decomposition of the guided wave scatter detection signal of the square plate defect of the above steel plate is decomposed, and only one dominant natural mode function IMF1 is decomposed, and other intrinsic modal functions that are similarly dominant are not present. It shows that there is only one

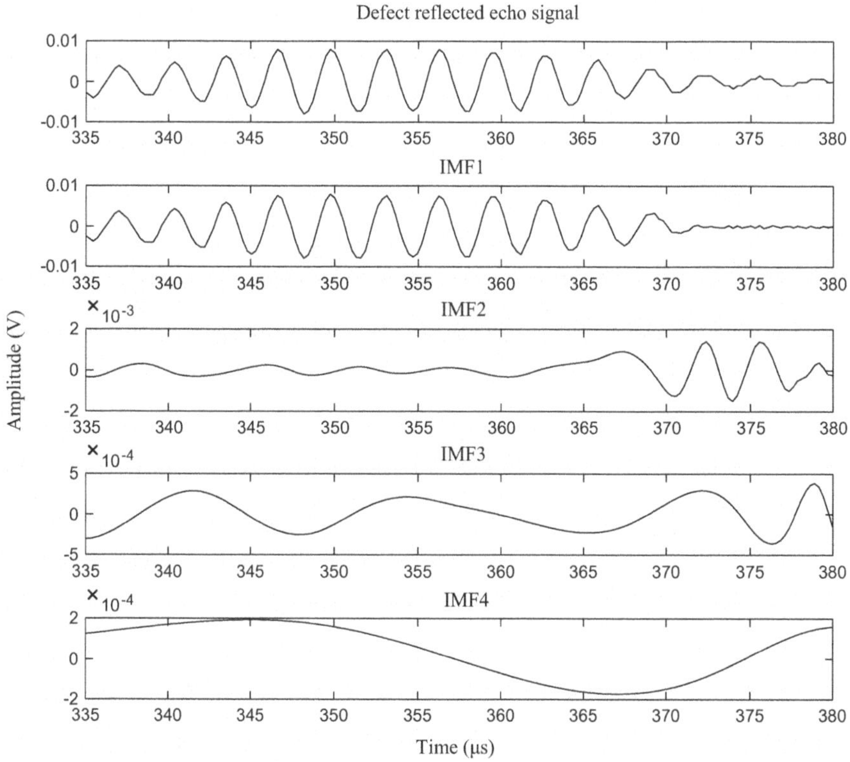

Fig. 3.21 Intrinsic modal function IMF solution result of guided wave scattering detection signal of square steel defect

kind of modal guided wave in the square wave defect detection signal of this steel plate.

The analysis result of the EMD modal identification method on the detected wave scatter detection signal of the steel plate defect shows that after the SH0 modal guided wave and the defect scatter effect, there is no other type of guided wave except the SH0 modal guided wave in the signal. The above test results have an important reference for the establishment of a guided wave scattering imaging model of magneto-acoustic arrays. To a certain extent, the method of imaging the guided wave scattering defect based on the above experimental setup does not need to consider the modal transformation process in the guided wave scattering process. Since the SH wave guided reception EMAT can only receive the SH guided wave, in principle, only the SH0 modal guided wave will exist at this operating frequency. The EMD mentioned above modal identification method further analyzes the result of the analysis of the guided wave scattering detection signal of the steel plate defect by using the experimental method.

The EMD modal identification method is placed on the preprocessing position of the guided wave scattering detection signal by the magnetic acoustical array guided wave scattering imaging method, and used to judge and confirm whether the collected guided wave scattering detection signal contains other modes of guided wave, A pure and single guided wave mode is provided for the guided wave detection signal TOF or amplitude extraction and the guided imaging method of the magneto-acoustic array, so that the guided wave imaging method of the magneto-acoustic array does not need to consider the modal transformation process after the guided wave and the defect. This simplifies the analysis and processing of guided wave detection signals.

In the following, based on the EMD modal identification results of the guided wave scatter detection signal of square steel defects, the time–frequency energy density extraction TOF extraction method proposed in this chapter was used to extract the square wave guided wave scatter detection signal. According to the theoretical steps of the time–frequency energy density precipitation TOF extraction method, the short-time Fourier transform is first applied to the scattered wave detection signal of square steel defect of the steel mentioned above plate to obtain its time–frequency two-dimensional distribution. The above steel plate square defect guided wave scattering detection signal data contain a total of 1727 data points. In the discrete short-time Fourier transform, a Hamming window function is used to intercept the original guided wave detection signal, and the window function has a width of 431. That is one fourth of the data points of the original guided wave detection signal. According to the step (1) of the time–frequency energy density precipitation TOF extraction method, a time–frequency energy density value of a defect guided wave scattering detection signal in a two-dimensional time–frequency distribution is calculated and plotted in a two-dimensional time–frequency distribution. As shown in Fig. 3.22.

From Fig. 3.22, it can be seen that the energy of this defect guided wave scattering detection signal is mainly concentrated in the vicinity of the center frequency of 320 kHz. By performing step (2) to step (6) of the proposed time–frequency energy density precipitation TOF extraction method, a time–frequency energy density dis-

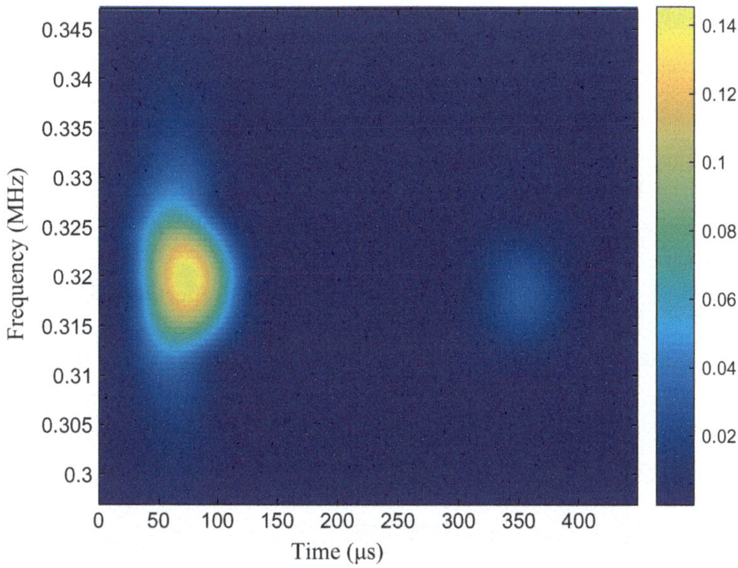

Fig. 3.22 Two-dimensional distribution of time–frequency energy density of guided wave scattering detection signal of square steel defect

tribution curve in the time domain based on the center frequency is obtained as shown in Fig. 3.23.

In Fig. 3.23, according to step (7) to step (8) of the time–frequency energy density precipitation TOF extraction method, the peak time of the initial pulse wave and the defect reflection echo on the time-domain energy density curve is first extracted. At 74 and 354 μs, respectively, the time for the re-evaluation of the reflected echo from the initial pulse wave was 280 μs. Therefore, using the time–frequency energy density precipitation TOF extraction method, the TOF extraction value of the square defect guided wave scatter detection signal is 280 μs.

According to the propagation path of the SH0 guided wave in Fig. 3.19, the propagation distance of the SH0 guided wave from being excited by the controllable EMAT to being received by the omnidirectional EMAT is 0.9 m. According to the theoretical group velocity of the guided wave at the P10 operating point SH0 in the dispersion curve, the theoretical propagation time of the SH0 guided wave can be obtained. The theoretical propagation time of SH0 guided wave is compared with the TOF extracted by time–frequency energy density extraction TOF extraction method, and the relative error between the theoretical TOF and the extraction TOF is calculated using the relative error defined by Eq. (3.9). The results are shown in Table 3.8.

According to Table 3.8, the time–frequency energy density precipitation TOF extraction method proposed in this chapter has a relatively small relative error in the extraction time of the guided wave scattering detection signal of the steel plate

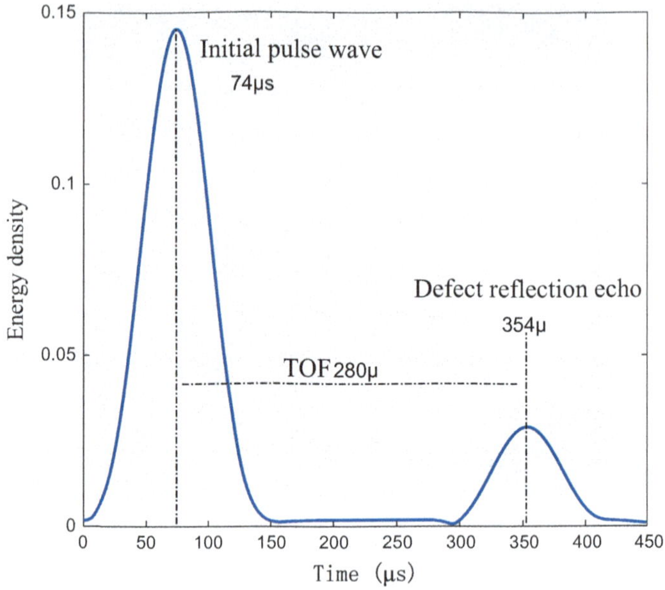

Fig. 3.23 Energy density distribution curve in the time domain for guided wave scattering detection signals of square steel defects

Table 3.8 Comparison of theoretical TOF and extraction time of guided wave scatter detection signal for steel plate square defect

SH guided mode	Theoretical propagation speed (m/s)	Transmission distance (cm)	Theoretical propagation time (μs)	TEDP TOF extraction result (μs)	The relative error of TEDP TOF (%)
SH0	3200	90	281.25	280	0.44

defect, which is only 0.44%, showing a high TOF extraction accuracy. The experimental method is used to prove that the time–frequency energy density precipitation TOF extraction method has high extraction accuracy for guided wave scatter detection signal TOF of steel plate defect and provides high-precision guided wave TOF information for the guided wave scattering imaging method of magneto-acoustic array. It is helpful to improve the accuracy of guided wave scattering and imaging.

References

1. N.E. Huang, Z. Shen, S.R. Long et al., The empirical mode decomposition and the Hilbert spectrum for nonlinear and non-stationary time series analysis. Proc. R. Soc. A Math. Phys. Eng. Sci. **454**(1971), 903–995 (1998)
2. Y. Zhang, S.L. Huang, S. Wang, W. Zhao, Time-frequency energy density precipitation method for time-of-flight extraction of narrowband lamb wave detection signals. Rev. Sci. Instrum. **87**(5), 054702, 1–8 (2016)
3. S.L. Huang, Y. Zhang, S. Wang, W. Zhao, Multi-mode electromagnetic ultrasonic lamb wave tomography imaging for variable-depth defects in metal plates. Sensors **16**(5), 628, 1–10 (2016)
4. Y. Zhang, S.L. Huang, S. Wang, Z. Wei, W. Zhao, Recognition of overlapped lamb wave detecting signals in aluminum plate by EMD-based STFT flight time extraction method. Int. J. Appl. Electromagnet. Mech. **52**(3–4), 991–998 (2016)
5. Y. Zhang, S. Wang, Q. Wang, S.L. Huang, W. Zhao, Profile imaging of actual defect in steel plate based on electromagnetic ultrasonic SH guided wave scattering. Insight
6. Y. Zhang, S.L. Huang, W. Zhao, S. Wang, Q. Wang, Electromagnetic ultrasonic guided wave long-term monitoring and data difference adaptive extraction method for buried oil-gas pipelines. Int. J. Appl. Electromagnet. Mech.
7. S. Wang, S.L. Huang, Y. Zhang, W. Zhao, Modeling of an omni-directional electromagnetic acoustic transducer driven by the Lorentz force mechanism. Smart Mater. Struct. **25**(12), 125029 (2016)
8. S. Wang, S.L. Huang, Y. Zhang, W. Zhao, Multiphysics modeling of a Lorentz force-based meander coil electromagnetic acoustic transducer via steady-state and transient analyses. IEEE Sens. J. **16**(17), 6641–6651 (2016)

Chapter 4
Guided Wave Electromagnetic Ultrasonic Tomography

4.1 Electromagnetic Ultrasonic Straight-Ray Lamb Wave Cross-Hole Tomography Imaging Method

4.1.1 Fundamental Principles of Lamb Wave Cross-Hole Tomography with Electromagnetic Ultrasound

The defect of the aluminum plate concerned in this chapter is mainly the plate thickness loss-type defect caused by corrosion, and the plate thickness change caused by such defects is relatively flat. When Lamb waves encounter such defects, the amplitudes of other modal waves generated by modal transformation are relatively small. Therefore, the effects caused by the modal transformation of Lamb waves are ignored in the following studies.

The basic principle of cross-hole tomography with Lamb wave electromagnetic radiography is shown in Fig. 4.1. The excitation and reception EMATs were placed on both sides of the rectangular imaging area on the aluminum plate. The rectangular imaging area is evenly divided into $M_s \times N_s$ small grids, each of which is a pixel, and it is considered that the parameters such as aluminum plate thickness and Lamb wave velocity in each pixel grid are fixed. The ultimate goal of imaging is to solve the thickness of the aluminum plate in each pixel grid.

In the imaging process, Lamb wave detection is first performed using all available EMAT transceiver combinations to obtain the detected waveform. Then, the obtained detection waveform is used to extract the TOF of the Lamb wave ray that connects each pair of EMATs, and the actual TOF of the *i*th Lamb wave ray is recorded as T_i ($i = 1, 2, \ldots, P^2$; P is the number of EMATs for excitation, also used for receiving EMAT number). After all the T_i is obtained, T_i needs to be input to the corresponding image reconstruction algorithm, and the thickness of the aluminum plate in each grid is calculated through a certain number of iterations.

Based on the straight-ray model, the Lamb wave rays connecting each pair of transceiving EMATs pass through the grids from different directions along a straight

S. Huang et al., *Theory and Methodology of Electromagnetic Ultrasonic Guided Wave Imaging*, https://doi.org/10.1007/978-981-13-8602-2_4

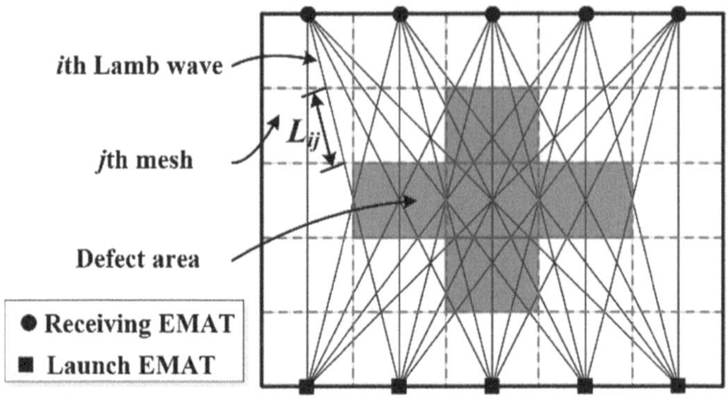

Fig. 4.1 Basic principle of Lamb wave cross-hole tomography with electromagnetically direct ray

line. From the dispersion curve of the Lamb wave shown in Fig. 2.4, the group velocity of each modal Lamb wave is a function of the plate thickness at a certain operating frequency. Therefore, when a Lamb wave ray encounters a defective (such as thinning) mesh, the Lamb wave group velocity will change with the thickness of the aluminum plate, and the Lamb wave group velocity within each mesh can be calculated accordingly. Specifically, if the slowness (the reciprocal of the group velocity) of the modal Lamb wave used in the jth grid is recorded as S_j, and the length of the ith Lamb wave ray in the jth grid is recorded as $L_{ij.}$ According to the geometric relations, T_i, L_{ij}, and S_j should satisfy the following relationships:

$$T_i = \sum_{j=1}^{M_s \times N_s} L_{i\,j} \times S_j \quad \left(i = 1, 2, \ldots, P^2\right) \tag{4.1}$$

Equation (4.1) is a linear system of equations. Each equation in the system of equations represents a Lamb wave ray. Each unknown S_j represents the desired slowness within a pixel grid. Under the control of the equation, the measured TOF T_i of each ray constrains S_j within each pixel grid it passes through; and S_j in each pixel grid is simultaneously constrained by the measured TOF T_i of all rays passing through it. Therefore, by solving Eq. (3.1), S_j in each grid can be calculated. Further, using the Lamb wave dispersion relation shown in Fig. 2.4, the corresponding group velocity cg_j, phase velocity cp_j, and plate thickness d_j in each grid can be obtained. After the parameters such as the size of the imaging area, the transceiving EMAT position, and the pixel grid size and number in Fig. 3.1 are determined, the matrix (L_{ij}) in Eq. (4.1) can be calculated based on the corresponding geometric relationships.

Equation shown in (4.1) are usually large-scale and need to be solved using iterative methods. Also, the system of equations is usually an underdetermined or overdetermined system of equations, and since the density of Lamb wave ray in the imaging region is not uniform, no ray passes in a large number of pixel grids, and the matrix (L_{ij}) contains a large number of zero elements. Therefore, the solution algorithm of this equation group needs to have good stability and convergence, and it needs to have high solution efficiency. In previous studies, there are two major algorithms for solving the equations: ART and SIRT. Compared with ART, SIRT has better iterative convergence, is less sensitive to TOF errors, and has lower requirements on the accuracy of initial iterations [1]. Therefore, the following studies all use SIRT to perform Lamb wave cross-hole tomography.

The SIRT-based equation solving (i.e., image reconstruction) process is: First, a set of slow initial values $S_j^{(0)}$ ($j = 1, 2, \ldots, M_s \times N_s$) is given. $S_j^{(0)}$ is then substituted into Eq. (3.1) to calculate the corresponding $T_i^{(0)}$ ($i = 1, 2, \ldots, P^2$). Subsequently, $S_j^{(0)}$ is corrected by using the deviation between the iterative output value $T_i^{(0)}$ and the measured TOF T_i to obtain a new set of slowness values $S_j^{(1)}$. This continues, and the iterative process is repeated continuously. The relative error $|T_i^{(k)} - T_i|/T_i \times 100\%$ between $T_i^{(k)}$ and T_i after k iterations is less than the preset threshold T_{TH} so that the solution of the system of equations is obtained. In each iteration, the calculation formula for correcting the slowness value in each grid by using the deviation between $T_i^{(k)}$ and T_i is:

$$S_j^{(k+1)} = S_j^{(k)} + \mu_{\mathrm{I}} \frac{\sum_{i=1}^{P^2} \left(L_{ij} / \sum_{j=1}^{M_s \times N_s} L_{ij} \right) \left(T_i - T_i^{(k)} \right)}{\sum_{i=1}^{P^2} L_{ij}} \tag{4.2}$$

where μ_{I} ($0 < \mu_{\mathrm{I}} \leq 1$) is a relaxation factor, and its role is to ensure the stability of iterative calculations.

4.1.2 Improved Electromagnetic Ultrasound Straight-Ray Lamb Wave Cross-Hole Tomography Imaging Method

The modal control capabilities of conventional EMATs that have been used to transmit and receive Lamb waves are limited. When using the traditional EMAT for straight-ray Lamb wave cross-hole tomography, the EMAT detection waveform usually contains more modal Lamb waves. In the PWVD of the detected waveform, the energy density distribution of each mode wave will overlap. This will seriously affect the accuracy of PWVD-based TOF extraction, thus affecting the performance of tomography.

In response to the above problems, this section attempts to propose an improved electromagnetic ultrasonic direct ray Lamb wave cross-hole tomography imaging method (from now on referred to as "improved direct ray imaging method"). The aim is to improve the accuracy of electromagnetic ultrasonic Lamb wave TOF extraction by introducing a new EMAT capable of transmitting and receiving pure single-mode Lamb waves, thereby improving imaging performance. In the improved direct ray imaging method, based on the direct ray cross-hole projection method shown in Fig. 3.1, Lamb wave transmission and reception is performed using the new EMAT developed in the work of this chapter, and the corresponding TOF T_i ($i = 1, 2, \ldots, P^2$) is extracted. Based on SIRT, using Eqs. (3.1) and (3.2) to solve the S_j ($j = 1, 2, \ldots, M_s \times N_s$) of Lamb waves in each mesh of the imaging region. On this basis, using the slowness definition $cg_j = 1/S_j$ and the Lamb wave dispersion relationship in Fig. 2.4, the Lamb wave group velocity cg_j, the plate thickness d_j, and the Lamb wave phase velocity cp_j in each grid are calculated.

Huthwaite [2] and Belanger [3] have studied in detail how to select the mode and operating frequency of the Lamb wave used in single-mode Lamb wave tomography. According to their research results, the selection of the modal and operating frequency of the Lamb wave should follow the following principles: ① To minimize the Lamb wave mode in the aluminum plate at the selected operating frequency to avoid interference with the appearance of the modal Lamb wave; ② at the selected operating frequency, the dispersion curve of the selected mode Lamb wave with the thickness of the plate has a large slope near the thickness of the healthy plate, so that the Lamb wave is sensitive to the variation of the plate thickness; ③ Lamb waves of low-order modes should be selected as much as possible to ensure that the selected mode Lamb waves can propagate in the defect area with less residual thickness. In the improved direct ray imaging method, the choice of the Lamb wave mode and operating frequency follow the above principles.

The specific steps of the improved direct ray imaging method are as follows:

(1) Using the selected new EMAT to emit the Lamb wave of the modality used at a specific operating frequency, simultaneously using all the new EMATs for the reception, and recording the detection waveform of each EMAT.
(2) Using the proposed PWVD TOF extraction method and extracting the TOF T_i of the modal Lamb wave used from each detected waveform.
(3) If all the new EMATs have been subjected to Lamb wave emission, proceed to step (4). Otherwise, select a new EMAT for excitation and return to step (1).
(4) Calculating each element in the matrix (L_{ij}).
(5) Using all T_i and L_{ij} as input, based on SIRT and using Eqs. (3.1) and (3.2) to solve the distribution of the slowness S_j of the modal Lamb wave used.
(6) Using $cg_j = 1/S_j$ and the Lamb wave dispersion relationship shown in Fig. 2.4, the distributions of cg_j, cp_j, and d_j of the modal Lamb waves used in each grid are calculated.

The test was carried out on a 3-mm-thick aluminum plate to verify the TOF extraction performance of the new EMAT and the performance of the improved direct ray imaging method. The EMAT excitation and receiver used in the test is the same as in Sect. 4.2.3.

Compared with the traditional EMAT, the new EMAT developed in this chapter can effectively improve the accuracy of electromagnetic ultrasonic Lamb wave TOF extraction. On this basis, in order to verify the effect of the improvement of the TOF mentioned above extraction accuracy on the improvement of tomographic performance, that is, to verify the performance of the improved direct ray imaging method, the imaging method was used to perform a tomographic test on a 3-mm-thick aluminum plate. For comparison, a tomographic test was performed under the same test

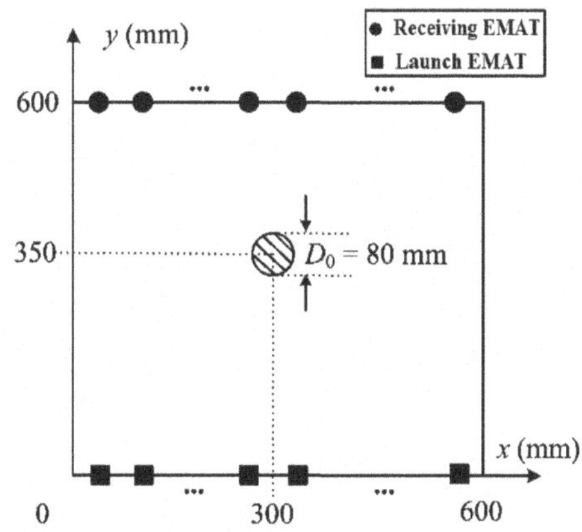

Fig. 4.2 Schematic diagram of tomographic test parameters

conditions using a conventional direct ray imaging method. The conventional direct ray imaging method differs from the improved direct ray imaging method in that the Lamb wave TOF is acquired using a conventional EMAT.

The imaging test was carried out in a rectangular area of 600 mm × 600 mm on an aluminum plate, which was evenly divided into 180 × 180 small grids. The round-hole-shaped artificial defect has a depth of 1.5 mm, and its diameter D_0 and center position (x_0, y_0) are shown in Fig. 4.2. Twelve excitation EMATs and 12 reception EMATs were evenly arranged on both sides of the imaging area. The iterative solution threshold T_{TH} of the SIRT algorithm is set to 0.05%, and the relaxation factor is taken as $\mu_I = 0.1$.

The tomographic results are obtained using the conventional direct ray imaging method and the improved direct ray imaging method, that is, the distribution of the A0 wave slowness S_j. Compared to the tomographic results obtained using conventional direct ray imaging methods, the tomographic results obtained using the improved direct ray imaging method contain less reconstructed artifacts and higher imaging quality. This shows that the latter contains a smaller calculation error ΔS_j than the former. Compared with the traditional direct ray imaging method, the improved direct ray imaging method for tomography can improve the accuracy of narrowband electromagnetic ultrasonic Lamb wave TOF extraction, effectively reduce the slowness calculation error, and suppress the occurrence of reconstructed artifacts, thereby improving the quality of tomographic image reconstruction.

The measured A0 wave slowness curves at $y = 350$ mm were extracted from the imaging results shown in Fig. 4.3a and b, respectively, as shown by the solid lines in Fig. 3.11a and b. Also, according to the actual plate thickness distribution, using the Lamb wave dispersion relationship and the equation $cg_j = 1/S_j$, the theoretical A0 wave slowness curve at $y = 350$ mm in the aluminum plate can be calculated. The theoretical slowness of the 290 kHz A0 wave used in the healthy and defective areas of the aluminum plate used in the test was 306.3 and 339.4 μs/m, respectively. Compared with the measured S_j curve based on the traditional direct ray imaging method, the measured S_j curve obtained based on the improved direct ray imaging

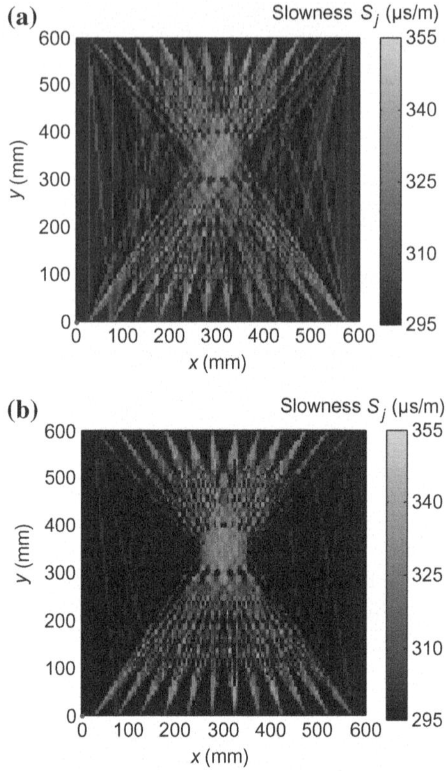

Fig. 4.3 Tomography results using two direct ray imaging methods: **a** traditional direct ray imaging method and **b** improved direct ray imaging method

method contains less noise and is more consistent with the theoretical S_j curve. Therefore, using the latter helps achieve more accurate defect feature information extraction (Fig. 4.4).

In order to compare the accuracy of the two imaging methods to extract defect feature information, the diameter of the defect is estimated by the following methods and steps:

(1) Roughly determine the position of the defect area by comparing the measured S_j curve with a preset threshold.
(2) Find the intersection of the main rising and falling edges of the measured S_j curve at the edge of the defect with the straight line $S_j = 306.3$ μs/m (the theoretical slowness value of the healthy region).
(3) Calculate the distance between the two intersection points, that is, obtain the estimated defect diameter.

Using the measured S_j curve, the estimated defect diameter is about 107 mm, and the absolute and relative errors of the estimate are 27 mm and 33%, respectively. Using the measured S_j curve, the estimated defect diameter is about 96 mm, and the absolute and relative errors of the estimate are 16 mm and 20%, respectively.

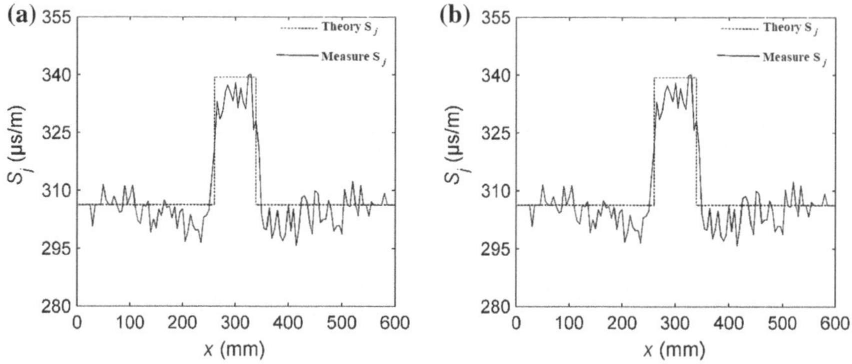

Fig. 4.4 A0 wave S_j curve at $y = 350$ mm extracted from the imaging results of two direct ray imaging methods: **a** conventional direct ray imaging method and **b** improved direct ray imaging method

The above test results show that compared with the traditional direct ray imaging method, the improved direct ray imaging method can effectively improve the accuracy of defect feature information extraction, and the reduction of the defect diameter estimation error is: $|16 - 27|/27 = 41\%$.

4.1.3 Electromagnetic Ultrasonic Multimode Direct Ray Lamb Wave Transforaminal Tomography Method

By improving the accuracy of the TOF extraction, the error of the slow S_j calculation in the tomographic result can be reduced to some extent, thereby suppressing the noise in the measured S_j curve. However, the above errors and noises are often difficult to suppress completely. It can be seen from Eq. (3.4) that when the accuracy of the TOF extraction and matrix (L_{ij}) calculation is constant, the noise amplitude can be approximated as a constant. Therefore, in the measured S_j curve obtained by imaging, when the magnitude of the S_j change caused by the defect is small, even when compared with the noise amplitude, the characteristic of the S_j change is masked by the noise. This will increase the difficulty and error in extracting the defect feature information from the measured S_j curve.

The above problems are particularly prominent when the aluminum sheet contains a depth defect. For example, a variable-depth circular-hole-shaped defect that often occurs in actual detection: In the healthy area of the aluminum plate, the plate thickness is d_0; in the circular-hole-shaped defect 1, the plate thickness of most of the area is d_1. In the round-hole-shaped defect 2, the plate thickness further becomes d_2. In the measured S_j curve obtained by imaging, the S_j change caused by the defect will be more complicated than the S_j change caused by the simple circular-hole-shaped defect, which will undoubtedly further increase the difficulty of completely

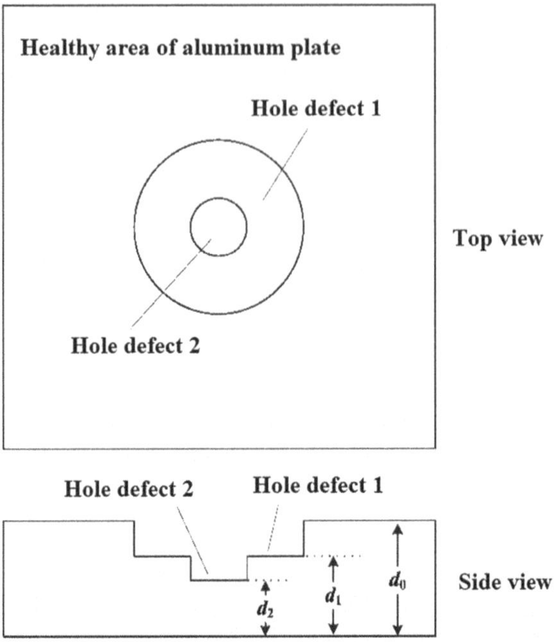

Fig. 4.5 Variable-depth circular-hole-shaped defects in aluminum plates

extracting the defect information from the noise-containing measured S_j curve. For the case where the aluminum plate contains a variable-depth defect, the base plate thickness d (from now on referred to as "base plate thickness") near a certain defect in the aluminum plate is defined for convenience of discussion—the thickness of the aluminum plate in the area outside the defect in the vicinity of a certain defect. For example, the base plate thickness near the defect 1 is d_0, and the base plate thickness near the defect 2 is d_1.

An effective method for solving the above problem is: By appropriately selecting the mode, and operating frequency of the Lamb wave used for imaging, the rate of change of the Lamb wave theory S_j with the thickness variation is large for any d. Therefore, for any d, the amplitude of S_j change in the measured S_j curve caused by the defect is larger and much larger than the noise amplitude (Fig. 4.5).

Define the sensitivity of a modal Lamb wave to plate thickness variation at any working frequency f and base plate thickness d (from now on referred to as "sensitivity"): the rate of change of Lamb wave S_j with plate thickness at any f and d. This sensitivity is referred to as $S_{\text{sen}}(f, d)$. It can be seen from the above definition that the sensitivity $S_{\text{sen}}(f, d)$ of a certain mode Lamb wave can be expressed by the absolute value of the partial derivative of d by the modal Lamb wave theory S_j, that is:

$$S_{\text{sen}}(f, d) = \left| \frac{\partial S_{j\_\text{the}}(f, d)}{\partial d} \right| \tag{4.3}$$

where $S_{j\_\text{the}}(f, d)$ is the theory S_j of the modal Lamb wave.

The key to implementing the above method is that by appropriately selecting the mode and operating frequency of the Lamb wave used for imaging, the selected mode Lamb wave has greater sensitivity for any d. It can be seen from the Lamb wave dispersion characteristics that the sensitivity of each mode Lamb wave changes with the change of d. Therefore, if only a single-mode Lamb wave is used for tomography, it is often difficult to ensure that the single-mode Lamb wave has high sensitivity at any d.

The implementation of multi-modal direct-ray Lamb wave cross-hole tomography can complement the Lamb waves of different modes, so that there is always a certain mode Lamb wave with higher sensitivity under different d. Thereby solving the problems in the above single mode Lamb wave transforaminal tomography. However, most of the existing researches are limited to the comparison of different modal Lamb waves for tomography. Moreover, in these comparative experiments, the researchers still only use one of the modal Lamb waves for tomography. Therefore, there is currently no practical multi-modal direct ray Lamb wave cross-hole tomography method.

Given this, this chapter attempts to study the variation of the sensitivity of the common modal Lamb wave with the working frequency and the thickness of the base plate based on the Lamb wave dispersion characteristics through analysis and calculation. Based on the variation law, the principle of cooperation between Lamb waves is given, and a new EMAT-based electromagnetic multi-modal direct ray Lamb wave cross-hole tomography imaging method is proposed.

It can be known from the Lamb wave dispersion characteristics that the high-order mode Lamb wave whose order is greater than or equal to 1 does not exist in the low-frequency–thickness product region. Therefore, when the operating frequency of the Lamb wave is constant, this high-order mode Lamb waves cannot propagate in the defect region where the remaining thickness is thin, which makes their application greatly limited. Therefore, in the study of multimodal Lamb wave cross-hole tomography, this chapter focuses on low-order Lamb waves of A0 and S0 modes. Also, in order to avoid interference caused by multimodal characteristics as much as possible, the working region of the Lamb wave is selected in a low-frequency–thickness region where the high-order mode Lamb wave is relatively small. Next, through the analysis and calculation, the sensitivity of the two modal Lamb waves in the low-frequency–thickness product region varies with the operating frequency f and the base plate thickness d.

Firstly, using the Lamb wave dispersion equation, the dispersion curve of the group velocity of the A0 and S0 mode Lamb waves in the aluminum plate in the low-frequency–thickness region is calculated. The longitudinal wave velocity and shear wave velocity in the aluminum plate are 6300 and 3300 m/s, respectively. On the left and right sides of the A_c point, the group velocity of the A0 wave increases monotonically and decreases with the frequency-constant product, respectively. Therefore, in order to ensure a one-to-one correspondence between the A0 wave group velocity and the plate thickness, the operating frequency–thickness product of the A0 wave needs to be smaller than the frequency–thickness product x_a of the A_c point. It is defined that A_c and x_a are the operating cutoff point and the cutoff frequency–thickness prod-

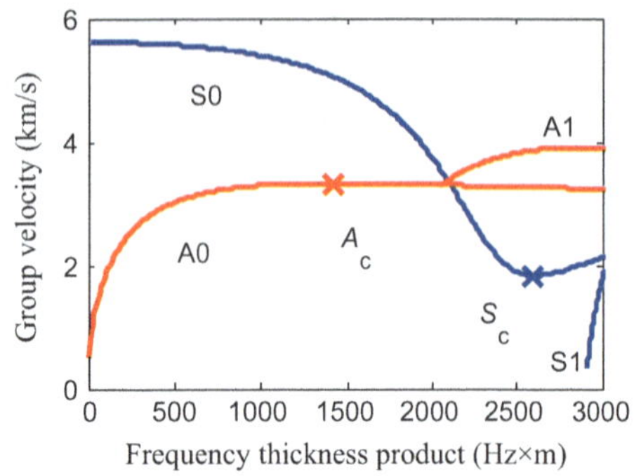

Fig. 4.6 Lamb wave group velocity dispersion curve in an aluminum plate

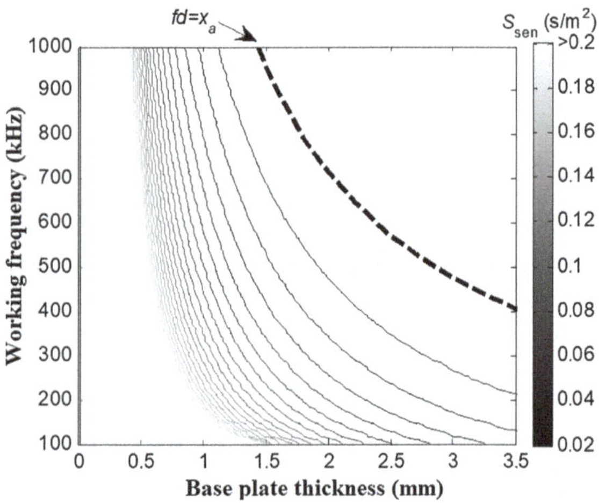

Fig. 4.7 Two-dimensional distribution of the amplitude of the A0 wave $S_{\text{sen}}(f, d)$ on f–d

uct of the A0 wave, respectively. Similarly, the S_c point and its frequency–thickness product x_s are the operating cutoff point and the cutoff frequency product of the S0 wave, respectively.

Then, using the dispersion curve and the slowness definition, the variation curves of the A0 and S0 wave sensitivities $S_{\text{sen}}(f, d)$ with the base plate thickness d under different operating frequencies f are calculated. Thereby, a two-dimensional distribution of the amplitudes of the A0 and S0 waves $S_{\text{sen}}(f, d)$ on the f–d plane is obtained. Consider that a higher f will cause the Lamb wave to decay too fast, and a lower f will cause the Lamb wave to have a lower ability to recognize defects. The Lamb operating frequency used in this chapter ranges from 100 to 1000 kHz. In Fig. 3.14, the different contour colors represent different $S_{\text{sen}}(f, d)$. $S_{\text{sen}}(f, d)$ corresponding to each contour line is equal; the difference between $S_{\text{sen}}(f, d)$ corresponding to two adjacent contour lines is a fixed value (Figs. 4.6, 4.7 and 4.8).

The A0 wave $S_{\text{sen}}(f, d)$ has the following laws with the distribution of f and d:

(1) An increase in d and f results in a decrease in $S_{\text{sen}}(f, d)$ of the A0 wave.

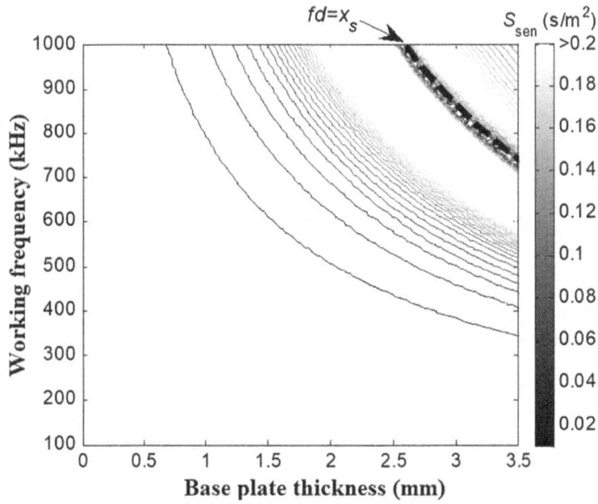

Fig. 4.8 Two-dimensional distribution of the amplitude of the S0 wave $S_{sen}(f, d)$ on f–d

(2) For any given f, when d is small, the sensitivity of the A0 wave is high, and thus, it is suitable for tomography. When d is large, the sensitivity of the A0 wave is low, and thus, it is not suitable for tomography.

(3) If the A0 wave $S_{sen}(f, d)$ corresponding to a certain contour line is set as the sensitivity threshold value S_{TH}, it is also defined that the region where the A0 wave $S_{sen}(f, d)$ is larger than the S_{TH} on the f–d plane is the sensitive region of the A0 wave. The area where the A0 wave $S_{sen}(f, d)$ is smaller than the S_{TH} (the upper right area of the selected contour line) is the insensitive area of the A0 wave. As can be seen from Fig. 3.14, as f increases, the width of the sensitive region of the A0 wave gradually narrows in the d direction. Therefore, in tomography, in order to ensure that the sensitive area of the A0 wave is sufficiently wide, the selected operating frequency f should not be too large.

(4) The broken line is a curve in which the frequency–thickness product is equal to x_a. Since the frequency–amplitude product of the A0 wave in operation must be lower than x_a, the working area of the A0 wave must be located at the lower left of the dotted line.

The S0 wave $S_{sen}(f, d)$ has the following laws with the distribution of f and d:

(1) The broken line is a curve in which the frequency–thickness product is equal to x_s. Since the frequency–thickness product of the S0 wave in operation must be lower than x_s, the working area of the S0 wave must be located at the lower left of the dotted line.

(2) In the working area, d and f decrease will cause the $S_{sen}(f, d)$ of the S0 wave to decrease.

(3) In the working area, for any given f, when d is large, the S0 wave has high sensitivity and is, therefore, suitable for tomography. When d is small, the S0 wave is less sensitive and thus is not suitable for tomography.

(4) If the S0 wave $S_{sen}(f, d)$ corresponding to a certain contour line is set as the sensitivity threshold value S_{TH}, it is also defined that the region where the S0 wave $S_{sen}(f, d)$ is larger than the S_{TH} on the f–d plane is the sensitive region of the S0 wave. The region where the S0 wave $S_{sen}(f, d)$ is smaller than the S_{TH} (the lower left region of the selected contour line) is the insensitive region of the S0 wave. As f decreases, the width of the sensitive region of the S0 wave gradually narrows in the d direction. Therefore, in tomography, in order to ensure that the sensitive area of the S0 wave is sufficiently wide, the selected operating frequency f should not be too small.

In summary, the A0 wave is sensitive to plate thickness change only when d is small; the S0 wave is sensitive to plate thickness change only when d is large. It can be seen that using a single A0 or S0 mode Lamb wave for direct ray transforaminal tomography, it is difficult to ensure that the Lamb wave has high sensitivity at any d. Therefore, according to the law of A0 and S0 wave sensitivity with the change of f and d, the principles and methods of their cooperation are studied, and the corresponding multi-modal direct ray Lamb wave cross-hole tomography imaging method is established.

From the A0 mentioned above and S0 mode Lamb wave sensitivity changes with f and d, it can be known that the following principles should be followed when using the two waves to perform multi-modal direct ray Lamb wave cross-hole tomography:

(1) The frequency–thickness products of the A0 wave and the S0 wave used must be lower than x_a and x_s, respectively.
(2) After setting the sensitivity threshold to determine the sensitive areas of the A0 wave and the S0 wave, the working areas of the A0 wave and the S0 wave must be located in the respective sensitive areas.
(3) The A0 wave is mainly used to identify the variation of the thickness caused by the defect in the small case of d, and the S0 wave is mainly used to identify the variation of the thickness caused by the defect in the case of d.
(4) The A0 wave and the S0 wave do not need to be imaged under the same f, but their working areas under the respective f must be partially overlapped to ensure that for any d, at least one wave of the working area can be covered. Specifically, the upper limit d value of the A0 wave working area should be greater than the lower limit d value of the S0 wave working area.
(5) Finally, according to the relationship between the measured plate thickness and the working area of the Lamb wave, the tomographic results of the two waves are superimposed and combined according to a certain weight.

This chapter is based on the law that the sensitivity of A0 wave and S0 wave varies with f and d, and the principle that should be followed when these two waves cooperate (from now on referred to as "coordination principle"). An electromagnetic ultrasonic multi-modal direct ray Lamb wave cross-hole tomography method (from now on referred to as "multimodal imaging method") is proposed. Firstly, based on the principle of coordination, the working frequency and working area of A0 wave and S0 wave are calculated. Subsequently, the new EMAT with good modal control

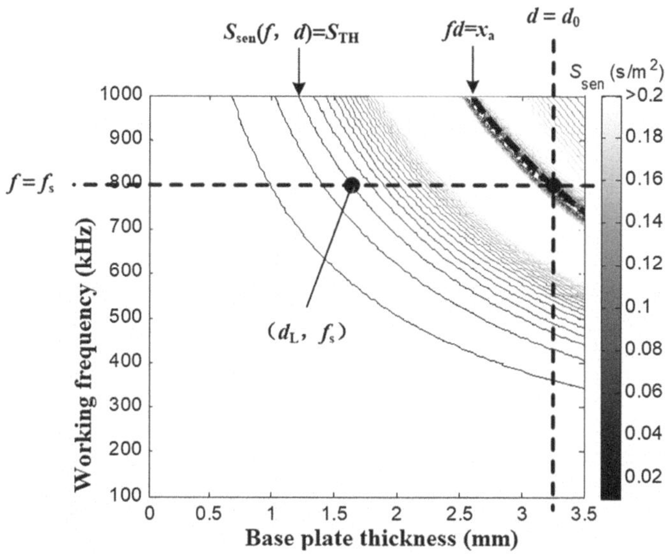

Fig. 4.9 Schematic diagram of calculating the operating frequency and working area of the S0 wave

capability developed in the work of this chapter is used to transmit and receive the A0 wave and the S0 wave, respectively, to obtain the TOF. Then, using the acquired TOF, based on the improved direct ray imaging method, A0 wave and S0 wave cross-hole tomography are, respectively, performed; finally, the tomographic results of the two waves are superimposed and fused according to a certain weight.

In the proposed multimodal imaging method, the process of calculating the operating frequency and working area of the A0 wave and the S0 wave based on the matching principle is: ① Calculate the working frequency of the A0 wave; ② calculate the working area of the A0 wave; ③ calculate the working area of the S0 wave; ④ calculate the operating frequency of the S0 wave. The specific calculation steps are as follows:

(1) Set the sensitivity threshold S_{TH} of the A0 wave and the S0 wave.
(2) In the S0 wave $S_{sen}(f, d)$ distribution map shown in Fig. 4.9, find the curve $fd = x_s$; then, the working area of the S0 wave should be at the lower left of the curve.
(3) In Fig. 4.9, find the line $d = d_0$, where d_0 is the thickness of the aluminum plate in the healthy area; then, the working area of the S0 wave should be on the left side of the line.
(4) In Fig. 4.9, calculate the intersection coordinates (d_0, f_s) of the curve $fd = x_s$ and the straight line $d = d_0$, and take f_s as the operating frequency of the S0 mode wave.

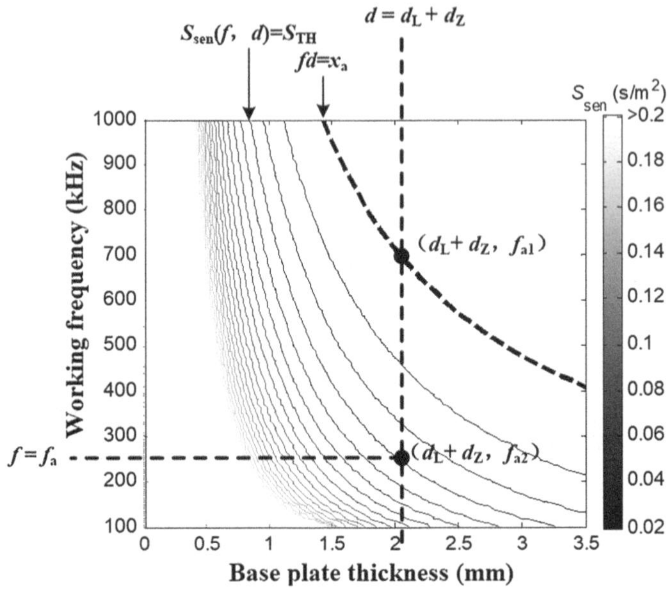

Fig. 4.10 Schematic diagram of calculating the operating frequency and working area of the A0 wave

(5) In Fig. 4.9, find the contour line $S_{sen}(f, d) = S_{TH}$; then, the working area of the S0 wave should be located at the upper right of the contour line, that is, the sensitive area of the S0 wave.

(6) In Fig. 4.9, calculate the intersection point (d_L, f_s) of the straight line $f = f_s$ and the contour line $S_{sen}(f, d) = S_{TH}$, and obtain the lower limit value d_L of the working area of the S0 wave at the operating frequency f_s; thereby, the working region $[d_L, d_0]$ of the S0 wave at the operating frequency f_s is obtained.

(7) In the A0 wave $S_{sen}(f, d)$ distribution map shown in Fig. 4.10, set the dead zone width $d_Z(d_Z > 0)$ of the A0 wave and S0 wave working area to obtain the working area of the A0 wave $[0, d_L + d_Z]$.

(8) In Fig. 4.10, find the curve $fd = x_a$; then, the A0 wave working area should be at the lower left of the curve.

(9) In Fig. 4.10, the intersection $(d_L + d_Z, f_{a1})$ of the curve $fd = x_a$ and the straight line $d = d_L + d_Z$ is calculated.

(10) In Fig. 4.10, find the high line $S_{sen}(f, d) = S_{TH}$; then, the working area of the A0 wave should be located at the lower left of the contour, i.e., the sensitive area of the A0 wave.

(11) In Fig. 4.10, the intersection $(d_L + d_Z, f_{a2})$ of the contour line $S_{sen}(f, d) = S_{TH}$ and the straight line $d = d_L + d_Z$ is calculated.

(12) Take a relatively small one of f_{a1} and f_{a2} as the operating frequency f_a of the A0 mode wave.

In the proposed multimodal imaging method, after obtaining the distribution of S_j by A0 wave tomography and S0 wave tomography, respectively, the image is superimposed and merged according to the following steps.

(1) Using the Lamb wave dispersion relationship, the two sets of S_j distribution are converted into plate thickness distribution.
(2) According to a certain weight, the two sets of thickness distribution results are superimposed and merged. The specific method is: For a certain pixel grid, the grid thickness obtained by A0 wave tomography is d_a. The grid thickness obtained by S0 wave tomography is d_a. If $d_a \in [0, d_L + d_Z]$ and $d_s \in [d_L, d_0]$, or $d_a \notin [0, d_L + d_Z]$ and $d_s \notin [d_L, d_0]$, then the stacked grid thickness is $d_{add} = 0.5d_a + 0.5d_s$; if $d_a \in [0, d_L + d_Z]$ and $d_s \notin [d_L, d_0]$, then $d_{add} = d_a$; if $d_a \notin [0, d_L + d_Z]$ and $d_s \in [d_L, d_0]$, then $d_{add} = d_s$.

To verify the imaging performance of the proposed multimodal imaging method for variable-depth defects, tomographic experiments were performed on 3-mm-thick aluminum plates. The specific test procedure for tomography using the proposed multimodal imaging method is to first determine the respective operating frequencies and working areas of the A0 and S0 waves. Then, the A0 wave and the S0 wave are separately used to perform tomography at the selected working frequency, and the results are analyzed and compared. By this, the tomographic results of the A0 wave and the S0 wave are weighted, superimposed, and merged. The corresponding multimodal tomography results were obtained and compared with the results of a single A0 wave or S0 wave tomography.

The tomographic test was carried out in a rectangular area of 600 mm × 600 mm on an aluminum plate, which was evenly divided into 120 × 120 small grids. A variable-depth circular-hole-shaped artificial defect on an aluminum plate can be considered to consist of two simple circular-hole-shaped defects. Defect 1 has a diameter D_1, a depth d_{01}, and a center position (x, y) of 120 mm, 1 mm, and (310 mm, 350 mm), respectively. Defect 2 is located inside defect 1, and its diameter D_2, depth d_{02}, and center position (x, y) are 60 mm, 2 mm, and (300 mm, 350 mm), respectively. It can be seen that the thickness of the base plate near the defect 1 is 3 mm; the thickness of the base plate near the defect 2 is 2 mm (Fig. 4.11).

It can be seen from the theoretical analysis that the A0 wave has a higher sensitivity only when the base plate thickness is small. Therefore, the sensitivity to the variation of the thickness caused by the defect 2 is relatively high, and the sensitivity to the variation of the thickness caused by the defect 1 is low; on the contrary, the S0 wave is sensitive only when the thickness of the base plate is large. Therefore, its sensitivity to the change in the thickness of the sheet caused by the defect 1 is relatively high, and the sensitivity to the change in the thickness of the sheet caused by the defect 2 is low.

In the test, the calculated A0 wave and S0 wave operating frequencies were 270 and 700 kHz, respectively. The parameters of the new EMAT for transmitting and receiving A0 waves and S0 waves are shown in Table 3.4. When imaging with each modal Lamb wave, 14 excitation EMATs, and 14 reception EMATs were uniformly arranged on both sides of the imaging area, respectively. The iterative solution thresh-

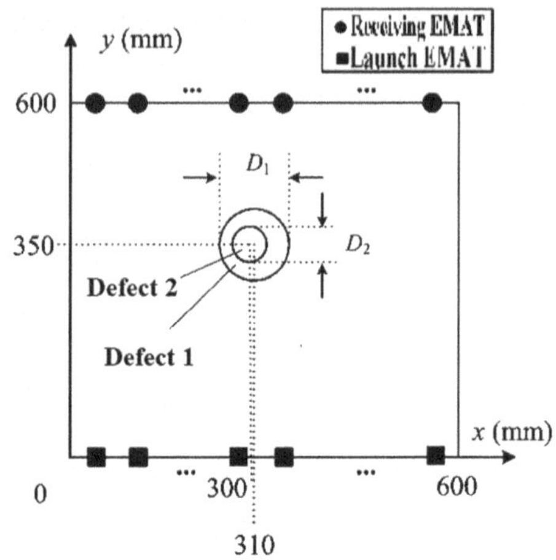

Fig. 4.11 Schematic diagram of tomographic test parameters

Table 4.1 Parameters of the new EMAT used in the test

	A0 wave EMAT	S0 wave EMAT
Working frequency (kHz)	270	700
l_c (mm)	4.2	3.6
Coil outer diameter (mm)	37.8	32.4
Coil inner diameter (mm)	12.6	10.8
Total number of coils	24	24
Excitation signal peak-to-peak value (V)	300	300
Number of excitation signal cycles	7	7

old T_{TH} of the SIRT is set to 0.05%, and the relaxation factor is taken as $\mu_I = 0.1$ (Table 4.1).

The tomographic results are obtained using the A0 wave and the S0 wave based on the improved direct ray imaging method alone, that is, the distribution of the A0 wave and the S0 wave slowness S_j. The dashed line is the actual contour of the two defects. The image reconstruction results show that when the A0 wave is used for tomography alone, the image of the reconstructed defect 2 and its outline are clear. In contrast, the image of the reconstructed defect one is integrated with the image of the healthy area, and the outer contour of the defect one reconstructed image is difficult to recognize. The image reconstruction results show that when the S0 wave is used for tomography alone, the image of the reconstructed defect 1 and its outline are clear. In contrast, the image of the reconstructed defect 2 is integrated with the image of the defect 1, and the contour of the reconstructed defect 2 images is difficult to recognize (Fig. 4.12).

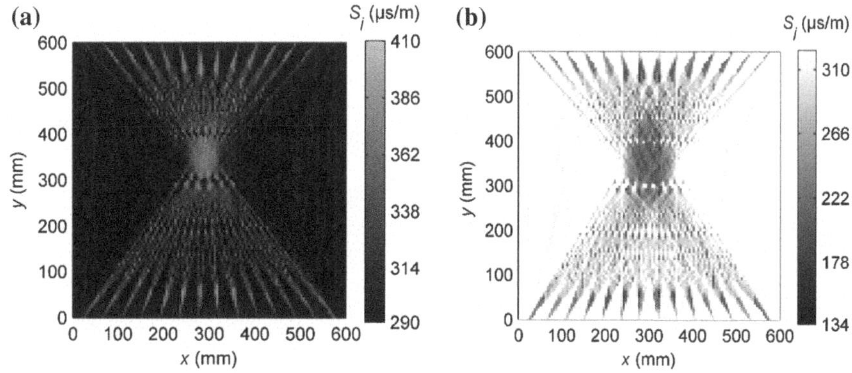

Fig. 4.12 Tomography results using the A0 and S0 waves alone based on the improved direct ray imaging method: **a** A0 wave; **b** S0 wave

To further analyze Fig. 3.19, the measured A0 wave and S0 wave S_j curve at $y = 350$ mm was extracted from the imaging results. Also, according to the actual plate thickness distribution, using the A0 wave and S0 wave dispersion relationship shown in Fig. 2.4 and the equation $cg_j = 1/S_j$, the theoretical A0 wave and S0 wave S_j curve at $y = 350$ mm in the aluminum plate can be calculated. The theoretical S_j of the A0 wave used in the test in the healthy area of the aluminum plate, the region where the defect one is located, and the region where the defect 2 is located are 308, 326, and 383 μs/m, respectively. The theoretical S_j of the S0 wave used in the healthy area of the aluminum plate, the region where the defect 1 is located, and the region where the defect 2 is located are 298, 197, and 181 μs/m, respectively. When tomography is performed using the A0 wave alone, the magnitude of the measured S_j change caused by the defect 2 is much larger than the noise amplitude. Moreover, in the region where the defect 2 is located, the measured S_j curve agrees well with the theoretical S_j curve, which makes the feature information of the defect 2 more accurately extracted. The magnitude of the measured S_j change caused by defect 1 is relatively small and close to the noise amplitude, which makes the feature information of defect 1 difficult to be accurately extracted. When the S0 wave is used for tomography alone, the magnitude of the measured S_j change caused by defect 1 is much larger than the noise amplitude. Moreover, in the region where defect 1 is located, the measured S_j curve agrees well with the theoretical S_j curve, which makes the feature information of defect 1 more accurately extracted. The measured S_j change caused by defect 2 is almost overwhelmed by noise and cannot be accurately identified, which makes the feature information of defect 2 difficult to extract accurately (Fig. 4.13).

Using the Lamb wave dispersion relationship, the S_j distribution is converted into a plate thickness distribution. When the A0 wave is used for tomography alone, the measured thickness variation amplitude in the region where the defect 2 is located agrees well with the theoretical value. In the area where defect 1 is located, the measured thickness variation amplitude is close to the noise amplitude and is even

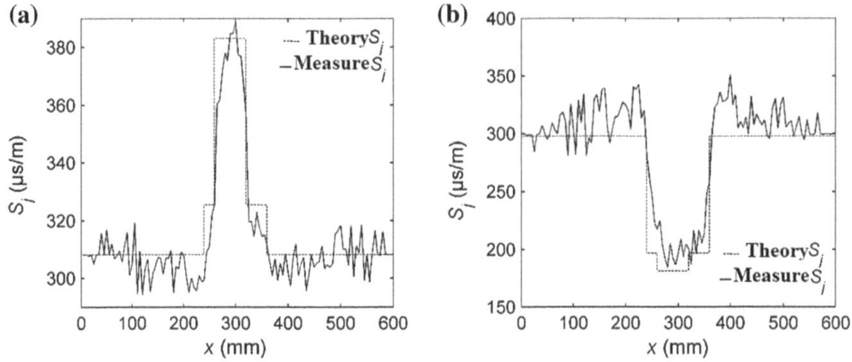

Fig. 4.13 S_j curve at $y = 350$ mm obtained by tomography using A0 and S0 waves alone: **a** A0 wave; **b** S0 wave

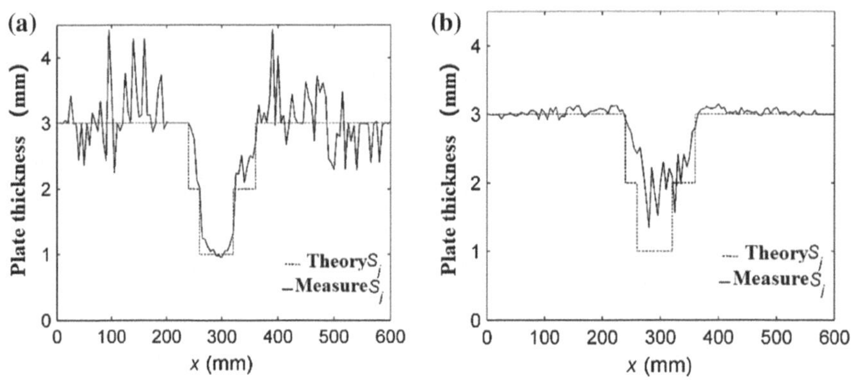

Fig. 4.14 Thickness distribution at $y = 350$ mm obtained by tomography using A0 and S0 waves alone: **a** A0 wave; **b** S0 wave

submerged by noise. When the S0 wave is used for tomography alone, the measured thickness variation amplitude is in good agreement with the theoretical value in the region where defect 1 is located. In the area where defect 2 is located, the measured thickness variation of the plate thickness is quite different from the theoretical value and is completely submerged by noise (Fig. 4.14).

The results obtained by imaging the variable-depth defect using the single-mode Lamb wave described above indicate that the feature information of the defect 2 can be extracted relatively accurately when the A0 wave is used for tomography alone. The feature information of defect 1 is almost overwhelmed by noise, and it is difficult to extract accurately; on the contrary, when the S0 wave is used for tomography alone, the feature information of defect 1 can be extracted relatively accurately. The feature information of defect 2 is almost overwhelmed by noise, and it is difficult to extract accurately. Therefore, the tomographic imaging using A0 wave

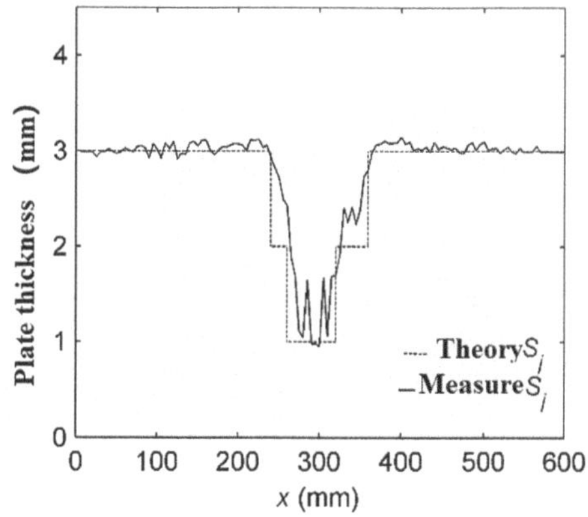

Fig. 4.15 Thickness distribution at $y = 350$ mm obtained by tomography using multimodal imaging

or S0 wave alone cannot accurately obtain the complete characteristic information of the circular-hole-shaped artificial defect with variable depth. The experimental results are consistent with the theoretical analysis results.

On this basis, according to the final step of the proposed multimodal imaging method, the plate thickness distribution obtained by A0 wave and S0 wave tomography is superimposed and fused according to the corresponding weights. Compared to the plate thickness distribution obtained by imaging with a single-mode Lamb wave, in the results obtained using the multimodal imaging method, the amplitude of the plate thickness variation in each region is significantly larger than the noise amplitude. The measured defect profile is stepped and closer to the profile of the actual depth defect, and therefore, the feature information of the defect 1 and the defect 2 can be extracted more accurately (Fig. 4.15).

Using the Lamb wave dispersion relationship, the S_j distribution is converted into a plate thickness distribution. When the A0 wave is used for tomography alone, the reconstructed image of the defect 2 and its contour are relatively clear; however, there are many artifacts in the imaging result, and the reconstructed image of the defect 1 is almost submerged by the image. When the S0 wave is used for tomography alone, the reconstructed image of the defect 1 and its contour are relatively clear, while the reconstructed image of the defect 2 is severely distorted and the outline is blurred. The above results show that the tomographic imaging using A0 wave or S0 wave alone cannot accurately obtain the clear and complete reconstructed image of the round-hole-shaped artificial defect with variable depth. The experimental results are consistent with the theoretical analysis results (Fig. 4.16).

On this basis, according to the final step of the proposed multimodal imaging method, the plate thickness distribution obtained by A0 wave and S0 wave tomography is superimposed and fused according to the corresponding weights. Compared to the tomography using a single-mode Lamb wave, the tomographic images obtained by the multimodal imaging method have relatively few artifacts, and the reconstructed images of defects 1 and 2 and their contours are relatively clear. Moreover,

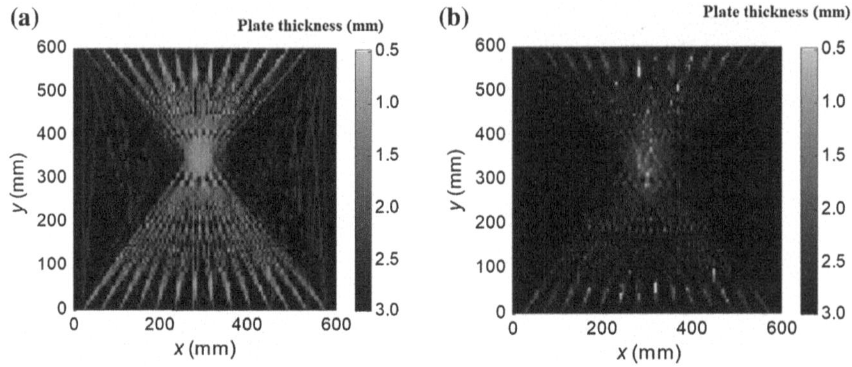

Fig. 4.16 Plate thickness distribution obtained by tomography using A0 and S0 waves alone: **a** A0 wave; **b** S0 wave

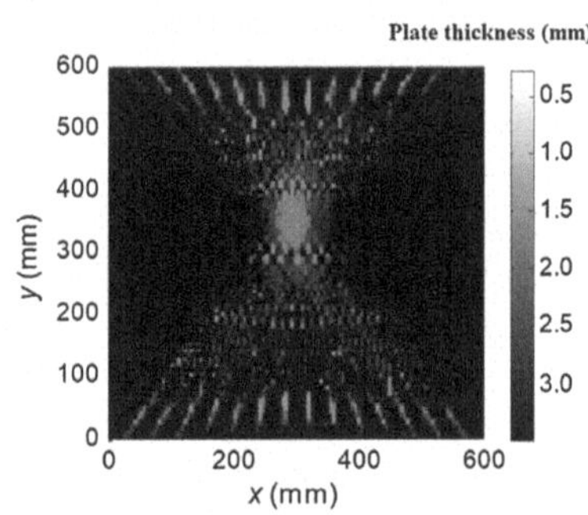

Fig. 4.17 Thickness distribution obtained by tomography using multimodal imaging

closer to the actual situation, therefore, the complete distribution of the variable-depth circular-hole-shaped artificial defects can be observed more clearly and accurately (Fig. 4.17).

4.2 Electromagnetic Ultrasonic Bending Ray Lamb Wave Cross-Hole Tomography Imaging Method

The propagation path of the Lamb wave ray used in the cross-hole tomography and the matrix (L_{ij}) corresponding to the propagation path are all solved based on the direct ray model [4, 5]. Under actual conditions, when the Lamb wave rays propagating in the aluminum plate encounter defects, the direction of propagation will be deflected, and the Lamb wave rays will appear to propagate along the curved path. Therefore, there is a certain deviation between the Lamb wave ray propagation path and the corresponding matrix (L_{ij}) solved based on the direct ray model and the actual situation. It can be seen from Eq. (3.1) that in the Lamb wave cross-hole tomography results, the accuracy of the matrix (L_{ij}) solution is the same as the accuracy of the TOF extraction, which also determines the accuracy of the slowness S_j calculation. It can be seen that in the electromagnetic ultrasonic direct ray Lamb wave cross-hole tomography imaging process, the inaccuracy of the matrix (L_{ij}) solution will lead to a certain error in the S_j distribution of the imaging, which will affect the imaging performance.

One of the methods to solve the above problem is to introduce Lamb wave test ray tracing (RT) technology in tomography [6], that is, to perform electromagnetic ultrasonic bending ray Lamb wave cross-hole tomography. The Lamb wave test radiation RT technique was used to calculate the defect distribution based on electromagnetic ultrasonic direct ray Lamb wave cross-hole tomography. Based on the mechanism of interaction between Lamb wave and defect, the curved Lamb wave ray propagation path closer to the actual situation is recalculated, and the solution result of the matrix (L_{ij}) is corrected. The Lamb wave cross-hole tomography was performed again using the modified matrix (L_{ij}).

However, there have been researching results on the Lamb wave test radiation RT technology, and there is a lack of systematic explanation about the mechanism of the influence of defect distribution on the Lamb wave propagation path. The bilinear interpolation (BI)-based Lamb wave test radiation RT algorithm used in solving the Lamb wave phase velocity gradient and implementing RT is less accurate, and the phenomenon of ray search failure often occurs, and the reliability is low. It is difficult to effectively use the electromagnetic ultrasonic bending ray Lamb wave cross-hole tomography. Given this, this chapter combines the derivation process of the extrapolation formula of the Lamb wave test radiation RT. The influence of the Lamb wave phase velocity distribution caused by the defect on the Lamb wave propagation path, the importance of improving the Lamb wave phase velocity gradient accuracy, and the limitation of the BI-based Lamb wave test radiation RT algorithm are analyzed. Aiming at the problems of the algorithm, an improved Lamb wave test radiation RT algorithm is proposed. On this basis, a new electromagnetic ultrasonic bending ray Lamb wave cross-hole tomography method is proposed.

4.2.1 BI-Based Lamb Wave Test Radiation RT Algorithm

When implementing Lamb wave test radiation RT in aluminum plate, firstly, according to the dispersion relationship of Lamb wave, the defect distribution or slowness distribution of straight-ray Lamb wave cross-hole tomography output is converted into the corresponding phase velocity distribution. Then, according to the gradient of the phase velocity at each position in the aluminum plate, the advancing direction of the Lamb wave is gradually extrapolated.

In the following, the derivation process of the Lamb wave test radiation RT extrapolation formula is combined with the Lamb wave test phase to analyze the influence of the Lamb wave phase velocity distribution on the Lamb wave propagation path and the accuracy of improving the Lamb wave phase velocity gradient. On this basis,

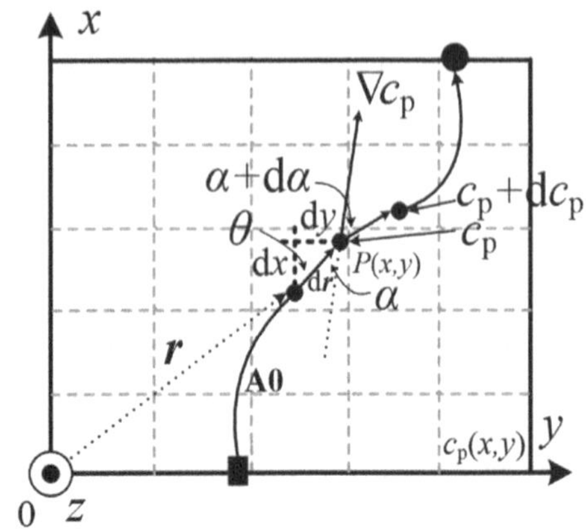

Fig. 4.18 Test radiation RT algorithm extrapolation schematic

the problems in the BI-based Lamb wave test radiation RT algorithm (from now on referred to as "traditional test radiation RT algorithm") are analyzed.

Based on the application background of Lamb wave cross-hole tomography, the EMAT array model and meshing method involved in the Lamb wave test radiation RT algorithm are the same as those in Lamb wave cross-hole tomography. The plate thickness d, the Lamb wave phase velocity c_{p}, and the Lamb wave group velocity c_{g} in each mesh are regarded as fixed. The Lamb wave test radiation RT process uses the gridded cp distribution obtained by straight-ray Lamb wave cross-hole tomography as a known quantity and calculates the true propagation path of Lamb waves between each pair of EMATs (Fig. 4.18).

For ease of research, a three-dimensional Cartesian coordinate system is established in the imaging region. Record $\boldsymbol{a}_x$, $\boldsymbol{a}_y$, $\boldsymbol{a}_z$ as the unit vector in the x, y, and z directions. The point $P(x, y)$ on the Lamb wave ray i has a position vector of $\boldsymbol{r} = x\boldsymbol{a}_x + y\boldsymbol{a}_y$. Here, the tangential micro-increment of ray i is: d d$\boldsymbol{r} = \mathrm{d}x\boldsymbol{a}_x + \mathrm{d}y\boldsymbol{a}_y$. The angle between d$\boldsymbol{r}$ and the x-axis is denoted as θ. The angle α of incidence that defines the ray to point P is the angle between d$\boldsymbol{r}$ and ∇c_{p}. ∇c_{p} is the gradient of the Lamb wave c_{p} at point P, and $\nabla c_{\mathrm{p}} = (\partial c_{\mathrm{p}}/\partial x)\boldsymbol{a}_x + (\partial c_{\mathrm{p}}/\partial y)\boldsymbol{a}_y$.

To perform the Lamb wave test RT, you first need to derive the corresponding extrapolation formula. The extrapolation formula should reflect the influence of the c_{p} distribution on the propagation path of the Lamb wave, that is, the relationship between the incident angle α, the c_{p} distribution, and the exit angle (α+dα) at the point P. Therefore, an extrapolation formula reflecting the relationship between α, c_{p}, and (α+dα) needs to be derived. The above extrapolation formula is derived based on the Snell theorem:

$$\sin(a + \mathrm{d}a)c_{\mathrm{p}} = \sin(a)\,(c_{\mathrm{p}} + \mathrm{d}c_{\mathrm{p}}) \tag{4.4}$$

where dc_{p} is the increment of c_{p} before and after point P.

$$\cos a = \frac{\nabla c_{\mathrm{p}} \cdot \mathrm{d}\boldsymbol{r}}{\left|\nabla c_{\mathrm{p}}\right| |\mathrm{d}\boldsymbol{r}|} \tag{4.5}$$

$$\sin a = \frac{(\nabla c_{\mathrm{p}} \times \mathrm{d}\boldsymbol{r}) \cdot \boldsymbol{a}_z}{\left|\nabla c_{\mathrm{p}}\right| |\mathrm{d}\boldsymbol{r}|} \tag{4.6}$$

Combined $\mathrm{d}\alpha = \mathrm{d}\theta$, $\tan\theta = \mathrm{d}y/\mathrm{d}x$, $\mathrm{d}c_p = \nabla c_p \cdot \boldsymbol{dr}$, the calculation formula of the ray deflection angle $\mathrm{d}\alpha$ can be derived:

$$\mathrm{d}\alpha = \frac{1}{c_{\mathrm{p}}}\left(\frac{\partial c_{\mathrm{p}}}{\partial x}\frac{\partial y}{\partial x} - \frac{\partial c_{\mathrm{p}}}{\partial y}\right)\mathrm{d}x \tag{4.7}$$

Further derivation of the discretized RT extrapolation formula can be shown as follows. Let P point be the discrete point on the ray i obtained by extrapolation of the K step, $P(x, y) = P_K(x_K, y_K)$. The two points before and after P_K are $P_{K-1}(x_{K-1}, y_{K-1})$ and $P_{K+1}(x_{K+1}, y_{K+1})$, respectively. From the geometric relationship: $\mathrm{d}x = x_K - x_{K-1}$, $\mathrm{d}y = y_K - y_{K-1}$, $\partial y/\partial x = (y_K - y_{K-1})/\Delta x$ and $\mathrm{d}\alpha = \mathrm{d}\theta = (y_{K+1} - 2y_K + y_{K-1})/[(y_K - y_{K-1})^2 + (\Delta x)^2]$, where $\Delta x = \mathrm{d}x$ is a fixed extrapolation step. Then:

$$y_{K+1} = 2y_K - y_{K-1} + \left[\frac{\partial c_{\mathrm{p}}}{\partial x}\frac{(y_K - y_{K-1})}{\Delta x} - \frac{\partial c_{\mathrm{p}}}{\partial y}\right]\left[(y_K - y_{K-1})^2 + (\Delta x)^2\right] \tag{4.8}$$

Equation (4.8) is the extrapolation formula of Lamb wave test RT, where $\partial c_{\mathrm{p}}/\partial x$ and $\partial c_{\mathrm{p}}/\partial y$ (i.e., the x-direction component and the y-direction component of ∇c_{p}) are still unknown.

It can be seen from the above derivation process that the variation of the thickness caused by defects in the aluminum plate causes a change in the c_{p} distribution of the Lamb wave, and the change of the c_{p} distribution of the Lamb wave causes a change in the gradient ∇c_{p}, and ∇c_{p} determines the change of the Lamb wave propagation path. It can be seen that to achieve accurate Lamb wave RT, the key is to accurately solve the c_{p} of the Lamb wave. The Lamb wave RT is known as a gridded c_{p} distribution, and the values of $\partial c_{\mathrm{p}}/\partial x$ and $\partial c_{\mathrm{p}}/\partial y$ cannot be obtained by obtaining a partial derivative of c_{p}. Therefore, it is necessary to further establish an accurate solution method of ∇c_{p}.

4.2.1.1 BI-Based Gradient Solving Method

To solve the ∇c_{p} required for the Lamb wave test RT, use BI to calculate $\partial c_{\mathrm{p}}/\partial x$ and $\partial c_{\mathrm{p}}/\partial y$ [6]:

$$\begin{aligned}\frac{\partial c_{\mathrm{p}}}{\partial x} &= \frac{c_{\mathrm{p}}(x_K, y_K) - c_{\mathrm{p}}(x_{K-1}, y_K)}{\Delta x}\\ \frac{\partial c_{\mathrm{p}}}{\partial y} &= \frac{c_{\mathrm{p}}(x_K, y_K) - c_{\mathrm{p}}(x_K, y_{K-1})}{y_K - y_{K-1}}\end{aligned} \tag{4.9}$$

Get the complete RT extrapolation formula. A conventional test RT algorithm is thus obtained: for ray i, a given launch position $(0, y_0)$, a receive position (x_r, y_r), and an initial launch angle θ. The next point coordinates (x_{K+1}, y_{K+1}) are extrapolated from the coordinates (x_K, y_K) and (x_{K-1}, y_{K-1}) of the adjacent two points on the Δx and the ray. The ray length $w(i, K)$ extrapolated from the Kth step, the extrapolation point (x_K, y_K) coordinates, and the Lamb wave phase velocity $c_p(i, K)$ at (x_K, y_K) are recorded. The corresponding $c_g(i, K)$ is calculated by using the Lamb wave dispersion characteristics. Repeat the extrapolation process until $x_{K+1} \geq x_r$, using the following formula to calculate the calculated TOF of the ray i obtained using the traditional test RT algorithm:

$$T\mathrm{conv}(i) = \sum_K \frac{w(i, K)}{c_g(i, K)} \tag{4.10}$$

Let $\theta = \theta + \Delta\theta$ and repeat the above RT process until finding the minimum travel path connecting the EMAT.

The Lamb wave test RT process uses the gridded c_p distribution obtained by direct ray Lamb wave cross-hole tomography as an input. The resolution of the meshed c_p distribution is relatively low, while the extrapolation step size Δx of the conventional test RT algorithm is relatively small. Therefore, in most of the extrapolation steps of the traditional test RT algorithm, the three points involved (x_K, y_K), (x_{K-1}, y_K), and (x_K, y_{K-1}) are located in the same grid. Under this condition, the obtained $\partial c_p/\partial x$ and $\partial c_p/\partial y$ are both 0. The Lamb wave rays obtained by these extrapolation steps will propagate along a straight line until the rays reach the edge of the mesh, and the above three points are no longer in the same mesh.

When the Lamb wave ray reaches the edge of the mesh, at least two of the above three points are in the same mesh. The ∇c_p obtained by the traditional test RT algorithm at the edge of the mesh must be perpendicular to the edge of the mesh, which makes the refractive surface of the obtained Lamb wave ray necessarily parallel to the edge of the mesh.

In a defective aluminum plate, the actual Lamb wave c_p distribution is usually much more complicated than the meshed c_p distribution, and the actual c_p throughout the same grid is not necessarily a fixed value. Therefore, the actual Lamb wave rays usually do not propagate along a straight line in the same grid, and the actual line surface of the Lamb wave is not always parallel to the edge of the mesh. It can be seen that the meshed c_p distribution as an input has a lower resolution. The Lamb wave ∇c_p, the ray refraction surface, and the ray propagation path obtained by the conventional test RT algorithm are lower in accuracy than the actual situation.

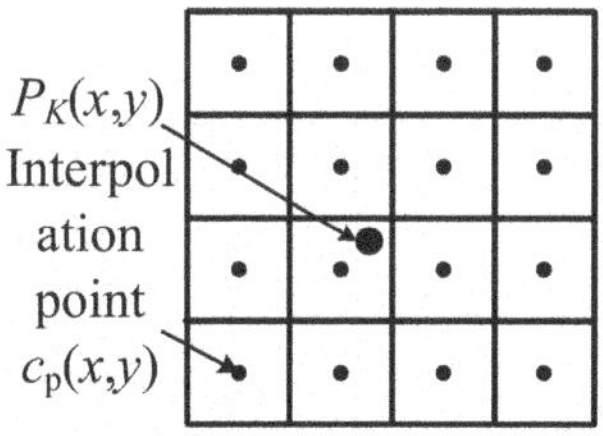

Fig. 4.19 Schematic diagram of TDCSI interpolation

4.2.2 Improved Lamb Wave Test Radiation RT Algorithm

Aiming at the low accuracy of Lamb wave c_p and ray path in the traditional test RT algorithm, it is considered that the Lamb wave ∇c_p distribution in the aluminum plate with defects is usually complicated in practical applications. In this paper, two-dimensional cubic spline interpolation (TDCSI) is used to perform surface fitting on the c_p distribution of Lamb waves to improve the resolution of cp distribution. In turn, the $\partial c_p/\partial x$ and $\partial c_p/\partial y$ required for extrapolation are accurately solved. In this method, the two-dimensional cubic spline function used to fit the c_p distribution is:

$$c_p'(x, y) = \sum_{m=0}^{3}\sum_{n=0}^{3} C_{mn}x^m y^n \tag{4.11}$$

where c_p' is the result of fitting c_p. The 16 unknown C_{mn} need to be solved by interpolation. As shown in Fig. 4.2, this paper extracts 16 interpolation points in the 4 × 4 grid area near the P_K point and substitutes $c_p(x, y)$ and corresponding coordinates at these interpolation points as known conditions to solve C_{mn}. The interpolation points are all at the center of the corresponding pixel grid. By deriving the formula, the values of $\partial c_p'/\partial x$ and $\partial c_p'/\partial y$ can be obtained. By using $\partial c_p'/\partial x$ and $\partial c_p'/\partial y$ instead of $\partial c_p/\partial x$ and $\partial c_p/\partial y$, the complete Lamb wave test RT extrapolation formula can be obtained (Fig. 4.19).

In this paper, the TDCSI method is used to improve the accuracy of ∇c_p. However, this method needs to solve the 16-element equations to obtain the coefficient C_{mn}, which makes the proposed algorithm more computational and longer. However, in order to achieve rapid scanning imaging of large-area aluminum plates, the Lamb wave test RT algorithm and the corresponding curved ray Lamb wave cross-hole tomography need to have a faster calculation speed. Given this, this paper adopts the adaptive block method, which aims to reduce the total calculation amount and calculation time of the proposed algorithm.

Lamb wave rays always propagate from bottom to top. Therefore, after the ray i is extrapolated by a certain step, there are only two possibilities for entering the direction of a pixel grid: from the lower left to the upper right; from the lower right to the upper left. Based on this, this paper proposes to perform Lamb wave test RT in blocks of 2 × 2 grid area. In the extrapolation process, the direction in which the ray enters the grid ABCD is predicted. If the ray enters the grid ABCD from the

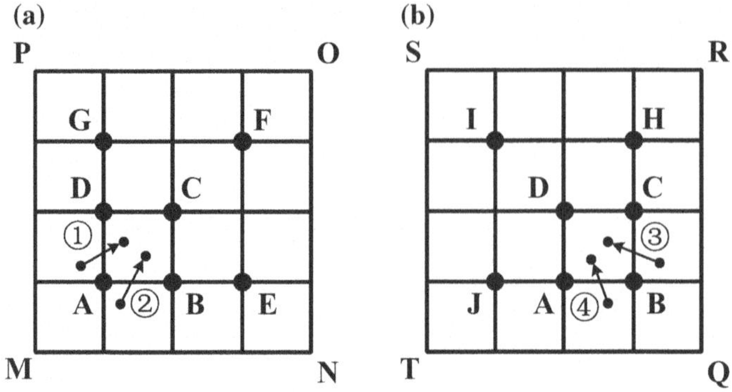

Fig. 4.20 Schematic diagram of the block method: **a** the interpolated region when the ray enters the grid from the lower left to the upper right; **b** the interpolated region when the ray enters the grid from the lower right to the upper left

lower left to the upper right as indicated by the arrow ① or ② in Fig. 4.3a, the 2 × 2 sub-mesh area AEFG is divided in the ray advancing direction. The MNOP is used as the corresponding interpolation region for solving C_{mn}, the c_p distribution is fitted once to solve the ∇c_p required for each step of AEFG extrapolation, and the ray extrapolation calculation in AEFG is completed. Similarly, if the ray arrow ③ or ④ enters the grid from the lower right to the upper left, the area BHIJ is divided, the RSTQ is used as the interpolation area for solving C_{mn}, the c_p distribution is fitted once, and the extrapolation in BHIJ is completed (Fig. 4.20).

On this basis, considering the problem of Lamb wave test RT in the aluminum-containing plate studied in this paper, the change of c_p of Lamb wave in an aluminum plate is concentrated only in the defect area. In the healthy area other than the defect, the c_p of the Lamb wave is almost unchanged, and the Lamb wave is approximately linearly propagated. Therefore, this paper proposes an adaptive method based on whether the 2 × 2 sub-grid region divided in the extrapolation process contains defects. Different methods are automatically selected for Lamb wave test RT: For the defect area, the Lamb wave test RT is performed using the algorithm based on the TDCSI method; for the healthy area, the ray path is quickly solved based on the direct ray model to reduce the calculation amount.

Using the blocking method, it is only necessary to solve a 16-element equation group for a sub-mesh region in the direction of the ray advancement, and the extrapolation of each step in the region can be completed. There is no need to solve a system of equations for each step in the region, so redundant calculations can be eliminated, saving computation time. Also, in the Lamb wave test RT process, in order to search for the minimum travel time of the EMAT, a large number of test shots are required at different launch angles, and each test shot will pass through a large number of healthy areas. Therefore, by using the adaptive method to identify the healthy sub-mesh region on the ray path and quickly solving the ray path based on the direct ray model, a large number of calculations caused by solving the equations,

∇c_p, and extrapolation can be avoided, thereby reducing the calculation amount and calculation time of the A0 wave RT.

Combining the TDCSI method and the adaptive block method, this paper proposes an improved Lamb wave test RT algorithm (from now on referred to as "improved test RT algorithm"). For the Lamb wave ray i between the ith pair of transceiving EMATs in the cross-hole model, the overall framework of the improved test RT algorithm is as follows:

(1) To judge the direction in which the ray is emitted from the previous 2 × 2 sub-mesh region and enters the grid ABCD in Fig. 4.3, and if it is from the lower left to the upper right, AEFG is extracted as a new sub-mesh region. Otherwise, extract BHIJ as the new sub-grid area.
(2) It is judged whether the boundary of the new sub-mesh region exceeds the boundary of the imaging region, and if so, it is translated into the imaging region so that the originally exceeded edge coincides with the boundary of the imaging region. The resulting 2 × 2 sub-mesh region is expanded into a 4 × 4 grid region. It is judged whether the boundary of the 4 × 4 grid area exceeds the boundary of the imaging area, and if so, it is translated into the imaging area so that the originally exceeded edge coincides with the boundary of the imaging area.
(3) Compare the c_p corresponding to each mesh in the 4 × 4 sub-grid area with the Lamb wave phase velocity in the healthy aluminum plate to determine whether the region contains defects. If yes, go to step (5) immediately; otherwise, go to step (4) immediately.
(4) Based on the direct ray model, the ray extrapolation is performed step by step in the 2 × 2 sub-grid region. Record the ray length $w(i, K)$ in the extrapolation of step K, the P_K point coordinate, and the phase velocity $c_p(i, K)$ of the A0 wave at P_K, and calculate the corresponding $c_g(i, K)$ according to the Lamb wave dispersion relationship. Step (9) is immediately executed until the new extrapolation point reaches or exceeds the 2 × 2 sub-grid area boundary.
(5) For the 4 × 4 grid area, the coefficient C_{mn} is solved.
(6) Solve the values of $c_p{}'$, $\partial c_p{}'/\partial x$, and $\partial c_p{}'/\partial y$ near the current extrapolation point P_K.
(7) Solve the coordinates of the new extrapolation point P_{K+1}, and record the ray length $w(i, K)$ extrapolated from the Kth step, the P_K point coordinate, and the phase velocity $c_p(i, K)$ of the A0 wave at the P_K. Moreover, calculate the corresponding $c_g(i, K)$ according to the Lamb wave dispersion relationship.
(8) It is judged whether P_{K+1} reaches the boundary of the 2 × 2 sub-mesh region. If yes, go to step (9) otherwise, go back to step (6).
(9) It is judged whether P_{K+1} has reached the boundary of the imaging area, and if not, return to step (1). If so, the extrapolation of the ray i is ended, and the measured travel time $T_{new}(i)$ of the ray i obtained using the proposed improved test RT algorithm is calculated according to the following equation:

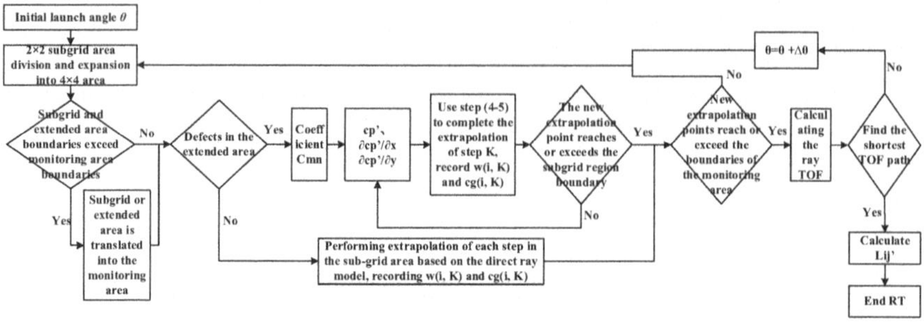

Fig. 4.21 Overall flowchart of the improved test RT algorithm

$$T\text{new}(i) = \sum_{K} \frac{w(i, K)}{c_g(i, K)} \tag{4.12}$$

(10) Let $\theta = \theta + \Delta\theta$, repeat steps (1) through (9) until you find the minimum travel path connecting the EMAT.

(11) For the propagation path of the finally obtained Lamb wave ray i, calculate the length L_{ij}' ($j = 1, 2, \ldots, M_s \times N_s$) of the ray i in the jth grid using the recorded $w(i, K)$ and P_K point coordinates.

The overall flow of the above improved test RT algorithm is shown in Fig. 4.21. It is worth noting that the 2 × 2 sub-mesh regions in step (5) is located at the center of the 4 × 4 interpolation region required to solve for the coefficient C_{mn}. The purpose of this setting is to avoid the accuracy of fitting in the sub-grid area due to large distortion at the edge of the fitted curved surface and low fitting accuracy. Also, it is known from the characteristics of the RT technology that when the top view of the defect is convex, the RT usually has relatively high accuracy. When the top view of the defect is concave, the accuracy of the RT is relatively low due to the large deflection angle of the radiation and the relatively complicated refractive process. Therefore, the improved test RT algorithm proposed in this paper is mainly for the common top view of the actual project, which is a convex defect, such as a round-hole-shaped defect.

In this paper, the performance of the improved test RT algorithm and the traditional test RT algorithm are compared by numerical simulation and experiment. In the numerical simulation, the two algorithms are written in MATLAB software, and the configuration of the computer platform used is Intel i5 (CPU 3.2 GHz, memory 8 GB). The size of the imaged area in the aluminum plate is 500 mm × 500 mm. The center is machined with a 1-mm-deep circular-hole-shaped artificial defect, and ten pairs of EMAT for excitation and reception are uniformly arranged on both sides, which are recorded as t1 to t10 and r1 to r10, respectively. The adjacent EMAT spacing is 50 mm. A0 wave in the aluminum plate used in the health and defect regions, where c_p were 2319 and 2039 m/s, and c_g were 3265 and 3109 m/s (Fig. 4.22).

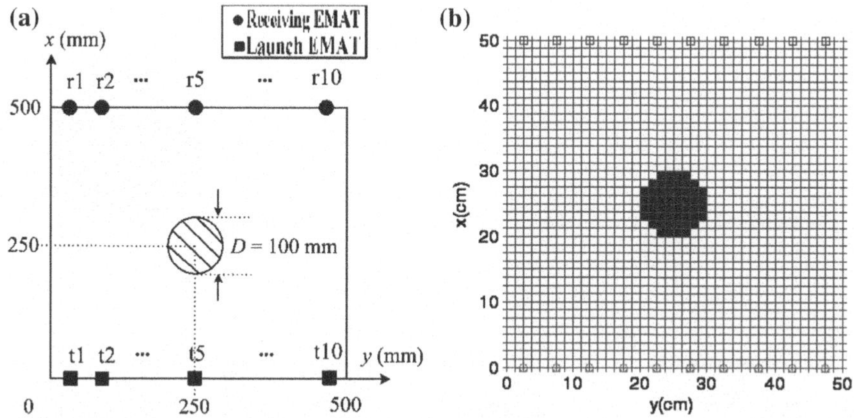

Fig. 4.22 Test setup and parameter diagram: **a** schematic diagram of test parameters; **b** schematic diagram of meshing of imaging area

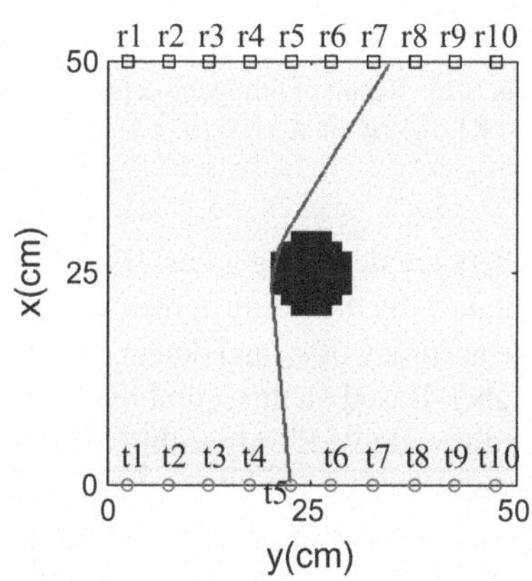

Fig. 4.23 A0 wave test RT numerical simulation results

In the numerical simulation, the imaging mentioned above area is evenly divided into 40 × 40 grids, the meshed A0 wave c_p distribution shown in Fig. 4.5b is obtained, and the shadow in the figure is the defect area. Considering that the step size Δx is too large, the accuracy of the RT is reduced, and if it is too small, the calculation amount is too large. To facilitate comparison and verification of the performance of the algorithm, Δx is 1/6 of the side length of the mesh.

Taking the gridded cp distribution as the input, let t5 be the emission position, and the initial emission direction and the y-axis positive angle are 95°. The improved test-time RT algorithm is used to simulate the A0 wave test RT (Figs. 4.23 and 4.24).

Taking k_4 as an example, the improved test RT algorithm is based on the TDCSI method to fit the distribution of the A0 wave c_p in k_4. Using the Lamb wave dispersion relationship shown in Fig. 2.4, the defect distribution in the aluminum plate can be obtained according to the fitted c_p distribution, wherein red is a healthy area, and the

Fig. 4.24 Numerical simulation results near the defect in Fig. 4.6

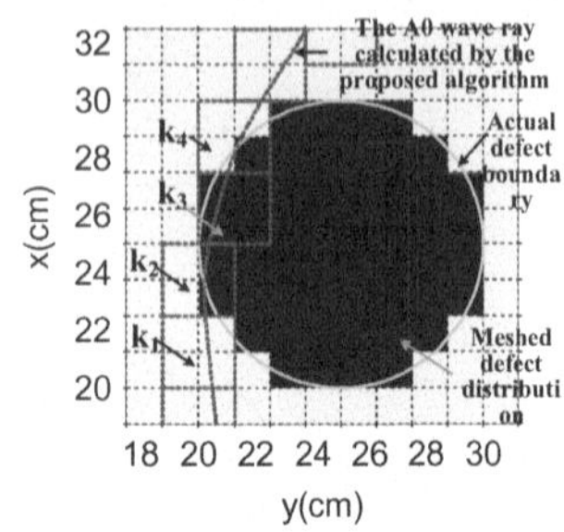

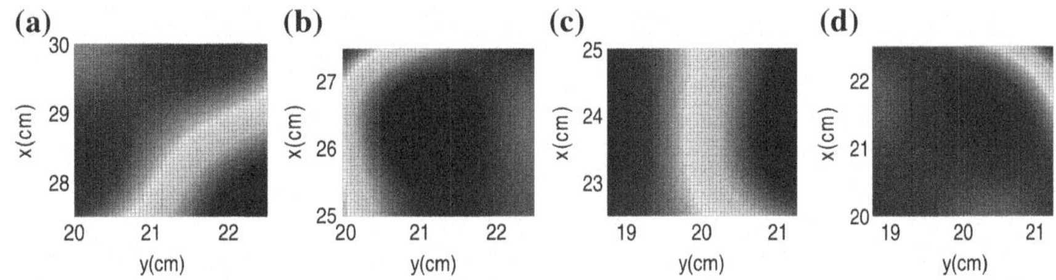

Fig. 4.25 Result of fitting the defect distribution in the divided sub-grid region using the improved test RT algorithm: **a** k_4; **b** k_3; **c** k_2; **d** k_1

others are defecting areas. By comparison, it can be found that the defect boundary obtained by the fitting agrees well with the actual defect boundary. This indicates that the accuracy of c_p and defect distribution fitted by the improved test RT algorithm is higher. Based on this, combined with the theoretical analysis of the limitations of the traditional test RT algorithm, the improved test RT algorithm is based on the above accurate fitting results to obtain the ∇c_p and ray paths, compared to the traditional test RT algorithm. The ∇c_p and ray propagation path are closer to the real situation and have higher accuracy. Similarly, the improved test RT algorithm fits the defect distributions in k_3, k_2, and k_1. The same conclusion can be obtained by comparing the obtained defect boundary with the actual defect boundary (Fig. 4.25).

By the qualitative analysis, the accuracy of the two algorithms is quantitatively compared. Taking t5 as the transmitting position and r1 to r10 as the receiving position, the numerical simulation is performed using the improved test RT algorithm and the traditional test RT algorithm (Fig.4.26). In the numerical simulation process, the calculated test shot RT algorithm and the traditional test shot RT algorithm calculate the travel time of the ray reaching the receiving position i as $T_{new}(i)$ and $T_{conv}(i)$ ($i = 1, 2, \ldots, 10$), respectively.

Figure 4.26 numerical simulation results were tested on an aluminum plate. The A0 wave is transmitted omnidirectionally using t5, and r1 to r10 are simultaneously received. Using the PWVD travel time extraction method, the measured travel time of the A0 wave ray arriving at the receiving position i is extracted from the test detection waveform and is recorded as $T_{exp}(i)$ ($i = 1, 2, \ldots, 10$). Comparison results of $T_{exp}(i)$, $T_{new}(i)$, and $T_{conv}(i)$ is shown in Fig.4.27a. In order to quantitatively compare the

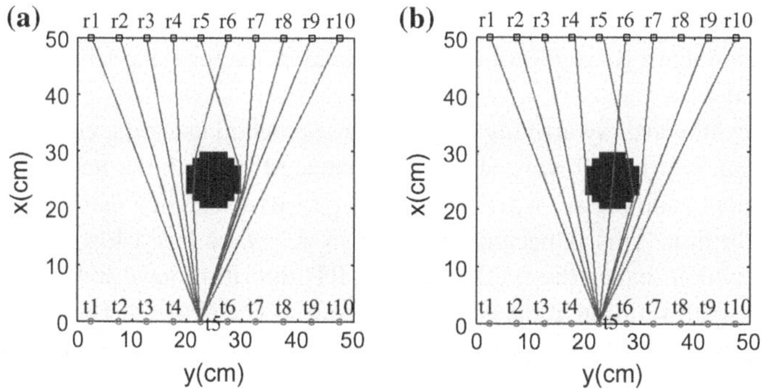

Fig. 4.26 Numerical simulation results of two algorithms: **a** improved test RT algorithm; **b** traditional test RT algorithm

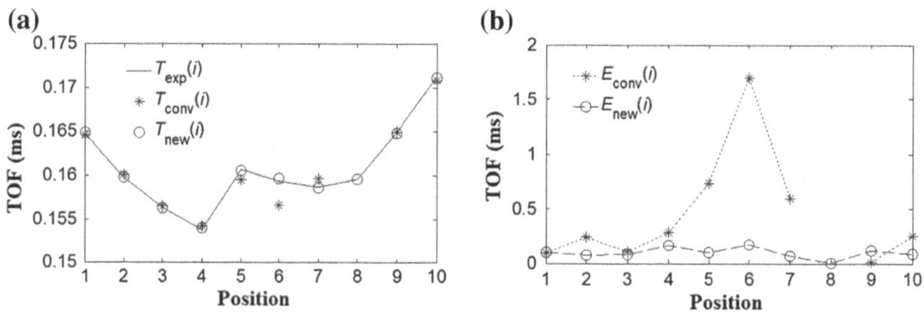

Fig. 4.27 Comparison of travel time with t5 as the launch position: **a** travel time based on test and numerical simulation; **b** comparison of travel time relative error obtained by numerical simulation using two algorithm numbers

accuracy of the two algorithms for test RT and calculate the travel time of the ray, the relative error of the travel time $T_{\text{new}}(i)$ and $T_{\text{conv}}(i)$ relative to the measured travel time $T_{\text{exp}}(i)$ is defined as:

$$E_{\text{new}}(i) = |T_{\text{exp}}(i) - T_{\text{new}}(i)|/T_{\text{exp}}(i) \tag{4.13}$$

$$E_{\text{conv}}(i) = |T_{\text{exp}}(i) - T_{\text{conv}}(i)|/T_{\text{exp}}(i) \tag{4.14}$$

The comparison between $E_{\text{new}}(i)$ and $E_{\text{conv}}(i)$ is shown in Fig. 4.27b.

It can be seen from the relative position of the EMAT that the A0 wave emitted from t5 to r8 will pass through the defect, while the other rays do not pass through the defect. Based on this, when the ray does not pass through the defect, the relative error of using the two algorithms to calculate the A0 wave travel time is small. When the ray passes through the defect, the relative error of the A0 wave travel time calculated by the traditional test RT algorithm is much larger than the calculated relative error of the improved test RT algorithm. It can be seen that the accuracy of the A0 wave ray travel time calculated by the improved test RT algorithm is higher than that of the conventional test RT algorithm, which indicates that the obtained ray path is also more accurate.

Also, the A0 wave ray arriving at r8 was not searched using the conventional test RT algorithm, but the A0 wave signal was detected by r8 in the test results on the aluminum plate, and the A0 wave ray reaching r8 was searched using the improved test RT algorithm. This is because the known A0 wave has a low c_p distribution resolution, which makes the traditional test RT algorithm have lower accuracy of ∇c_p and ray path solution at the edge of the defect. The above test results show that if the traditional test RT algorithm is used for bending ray tomography, there will be a phenomenon that the RT and tomography processes fail due to the failure of the ray search. The improved test RT algorithm has relatively high ray path solving accuracy, so it has high reliability. If it is used for bending ray tomography, it can effectively avoid the phenomenon of ray search failure. Thereby, the reliability of the bending ray tomography can be improved.

4.2.3 *Electromagnetic Ultrasonic Bending Ray Lamb Wave Cross-Hole Tomography Imaging Method Based on Test RT*

In order to improve the performance of tomography, and to avoid the failure of Lamb wave test RT and tomography, the improved test launch RT algorithm with higher reliability proposed in this paper is introduced into electromagnetic ultrasonic Lamb wave transforaminal tomography. A new electromagnetic ultrasonic bending ray Lamb wave cross-hole tomography method (from now on referred to as "bending ray imaging method") is proposed.

In the bending ray imaging method, first, the Lamb wave c_p distribution in the imaging region is calculated using the improved direct ray imaging method. Subsequently, using the c_p distribution as an input, the matrix (L_{ij}') is recalculated using the improved test RT algorithm. Finally, the modified matrix (L_{ij}') is used to solve the S_j distribution of Lamb waves in each mesh of the imaging region.

The specific steps of the curved ray imaging method are as follows:

(1) Using the selected new EMAT for excitation, emit the Lamb wave of the modality used at a specific operating frequency, simultaneously receive all the new EMATs for the reception, and record the detection waveform of each EMAT.
(2) Using the proposed PWVD TOF extraction method, extracting the TOF T_i of the modal Lamb wave used from each detected waveform.
(3) If all the new EMATs have been subjected to Lamb wave emission, proceed to step (4). Otherwise, select a new EMAT for excitation and return to step (1).
(4) Calculating a matrix (L_{ij}) based on a direct ray model.
(5) Using all of T_i and L_{ij} as input, using a modified direct ray imaging method to solve the distribution of the modal Lamb wave c_p used.

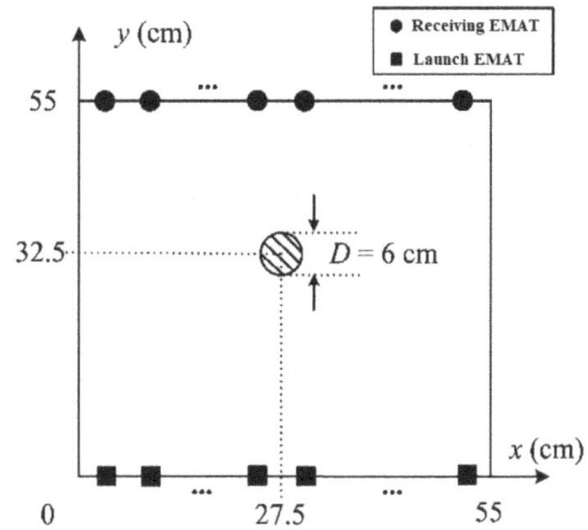

Fig. 4.28 Schematic diagram of the parameters of the single-defect aluminum plate imaging test

(6) Using the distribution of c_p as the input, using the improved test RT algorithm, and solving the matrix (L_{ij}').
(7) Using all T_i and L_{ij}' as input, the distribution of the slowness S_j of the modal Lamb wave used is solved based on SIRT.
(8) Using the equation $cg_j = 1/S_j$ and the Lamb wave dispersion relationship, the distributions of cg_j, cp_j, and d_j of the used modal Lamb waves in each grid are calculated.

To verify the performance of the curved ray imaging method, a tomographic test was performed on a 3-mm-thick aluminum plate using this method. For comparison, a tomographic test was performed under the same test conditions using a modified direct ray imaging method. The imaging test was carried out in a rectangular area of 550 mm × 550 mm on an aluminum plate, which was evenly divided into 110 × 110 small grids. The round-hole-shaped artificial defect has a depth of 1 mm, and its diameter D and center position are shown in Fig. 4.28. Twelve excitation EMATs and 12 reception EMATs were evenly arranged on both sides of the imaging area. The iterative solution threshold T_{TH} of the SIRT algorithm is set to 0.05%, and the relaxation factor is taken as $\mu_I = 0.1$. The extrapolation step for ray tracing is 1/6 of the grid side length.

First, tomography using the improved direct ray imaging method: 1 Using the geometric relationship, solve the A0 wave ray path based on the direct ray model, i.e., solve (L_{ij}). The dotted line in Fig. 4.28 is the contour of the actual defect; 2 based on the obtained straight ray path, the improved direct ray imaging method is used to solve the distribution of the A0 wave cg_j and the distribution of the A0 wave cp_j, wherein the obtained A0 wave cg_j distribution is shown in Fig.4.29.

Second, tomography was performed using a curved ray imaging method. Considering steps (1)–(5) of the curved ray imaging method, imaging is performed using the improved direct ray imaging method to obtain a distribution of the A0 wave c_p. When imaging is performed using the curved ray imaging method, the subsequent steps are performed directly using the above-described A0 wave c_p distribution obtained using

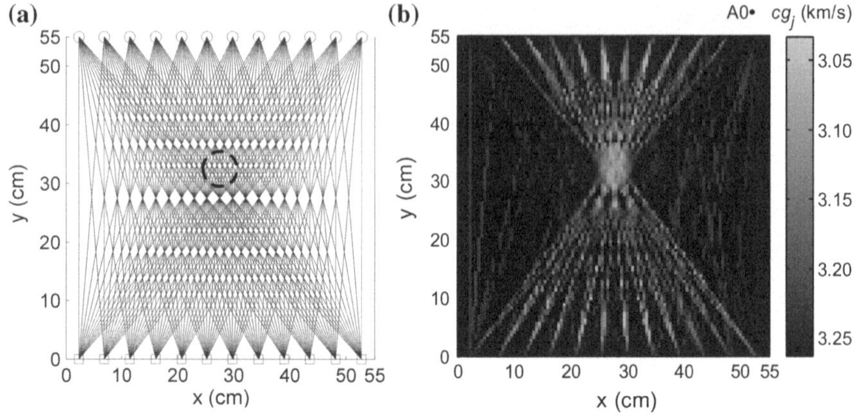

Fig. 4.29 Imaging results obtained using the improved direct ray imaging method: **a** A0 wave ray path based on the direct ray model; **b** A0 wave cg_j distribution

the improved direct ray imaging method. That is: 1 based on the above-obtained A0 wave cp_j distribution, using step (6) in the curved ray imaging method, that is, using the improved test RT algorithm to solve the A0 wave ray path, the dotted line in the figure is the outline of the actual defect; 2 based on the ray path, the A0 wave cg_j distribution is solved using steps (7) and (8) in the curved ray imaging method .

The A0 wave ray path connecting all EMAT pairs was searched using the improved test RT algorithm, and the resulting ray was bent as it passed through the defect. These curved ray paths are closer to the true A0 wave ray path than the straight ray path. This shows that the ray path and matrix (L_{ij}') obtained using the improved test RT algorithm have higher accuracy than the ray path and matrix (L_{ij}) obtained based on the direct ray model. Also, the reconstruction results contain more artifacts, which are mainly distributed along the ray path, while the reconstruction results are shown in Fig. 4.30b containing only a small number of artifacts. This indicates that the accuracy of the A0 wave cg_j distribution obtained using the curved ray imaging method is higher than that of the improved direct ray imaging method. It can be seen from the above results that the proposed bending ray tomography method improves the accuracy of the ray path solution by using the improved test RT algorithm, effectively improves the accuracy of the A0 wave cg_j distribution solution, and suppresses the generation of interference artifacts, thereby effectively improving the quality of image reconstruction.

According to the actual plate thickness distribution, the theoretical A0 wave cg_j curve at $y = 325$ mm in the aluminum plate can be calculated by using the Lamb wave dispersion relationship. The theoretical cg_j of the 290 kHz A0 wave used in the healthy and defective areas of the aluminum plate used in the test was 3265 and 3109 m/s, respectively. Compared to the measured cg_j curve obtained based on the improved direct ray imaging method, the measured cg_j curve obtained based on the curved ray imaging method contains less noise and is more consistent with the

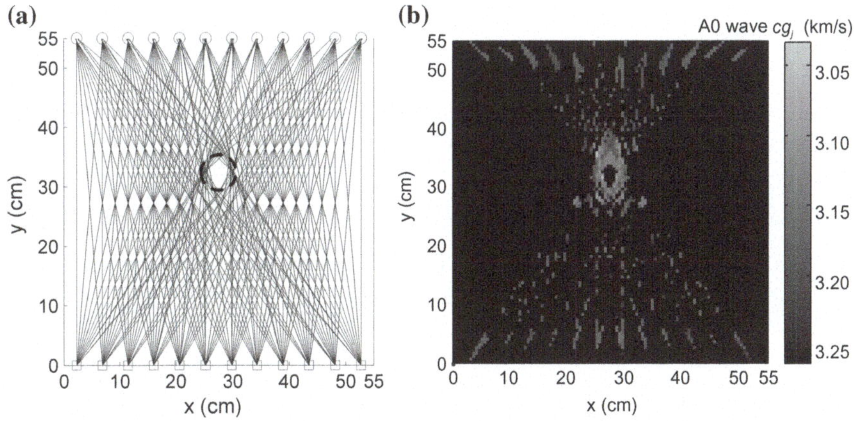

Fig. 4.30 Imaging results obtained using the curved ray imaging method: **a** A0 wave ray path obtained using the improved test RT algorithm; **b** A0 wave cg_j distribution

Fig. 4.31 A0 wave cg_j curve at $y = 325$ mm obtained using the improved direct ray imaging method

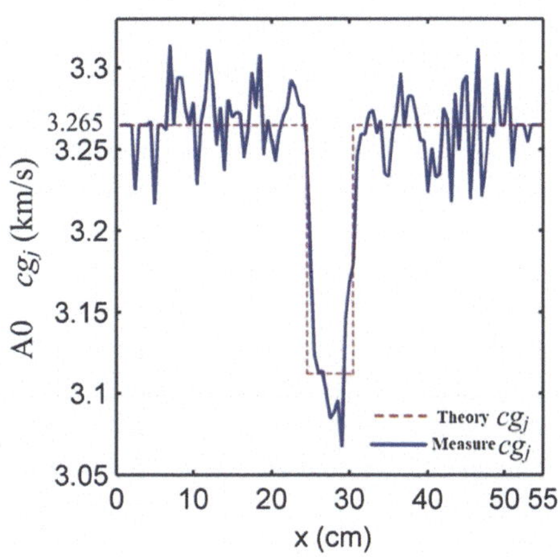

theoretical cg_j curve. Therefore, using the measured cg_j curve obtained based on the bending ray imaging method helps to more accurately determine the position of the defect boundary, thus contributing to more accurate estimation of the diameter of the defect (Figs. 4.31 and 4.32).

In order to compare the accuracy obtained by using two imaging methods for defect feature information extraction, the diameter of the defect is estimated according to the following methods and steps:

(1) Compare the measured cg_j curve with a preset threshold, and combine the defect distribution to roughly determine the position of the defect area.
(2) Find the intersection of the main rising and falling edges of the measured cg_j curve near the defect boundary with the straight line cg_j= 3265 m/s (the healthy region theory cg_j value).

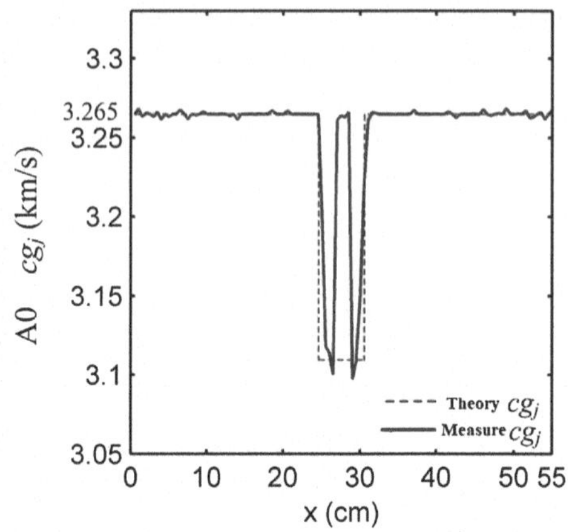

Fig. 4.32 A0 wave cg_j curve at $y = 325$ mm obtained using the bending ray imaging method

(3) Calculate the distance between the two intersection points, that is, obtain the estimated defect diameter.

The defect diameter estimated using the measured cg_j curve is approximately 78 mm, and the absolute and relative errors of the estimate are 18 mm and 30%, respectively. The defect diameter estimated using the measured S_j curve is approximately 69 mm, and the corresponding absolute and relative errors are 9 mm and 15%, respectively. The estimation results show that the accuracy of the defect feature information extraction can be further improved by using the curved ray imaging method compared to the improved direct ray imaging method. The reduction in the defect diameter estimation error is: $|9 - 18|/18 = 50\%$.

To further verify the performance of the curved ray imaging method, experiments were carried out on 3-mm-thick aluminum plates with double circular-hole-shaped artificial defects, and an improved direct ray imaging method was used as a control. The test was carried out in a rectangular area of 520 mm × 520 mm on an aluminum plate, which was evenly divided into 104 × 104 small grids. The depths of defect 1 and defect 2 are 1 and 1.5 mm, respectively. Twelve excitation EMATs and 12 reception EMATs were evenly placed on either side of the imaging area. The iterative solution threshold T_{TH} of the SIRT algorithm is set to 0.05%, and the relaxation factor is taken as $\mu_{\mathrm{I}} = 0.1$. The ray tracing extrapolation step is 1/6 of the grid side length. The test procedure is the same as the above-described single-defect aluminum plate test: 1 using the improved direct ray imaging method for tomography; 2 using the A0 wave c_{p} distribution obtained in step 1 to complete the subsequent step of the curved ray imaging (Fig. 4.33).

The A0 wave ray path connecting all EMAT pairs was searched using the improved test RT algorithm, and the resulting ray was bent as it passed through the defect. These curved ray paths are closer to the true A0 wave ray path than the straight ray path. When the improved direct ray imaging method is used, more artifacts appear in the imaging results of the double-defect aluminum plate compared to the imaging results of the single-defect aluminum plate. Also, in the imaging results of

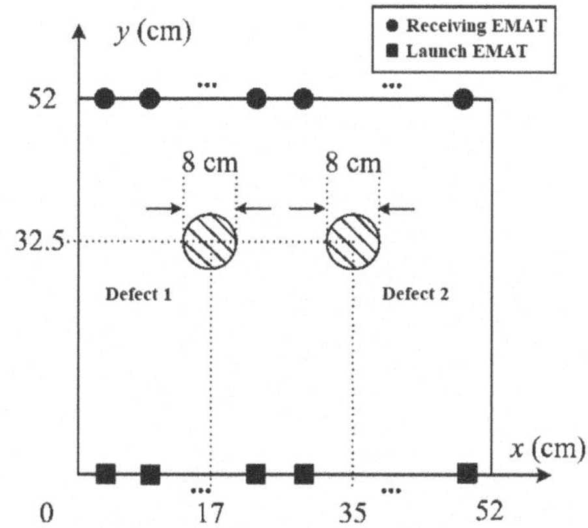

Fig. 4.33 Schematic diagram of the parameters of the double-defect aluminum plate imaging test

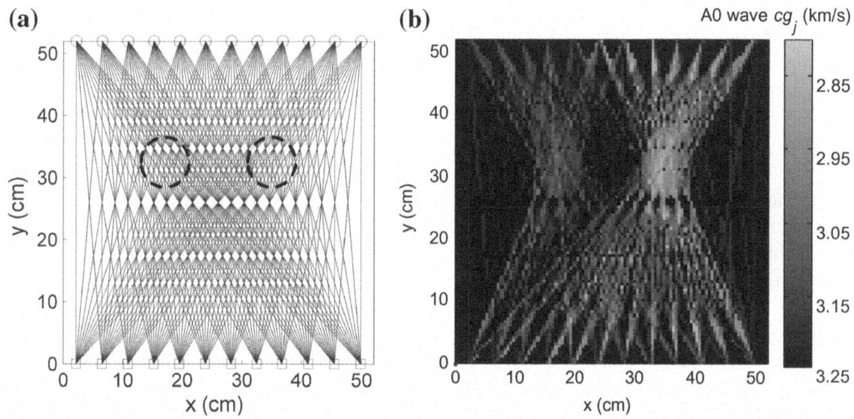

Fig. 4.34 Imaging results obtained using a modified direct ray imaging method for a double-defect aluminum plate: **a** A0 wave ray path based on the direct ray model; **b** A0 wave cg_j distribution

the double-defect aluminum plate, the defect 1 having a relatively shallow depth is almost drowned by the image. From the above results, it is known that the proposed bending ray tomography method can improve the accuracy of the ray path solution for the double-defect aluminum plate, effectively improve the accuracy of the A0 wave cg_j distribution solution, and suppress the generation of interference artifacts, thereby effectively improving the quality of image reconstruction (Figs. 4.34 and 4.35).

According to the actual plate thickness distribution, using the Lamb wave dispersion relationship [7–10], the theoretical A0 wave cg_j curve at $y = 325$ mm in the aluminum plate can be calculated. The theoretical cg_j of the 290 kHz A0 wave used in the health, defect 1, and defect 2 regions of the aluminum plate used for the test were 3265, 3109, and 2946 m/s, respectively. When imaging a double-defect aluminum plate, the measured cg_j curve based on the curved ray imaging method contains less

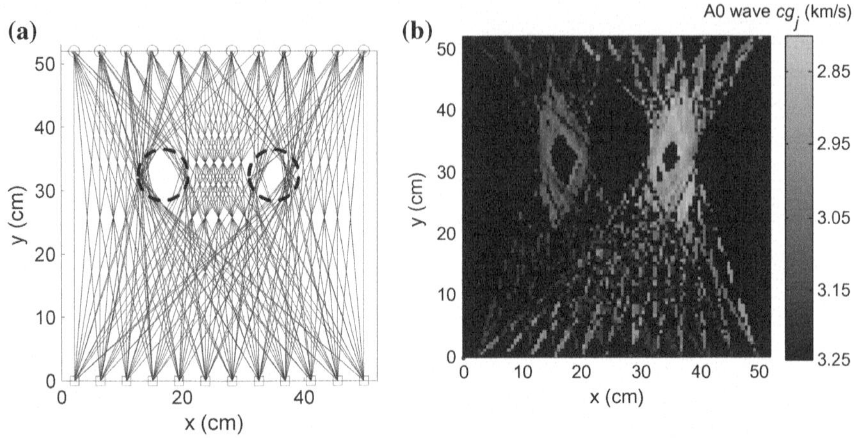

Fig. 4.35 Imaging results obtained using a curved ray imaging method for a double-defect aluminum plate: **a** A0 wave ray path obtained using the improved test RT algorithm; **b** A0 wave cg_j distribution

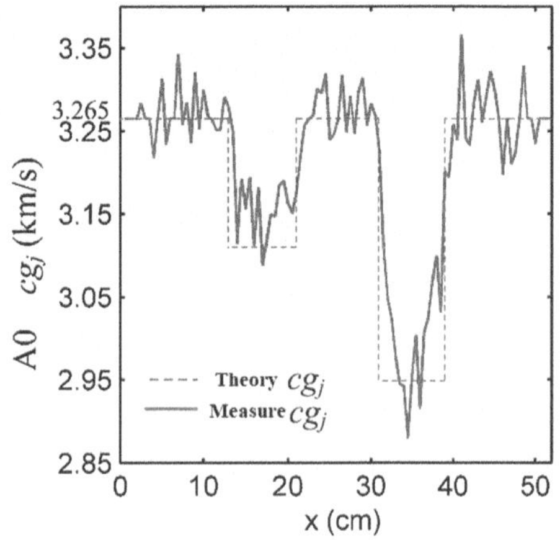

Fig. 4.36 A0 wave cg_j curve at $y = 325$ mm obtained using a modified direct ray imaging method for a double-defect aluminum plate

noise and is more consistent with the theoretical cg_j curve than the measured cg_j curve obtained based on the improved direct ray imaging method (Figs. 4.36 and 4.37).

The estimated defect 1 has a diameter of approximately 94 mm, and the absolute and relative errors of the estimate are 14 mm and 18%, respectively. The estimated defect 2 has a diameter of approximately 97 mm, and the corresponding absolute and relative errors are, respectively, 17 mm and 21%. The diameter of defect 1 estimated using the measured cg_j curve is approximately 89 mm, and the absolute and relative errors of the estimate are 9 mm and 11%, respectively. The estimated defect 2 has a diameter of approximately 86 mm, and the corresponding absolute and relative errors are 6 mm and 8%, respectively. The above estimation results

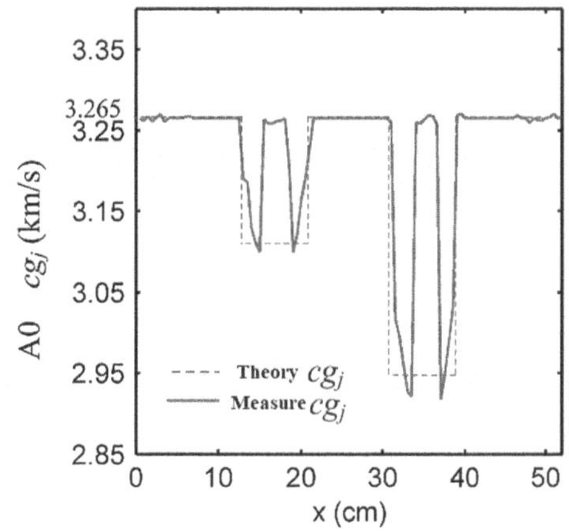

Fig. 4.37 A0 wave cg_j curve at $y = 325$ mm obtained using a bending ray imaging method for a double-defect aluminum plate

indicate that when the double-defect aluminum plate is imaged, it is compared to the improved direct ray imaging method. The bending ray imaging method can improve the accuracy of defect feature information extraction. The reduction of defect 1 and defect 2 diameter estimation errors are $|11 - 18|/18 = 39\%$ and $|8 - 21|/21 = 62\%$, respectively.

References

1. S. Wang, S. Huang, W. Zhao, Z. Wei, 3D modeling of circumferential SH guided waves in pipeline for axial cracking detection in ILI tools. Ultrasonics **56**, 325–331 (2015)
2. P. Huthwaite, R. Ribichini, P. Cawley et al., Mode selection for corrosion detection in pipes and vessels via guided wave tomography. IEEE Trans. Ultrason. Ferroelectr. Freq. Control **60**(6), 1165–1177 (2013)
3. P. Belanger, Feasibility of thickness mapping using ultrasonic guided waves [D] (Imperial College, 2009)
4. Z. Wei, S. Huang, W. Zhao, S. Wang, A large ratio data compression method for pipeline inspection based on EMAT-generated guided wave. Int. J. Appl. Electromagnet. Mech. **45**, 511–517 (2014)
5. S. Huang, Z. Wei, S. Wang, W. Zhao, A new omni-directional EMAT for ultrasonic lamb wave tomography imaging of metallic plate defects. Sensors **14**(2), 3458–3476 (2014)
6. E.V. Malyarenko, M.K. Hinders, Ultrasonic lamb wave diffraction tomography. Ultrasonics **39**, 269–281 (2001)
7. S. Wang, S. Huang, W. Zhao et al., Approach to lamb wave lateral crack quantification in elastic plate based on reflection and transmission coefficients surfaces. Res. Nondestr. Eval. **21**(4), 213–223 (2010)
8. Y. Zhang, S. Wang, S. Huang et al., Mode recognition of lamb wave detecting signals in metal plate using the hilbert-huang transform method. J. Sens. Technol. **5**, 1–14 (2015)

9. S. Huang, Y. Zhang, S. Wang et al., Multi-mode electromagnetic ultrasonic lamb wave tomography imaging for variable-depth defects in metal plates. Sensors **16**(5), 628, 1–10 (2016)
10. Y. Zhang, S. Huang, S. Wang et al., Recognition of overlapped lamb wave detecting signals in aluminum plate by EMD-based STFT flight time extraction method. Int. J. Appl. Electromagnet. Mech. **52**(3–4), 991–998 (2016)

Chapter 5
Guided Wave Electromagnetic Ultrasonic Scattering Imaging

5.1 Directional Emission-Omnidirectional Reception Guided Wave Scattering Imaging Using the Magnetic Acoustic Array

At present, there are few studies on the defect imaging using guided wave scattering method. Most of the defect detection based on guided wave scattering is limited to the location of the defects, and the defects studied are mostly artificial defects of standard contour shape. The researches on defect imaging utilized transmitted guided waves. However, (1) the analysis and processing of the guided wave detection signal are greatly affected by the scattering effect of the guided wave after encountering the defect, resulting in a large error in the characteristics of the guided wave, amplitude, and other guided wave signals extracted from the guided wave detection signal. This has a negative impact on the accuracy of defect imaging; (2) the defect size accounts for a small proportion of the transmitted ray, and the propagation delay caused by the variation of the wave speed caused by the thickness reduction in the defect is small. In particular, when the distance between the two transducers is relatively long or the guided wave propagation path is long, the propagation delay caused by the thickness reduction in the defect is smaller, which has a serious negative impact on the imaging accuracy of the defect; (3) for the defects with shallow depth and clear outline, the variation of the thickness of the material is not obvious, and the transmission guided wave imaging method based on the thickness map has lower accuracy for the contour of this type of defect. In addition, studies on the interaction between guided waves and defects show that the signal energy of the guided wave and the non-permeability type are mostly concentrated on the scattered wave, and the signal energy is mostly concentrated on the transmitted wave after the type of the defect. Most of the defects to be detected in the actual inspection project are non-permeable, and the depth is less than the percentage of the thickness of the structural member. The main reason is that the safety requirements of industrial structural parts are increasingly strict to achieve the purpose of preventing micro-duplication. However, the signal energy of

S. Huang et al., *Theory and Methodology of Electromagnetic Ultrasonic Guided Wave Imaging*, https://doi.org/10.1007/978-981-13-8602-2_5

the guided wave and such non-permeability defects are mostly concentrated on the scattered wave. Therefore, the use of guided wave scattering for defect imaging is necessary and reasonable for the actual defect detection engineering.

The existing guided wave scattering defect reconstruction model needs to use the guided wave detection signal when the plate to be tested is free of defects and the guided wave scattering signal when there is a defect, and it is necessary to perform a guided wave detection when the plate to be tested is free from defects. This is not possible in many actual inspection projects. Because the defect is already in its specific position when the board is to be tested, it is impossible to obtain the guided wave detection signal when the board to be tested is free from defects. It is difficult to apply this guided wave scattering defect reconstruction model in engineering practice. The comparison of the signals of the two guided wave detection signals also reduces the detection efficiency of the guided waves to some extent, and it is easy to introduce a variety of unknown interference factors. Including the consistency of the transducer position, the transducer excitation and the consistency of the detection parameters are not conducive to providing accurate guided wave detection data [1–7].

The contour and depth of the defect are important features of the defect itself and are an important basis for quantitative evaluation of defects and safety assessment of structural parts. In some engineering inspection situations, the depth of defects is more concerned, which is expressed as a quantitative evaluation of the wall thickness or thickness reduction of structural parts. However, there are some engineering inspection situations that pay more attention to defect contours, especially for the type of defects with shallow depth but a large area. The defect contour is the focus of attention. Also, the shallower type of defects is not easily evaluated by the traditional transmission guided wave detection imaging method. At this time, the defect profile can better reflect the real defect and the health of the structural parts. Relative to the defect depth, the defect profile can provide more information about the defect, which is the direct basis for judging the defect formation mechanism, solving the defect expansion rate, and degrading the defect type. The high-precision reconstruction of the actual complex defect profile is the key link for maintenance planning and life prediction of structural parts such as steel plates. It is of great significance for timely and effective detection of structural parts and ensuring the safe and stable operation of structural parts. Therefore, the directional emission-omnidirectional receiving magnetic acoustic array guided wave scattering imaging method is proposed in this chapter and the omnidirectional emission-omnidirectional receiving magnetic acoustic array guided wave scattering imaging method in the next chapter. Both are reconstruction and imaging models and methods for the defect profile of steel plates, and the quantification of defect depth is not considered in this chapter.

Based on the basic principle of guided wave scattering, when the guided wave encounters a strong degree of scattering of the defect, the boundary between the defect and the guided wave acts. That is, the interface between the structural member medium and the air medium is macroscopically represented as a contour of the defect. Therefore, the scattering angle of the guided wave and the distribution of the scattering field in the structural member are determined by the shape of the defect contour in the local region of the scattering, and the depth of the defect affects only the proportion of the guided wave in the incident guided wave. Therefore, the guided wave scattering detection method has a close correlation relationship with the defect contour, so it is logical and scientific to use the guided wave scattering detection method to detect the defect contour. The guided wave scatter imaging detection method is well suited for reconstruction and imaging of defect profiles.

5.1.1 Directional Emission-Omnidirectional Reception Magnetic Acoustic Array Guided Wave Scattering Imaging Model

This section is based on the guided wave scattering and guided wave scattering detection signals, combined with the direction-controllable emission EMAT to excite the known a priori conditions of the SH guided wave direction, considering the guided wave propagation and scattering paths. A directional transmission-omnidirectionally received magnetic acoustic array guided wave scattering defect contour imaging model was established.

The actual defect contours are different in shape, so it is desirable to establish a magneto-acoustic array guided wave scattering imaging model suitable for contour reconstruction of arbitrary shape defects, rather than only having high sensitivity for reconstruction of regular contour shape defects. For the proposed directional emission-omnidirectional receiving magnetic acoustic array guided wave scattering imaging model, consider the guided wave excitation, propagation, scattering, and receiving processes as shown in Fig. 5.1.

Figure 5.1 takes the directional emission-omnidirectional receiving magneto-acoustic array guided wave scattering imaging model as an example to consider the process of guided wave excitation, propagation, scattering, and reception. However, the basic principle is not limited by the form of the magneto-acoustic array transmission or reception, and the emphasis is on the establishment of the scattering model. As shown in Fig. 5.1, the direction controllable emission EMAT excites the incident ultrasonic guided wave in a certain direction. The incident ultrasonic guided wave propagates along the excitation direction, and after encountering the defect, it acts on the defect profile and scatters out to the outside of the defect. The scattered guided wave propagates through a certain distance and is received by the omnidirectional receiving EMAT. The dashed boundary of the shaded portion in Fig. 5.1 represents the boundary of the ultrasonic guided sound field distribution, so only the omnidirectional receiving EMAT that received the scattered guided wave is indicated. Moreover, the omnidirectional receiving EMAT that receives the scattered guided wave is located in the coverage area of the scattered guided sound field, and the omnidirectional receiving EMAT that does not receive the scattered guided wave is not indicated in Fig. 5.1.

The defect contour scattering guided wave sound field covers a wide range, and the omnidirectional receiving EMAT receives only the scattered guided waves of the local area occupied by the EMAT in the range of the fault scattering guided sound field coverage due to its size limitation. The scattered guided waves outside the local area occupied by the omnidirectional receiving EMAT itself cannot be received. Based on the above facts, in order to facilitate the quantitative modeling study, the propagation process of the scattered guided wave received by the omnidirectional receiving EMAT and its corresponding incident guided wave is abstracted into an incident guided beam and a scattered guided beam. As indicated by the solid black line with an arrow in Fig. 4.1, the direction of the arrow indicates the direction of propagation of the guided wave. Thus, the intersection of the incident guide beam and the scattered guide beam is a position point at which the incident guide beam changes direction, that is, a position point at which the incident guide beam is scattered. Moreover, the point must be located at a spatial location on the local defect profile that scatters the incident guide beam. Therefore, the point defining the above feature is a scattering point. As described above, the scattering point is a point on the defect profile that scatters the incident beam and feeds the resulting scattered beam into the omnidirectional receiving EMAT. It can be seen that the scattering point is the intersection of the incident guiding beam and the scattered guiding beam, and the scattering point must be located on the defect contour. Thus, if the specific positions of the incident guide beam and the scattered guide beam are determined, the specific position of the scattering point can be determined; at this time, the multi-angle omnidirectional guided wave incidence and scattering detection is performed on the defect. If a sufficient number of scattering points on the defect contour can be obtained, the

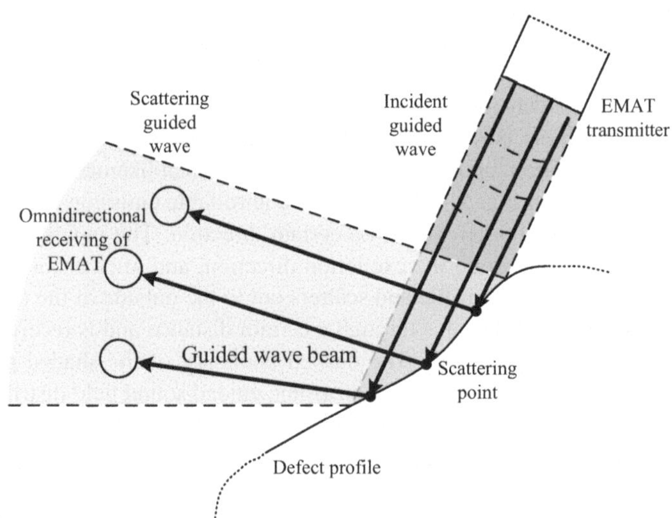

Fig. 5.1 Schematic diagram of guided wave excitation, propagation, scattering, and reception under directional emission conditions

scattering point image of the defect contour can be formed, and the defect contour curve or image can be obtained by fitting the position of the scattering point. The above idea is the basic principle and idea of the guided contour magneto-acoustic guided wave scattering imaging model proposed in this chapter. The defect contour is characterized by scattering points, and the defect contour is imaged by solving the position of the scattering point. The scattering points defined in this chapter establish the relationship between the defect profile and the guided wave scattering process. It is an important bridge connecting the defect contour and the input guided wave scattering detection signal, which is of great significance for the establishment of the guided acoustic scattering imaging model of the magneto-acoustic array.

The key to the guided contoured magnetic acoustic array guided wave scattering imaging model proposed in this chapter is the solution of the scattering point position. The directional emission-omnidirectional receiving magnetic acoustic array guided wave scattering imaging model is considered below, and the model and method for solving the scattering point position under directional emission are established.

The guided wave excitation, propagation, scattering, and reception processes shown in Fig. 5.1 are abstracted into a plane rectangular coordinate system, as shown in Fig. 5.2. In the plane rectangular coordinate system, the position of the direction-controllable emission EMAT is T, and direction of excitation of the ultrasonic waveguide is $\overrightarrow{TP}$. The position of the omnidirectional receiving EMAT receiving the scattered guided wave is R; the position of the scattering point is assumed to be P. Then, the incident guide beam is expressed as $\overrightarrow{TP}$, and the scattering guide beam is expressed as $\overrightarrow{PR}$. The distance between the direction-controllable emission EMAT position and the omnidirectional receiving EMAT position is $\left|\overrightarrow{RT}\right|$.

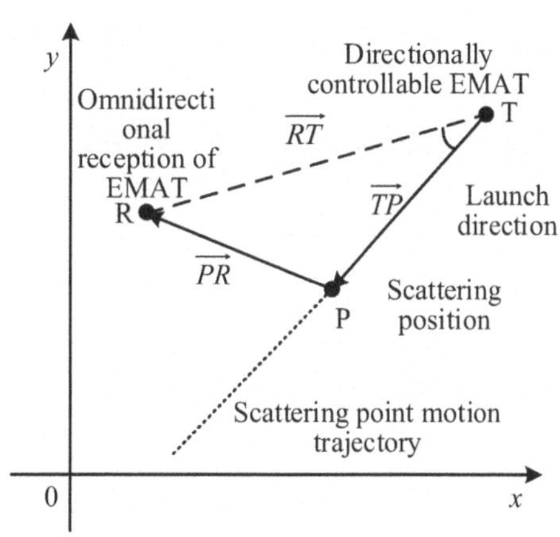

Fig. 5.2 Schematic diagram of the scatter target location model for directional emission-omnidirectional receiving magnetic acoustic array guided wave scattering imaging

In the model described in Fig. 5.2, the known quantities are the position T of the direction-controllable emission EMAT, the position R of the omnidirectional receiving EMAT, guide wave excitation direction $\overrightarrow{TP}$, the angle $\angle RTP$ between vector $\overrightarrow{RT}$ and vector $\overrightarrow{TP}$, the distance between the controlled EMAT and the omnidirectional receiving EMAT $\left|\overrightarrow{RT}\right|$, the propagation velocity of the ultrasonic guided wave in the medium v, the omnidirectional receiving EMAT receives the scattered guided wave detection signal TOF t_r. The amount to be determined is the position P of the scattering point. The a priori condition is that the scattering point position P is located in the direction of the guided wave emission of the direction-controllable emission EMAT; that is, the motion trajectory of the scattering point is the ray TP. According to the comprehensive demand and the prior condition, since the scattering point P must be located on the ray TP and the ray TP direction is known, only the length of the vector $\left|\overrightarrow{TP}\right|$ is required, and the position of the scattering point P can be uniquely determined. Therefore, in the model described in Fig. 4.2, the problem of solving the position of the scattering point P is transformed into the problem of solving the length.

The ultrasonic guided wave is first excited by the direction-controllable emission EMAT of the T position and is scattered by the scattering point P and received by the omnidirectional receiving EMAT of the R position. The propagation distance S experienced by it is the sum of the length of the vector $\overrightarrow{TP}$ and the length of the vector $\overrightarrow{PR}$, and $S = \left|\overrightarrow{TP}\right| + \left|\overrightarrow{PR}\right|$, and the relationship between the propagation distance and the propagation velocity v, and the scattered guided wave detection signal TOF t_r received by the omnidirectional receiving EMAT is as follows:

$$S = \left|\overrightarrow{TP}\right| + \left|\overrightarrow{PR}\right| = v * t_r \tag{5.1}$$

where $v * t_r$ is a known value. Therefore, in ΔTPR, according to the cosine theorem, $\left|\overrightarrow{PR}\right|$ can be expressed as $\left|\overrightarrow{TP}\right|$, $\left|\overrightarrow{RT}\right|$, and $\angle RTP$.

$$\left|\overrightarrow{PR}\right| = \sqrt{\left|\overrightarrow{TP}\right|^2 + \left|\overrightarrow{RT}\right|^2 - 2 * \left|\overrightarrow{TP}\right| * \left|\overrightarrow{RT}\right| * \cos(\angle RTP)} \tag{5.2}$$

In the expression of the above formula, only $\left|\overrightarrow{TP}\right|$ is an unknown quantity. Therefore, the left side of Eq. (5.1), that is, the propagation distance S experienced by the guided wave, can be expressed as $S(\left|\overrightarrow{TP}\right|)$

$$S(\left|\overrightarrow{TP}\right|) = \left|\overrightarrow{TP}\right| + \sqrt{\left|\overrightarrow{TP}\right|^2 + \left|\overrightarrow{RT}\right|^2 - 2 * \left|\overrightarrow{TP}\right| * \left|\overrightarrow{RT}\right| * \cos(\angle RTP)} \tag{5.3}$$

In order to solve the independent variable $\left|\overrightarrow{TP}\right|$ in the function $S(\left|\overrightarrow{TP}\right|)$, the results of the derivation $\frac{\mathrm{d}S(\left|\overrightarrow{TP}\right|)}{\mathrm{d}\left|\overrightarrow{TP}\right|}$ are as follows:

$$\frac{\mathrm{d}S(\left|\overrightarrow{TP}\right|)}{\mathrm{d}\left|\overrightarrow{TP}\right|} = 1 + \frac{\left|\overrightarrow{TP}\right| - \left|\overrightarrow{RT}\right| * \cos(\angle RTP)}{\sqrt{\left|\overrightarrow{TP}\right|^2 + \left|\overrightarrow{RT}\right|^2 - 2 * \left|\overrightarrow{TP}\right| * \left|\overrightarrow{RT}\right| * \cos(\angle RTP)}} \tag{5.4}$$

The following will prove that the derivative is never negative, assuming

$$\frac{\mathrm{d}S(\left|\overrightarrow{TP}\right|)}{\mathrm{d}\left|\overrightarrow{TP}\right|} < 0 \tag{5.5}$$

Then

$$\frac{\left|\overrightarrow{RT}\right| * \cos(\angle RTP) - \left|\overrightarrow{TP}\right|}{\sqrt{\left|\overrightarrow{TP}\right|^2 + \left|\overrightarrow{RT}\right|^2 - 2 * \left|\overrightarrow{TP}\right| * \left|\overrightarrow{RT}\right| * \cos(\angle RTP)}} > 1 \tag{5.6}$$

And

$$\begin{aligned} &\sqrt{\left|\overrightarrow{TP}\right|^2 + \left|\overrightarrow{RT}\right|^2 - 2 * \left|\overrightarrow{TP}\right| * \left|\overrightarrow{RT}\right| * \cos(\angle RTP)} \\ &\quad < \left|\overrightarrow{RT}\right| * \cos(\angle RTP) - \left|\overrightarrow{TP}\right| \end{aligned} \tag{5.7}$$

So

$$\left|\overrightarrow{RT}\right| * \cos(\angle RTP) - \left|\overrightarrow{TP}\right| > 0 \tag{5.8}$$

Thus, square processing on both sides of Eq. (5.7)

$$\begin{aligned} &\left|\overrightarrow{TP}\right|^2 + \left|\overrightarrow{RT}\right|^2 - 2 * \left|\overrightarrow{TP}\right| * \left|\overrightarrow{RT}\right| * \cos(\angle RTP) \\ &\quad < \left|\overrightarrow{RT}\right|^2 * \cos^2(\angle RTP) - 2 * \left|\overrightarrow{TP}\right| * \left|\overrightarrow{RT}\right| * \cos(\angle RTP) + \left|\overrightarrow{TP}\right|^2 \end{aligned} \tag{5.9}$$

Simplify Formula (5.9)

$$\left|\overrightarrow{RT}\right|^2 < \left|\overrightarrow{RT}\right|^2 * \cos^2(\angle RTP) \tag{5.10}$$

That is

$$1 < \cos^2(\angle RTP) \tag{5.11}$$

The above Formula (5.11) is obviously contradictory, so the hypothesis (5.5) is wrong. Therefore,

$$\frac{dS(\left|\overrightarrow{TP}\right|)}{d\left|\overrightarrow{TP}\right|} \geq 0 \tag{5.12}$$

Therefore, the function $S(\left|\overrightarrow{TP}\right|)$ is a monotonically increasing function for the independent variable $\left|\overrightarrow{TP}\right|$. Therefore, in Eq. (5.3), if a function value $S(\left|\overrightarrow{TP}\right|)$ that exists is given, the value $\left|\overrightarrow{TP}\right|$ obtained by Eq. (5.3) can be uniquely determined. The value of $S(\left|\overrightarrow{TP}\right|)$ can be provided by Eq. (5.1)

$$S(\left|\overrightarrow{TP}\right|) = v * t_r \tag{5.13}$$

In the actual detection project, the ultrasonic guided wave propagation velocity v is a known value, and the TOF t_r of the ultrasonic guided wave detection signal can be accurately extracted by the time–frequency energy density precipitation TOF extraction method proposed in Chap. 3. Therefore, the analytical equation by solving $\left|\overrightarrow{TP}\right|$ can be expressed as

$$\left|\overrightarrow{TP}\right| + \sqrt{\left|\overrightarrow{TP}\right|^2 + \left|\overrightarrow{RT}\right|^2 - 2 * \left|\overrightarrow{TP}\right| * \left|\overrightarrow{RT}\right| * \cos(\angle RTP)} = v * t_r \tag{5.14}$$

In the analytical equation shown by Eq. (5.14), $\left|\overrightarrow{TP}\right|$ is the quantity to be determined, and t_r, $\left|\overrightarrow{RT}\right|$, and $\angle RTP$ are input quantities. For a single detection process in which a controlled emission EMAT emits an ultrasonic guided wave at a certain angle, t_r can be obtained by taking a time-lapse extraction of the scattered guided wave detection signal. $\left|\overrightarrow{RT}\right|$ and $\angle RTP$ are determined according to the specific positional relationship between the direction-controllable emission EMAT and the omnidirectional receiving EMAT that receives the scattered guided waves during this single detection. Therefore, the solution of $\left|\overrightarrow{TP}\right|$ depends on the input of t_r, $\left|\overrightarrow{RT}\right|$, and $\angle RTP$ and the relationship of Eq. (5.14). Combining the above considerations, we can write Eq. (5.14) as

$$\begin{aligned} &F(\left|\overrightarrow{TP}\right| \mid t_r, \left|\overrightarrow{RT}\right|, \angle RTP) \\ &= \left|\overrightarrow{TP}\right| + \sqrt{\left|\overrightarrow{TP}\right|^2 + \left|\overrightarrow{RT}\right|^2 - 2 * \left|\overrightarrow{TP}\right| * \left|\overrightarrow{RT}\right| * \cos(\angle RTP)} - v * t_r \\ &= 0 \end{aligned} \tag{5.15}$$

Then

$$F(\left|\overrightarrow{TP}\right| \mid t_r, \left|\overrightarrow{RT}\right|, \angle RTP) = 0 \tag{5.16}$$

where $F(\left|\overrightarrow{TP}\right| \mid t_r, \left|\overrightarrow{RT}\right|, \angle RTP)$ is the relationship among $\left|\overrightarrow{TP}\right|$, $\left|\overrightarrow{RT}\right|$, t_r, and $\angle RTP$.

Analytic solution of $\left|\overrightarrow{TP}\right|$ can be obtained according to Eq. (5.16)

$$\left|\overrightarrow{TP}\right| = \frac{(v * t_r)^2 - \left|\overrightarrow{RT}\right|^2}{2 * v * t_r - 2 * \left|\overrightarrow{RT}\right| * \cos(\angle RTP)} \tag{5.17}$$

The guided wave excitation, propagation, scattering, and reception processes in the directional emission-omnidirectional reception detection test are shown in Fig. 4.1. In a test of directional guided waves, if an omnidirectional receiving EMAT receives a guided wave scatter detection signal, the time–frequency energy density precipitation TOF extraction method can be used to extract the TOF t_r of the guided wave scatter detection signal. According to the propagation velocity v of the guided wave in the test medium, the distance $\left|\overrightarrow{TP}\right|$ between the scattering point P and the direction-controllable emission EMAT in the directional emission guided wave detection test can be obtained by using the analytical Eq. (5.16). The position P of the scattering point can be determined according to the current emission direction and the distance obtained. The obtained scattering point position P is subjected to three smooth spline interpolations to obtain a contour curve of the defect to be detected.

This section defines the concept of incident guide beam, scattered guide beam, and scattering point by analyzing the process of guided wave excitation, propagation, scattering, and reception. The defect profile is described by the position of the scattering point. It is pointed out that the solution of the scattering point position is the key part of the guided wave scattering imaging model of the defect contour magneto-acoustic array, and the analytical solution model and method of the scattering point position under directional emission-omnidirectional reception are established. The uniqueness of the analytical model and method for solving the position of the scattering point is proved. The scatter point position resolution model and method proposed in this section are for a single detection process in which a controlled emission EMAT is emitted in a certain direction in the case of transmitting ultrasonic guided waves at an angle. In the following, an analytical solution model and method for the position of the scattering point in the case of directional emission-omnidirectional reception based on this subsection will be presented. A computational model for guided wave scatter imaging of defect-oriented directional emission-omnidirectional receiving magneto-acoustic arrays with controlled emission EMATs and their emission angles is established.

In the directional emission-omnidirectional receiving magnetic acoustic array guided wave scattering imaging method, a total of N direction-controllable emission EMATs are set as excitation transducers. The nth direction-controllable emission EMAT is labeled as T_n, where $n = 1, 2, \ldots, N$, N are positive integers. M omnidirectional receiving EMAT as receiving transducer, the mth omnidirectional receiving EMAT, is marked as Rm, where $m = 1, 2, \ldots, M$, M are positive integers; the controllable emission EMAT in each direction contains L emission angles, and the first emission angle is marked as θ_l, where $l = 1, 2, \ldots, L$, L are positive integers. In order to facilitate the unified calculation, the launch angle θ_l refers to the angle between the direction vector of the direction-controllable emission EMAT launch guide and the positive direction of the x-axis in the plane rectangular coordinate system. Therefore, the entire imaging detection process requires a total of $N \times L$ guided directional excitations, and all M omnidirectional receiving EMATs are prepared to receive the guided wave scatter detection signal each time the guided directional excitation is performed. The TOF information of the guided wave scatter detection signals received by all omnidirectional receiving EMATs during the whole imaging detection process is extracted, and the TOF data of $N \times L \times M$ scattered guided wave detection signals are obtained. According to the correspondence between the omnidirectional receiving EMAT serial number, the direction-controllable emission EMAT serial number, and the transmission angle, the TOF data of the extracted $N \times L \times M$ scattered guided wave detection signals constitute a TOF matrix $T(t_{mnl})$. $\angle RTP$ and TOF matrix $T(t_{mnl})$ have the same distribution law in omnidirectional receiving EMAT number, direction-controllable emission EMAT number, and transmission angle, so all $N \times L \times M$ angle data also form angle matrix $A(\alpha_{mnl})$.

$$T = \begin{bmatrix} t_{111} & t_{211} & \cdots & t_{m11} & \cdots & t_{M11} \\ t_{112} & t_{212} & \cdots & t_{m12} & \cdots & t_{M12} \\ \vdots & \vdots & \ddots & \vdots & \ddots & \vdots \\ t_{11l} & t_{21l} & \cdots & t_{m1l} & \cdots & t_{M1l} \\ \vdots & \vdots & \ddots & \vdots & \ddots & \vdots \\ t_{11L} & t_{21L} & \cdots & t_{m1L} & \cdots & t_{M1L} \\ t_{121} & t_{221} & \cdots & t_{m21} & \cdots & t_{M21} \\ t_{122} & t_{222} & \cdots & t_{m22} & \cdots & t_{M22} \\ \vdots & \vdots & \ddots & \vdots & \ddots & \vdots \\ t_{12l} & t_{22l} & \cdots & t_{m2l} & \cdots & t_{M2l} \\ \vdots & \vdots & \ddots & \vdots & \ddots & \vdots \\ t_{12L} & t_{22L} & \cdots & t_{m2L} & \cdots & t_{M2L} \\ \vdots & \vdots & \ddots & \vdots & \ddots & \vdots \\ t_{1n1} & t_{2n1} & \cdots & t_{mn1} & \cdots & t_{Mn1} \\ t_{1n2} & t_{2n2} & \cdots & t_{mn2} & \cdots & t_{Mn2} \\ \vdots & \vdots & \ddots & \vdots & \ddots & \vdots \\ t_{1nl} & t_{2nl} & \cdots & t_{mnl} & \cdots & t_{Mnl} \\ \vdots & \vdots & \ddots & \vdots & \ddots & \vdots \\ t_{1nL} & t_{2nL} & \cdots & t_{mnL} & \cdots & t_{MnL} \\ \vdots & \vdots & \ddots & \vdots & \ddots & \vdots \\ t_{1N1} & t_{2N1} & \cdots & t_{mN1} & \cdots & t_{MN1} \\ t_{1N2} & t_{2N2} & \cdots & t_{mN2} & \cdots & t_{MN2} \\ \vdots & \vdots & \ddots & \vdots & \ddots & \vdots \\ t_{1Nl} & t_{2Nl} & \cdots & t_{mNl} & \cdots & t_{MNl} \\ \vdots & \vdots & \ddots & \vdots & \ddots & \vdots \\ t_{1NL} & t_{2NL} & \cdots & t_{mNL} & \cdots & t_{MNL} \end{bmatrix}_{NL\times M}$$

$$A = \begin{bmatrix} \alpha_{111} & \alpha_{211} & \cdots & \alpha_{m11} & \cdots & \alpha_{M11} \\ \alpha_{112} & \alpha_{212} & \cdots & \alpha_{m12} & \cdots & \alpha_{M12} \\ \vdots & \vdots & \ddots & \vdots & \ddots & \vdots \\ \alpha_{11l} & \alpha_{21l} & \cdots & \alpha_{m1l} & \cdots & \alpha_{M1l} \\ \vdots & \vdots & \ddots & \vdots & \ddots & \vdots \\ \alpha_{11L} & \alpha_{21L} & \cdots & \alpha_{m1L} & \cdots & \alpha_{M1L} \\ \alpha_{121} & \alpha_{221} & \cdots & \alpha_{m21} & \cdots & \alpha_{M21} \\ \alpha_{122} & \alpha_{222} & \cdots & \alpha_{m22} & \cdots & \alpha_{M22} \\ \vdots & \vdots & \ddots & \vdots & \ddots & \vdots \\ \alpha_{12l} & \alpha_{22l} & \cdots & \alpha_{m2l} & \cdots & \alpha_{M2l} \\ \vdots & \vdots & \ddots & \vdots & \ddots & \vdots \\ \alpha_{12L} & \alpha_{22L} & \cdots & \alpha_{m2L} & \cdots & \alpha_{M2L} \\ \vdots & \vdots & \ddots & \vdots & \ddots & \vdots \\ \alpha_{1n1} & \alpha_{2n1} & \cdots & \alpha_{mn1} & \cdots & \alpha_{Mn1} \\ \alpha_{1n2} & \alpha_{2n2} & \cdots & \alpha_{mn2} & \cdots & \alpha_{Mn2} \\ \vdots & \vdots & \ddots & \vdots & \ddots & \vdots \\ \alpha_{1nl} & \alpha_{2nl} & \cdots & \alpha_{mnl} & \cdots & \alpha_{Mnl} \\ \vdots & \vdots & \ddots & \vdots & \ddots & \vdots \\ \alpha_{1nL} & \alpha_{2nL} & \cdots & \alpha_{mnL} & \cdots & \alpha_{MnL} \\ \vdots & \vdots & \ddots & \vdots & \ddots & \vdots \\ \alpha_{1N1} & \alpha_{2N1} & \cdots & \alpha_{mN1} & \cdots & \alpha_{MN1} \\ \alpha_{1N2} & \alpha_{2N2} & \cdots & \alpha_{mN2} & \cdots & \alpha_{MN2} \\ \vdots & \vdots & \ddots & \vdots & \ddots & \vdots \\ \alpha_{1Nl} & \alpha_{2Nl} & \cdots & \alpha_{mNl} & \cdots & \alpha_{MNl} \\ \vdots & \vdots & \ddots & \vdots & \ddots & \vdots \\ \alpha_{1NL} & \alpha_{2NL} & \cdots & \alpha_{mNL} & \cdots & \alpha_{MNL} \end{bmatrix}_{NL \times M} \tag{5.18}$$

The input quantity $\left|\overrightarrow{RT}\right|$ is only related to the position T_n of the direction-controllable emission EMAT and the position R_m of the omnidirectional receiving EMAT. Regardless of the direction of the controllable emission EMAT, all data $\left|\overrightarrow{RT}\right|$ are constructed similarly as described above. Distance matrix $D(d_{mnl})$ is:

$$D = \begin{bmatrix} d_{11l} & d_{21l} & \cdots & d_{m1l} & \cdots & d_{M1l} \\ d_{12l} & d_{22l} & \cdots & d_{m2l} & \cdots & d_{M2l} \\ \vdots & \vdots & \ddots & \vdots & \ddots & \vdots \\ d_{1nl} & d_{2nl} & \cdots & d_{mnl} & \cdots & d_{Mnl} \\ \vdots & \vdots & \ddots & \vdots & \ddots & \vdots \\ d_{1Nl} & d_{2Nl} & \cdots & d_{mNl} & \cdots & d_{MNl} \end{bmatrix}_{N\times M} \tag{5.19}$$

According to the analytical model $\left|\overrightarrow{TP}\right|$ of the solution of Eq. (5.16), the value TP_{mnl} calculated point by point is obtained by the following formula.

$$TP_{mnl} = \arg\left|\overrightarrow{TP}\right| \text{ s.t. } F(\left|\overrightarrow{TP}\right|, t_{mnl}, d_{mnl}, \alpha_{mnl}) = 0 \tag{5.20}$$

The TP_{mnl} calculated according to the above formula constitutes a matrix $TP_{NL\times M}$, and the matrix $TP_{NL\times M}$ is similar in form to the matrix $T(t_{mnl})$ and the matrix $A(\alpha_{mnl})$, and details are not described herein again. However, this matrix contains more invalid elements, mainly because of the single detection process in the case where the EMAT is controlled to emit ultrasonic waves in a certain direction. Not all omnidirectional receiving EMATs receive the scattered guided wave detection signal, and the omnidirectional receiving EMAT that does not receive the scattered guided wave detection signal corresponds to the m series serial number as an invalid element. The calculated value results $\left|\overrightarrow{TP}\right|$ only need to determine the direction of the controllable emission EMAT and its direction of the guided wave to determine the corresponding relationship. Therefore, in order to reduce the *TP* matrix redundancy, the *TP* matrix is rearranged according to the direction-controllable emission EMAT and its transmitted guided wave direction to obtain $TP_{N\times L}$.

$$TP = \begin{bmatrix} TP_{11} & TP_{12} & \cdots & TP_{1l} & \cdots & TP_{1L} \\ TP_{21} & TP_{22} & \cdots & TP_{2l} & \cdots & TP_{2L} \\ \vdots & \vdots & \ddots & \vdots & \ddots & \vdots \\ TP_{n1} & TP_{n2} & \cdots & TP_{nl} & \cdots & TP_{nL} \\ \vdots & \vdots & \ddots & \vdots & \ddots & \vdots \\ TP_{N1} & TP_{N2} & \cdots & TP_{Nl} & \cdots & TP_{NL} \end{bmatrix}_{N\times L} \tag{5.21}$$

After the *TP* matrix is obtained, the position coordinates of the scatter point can be solved by the positional controllable emission EMAT position coordinates, the guided wave emission angle, and the vector length $\overrightarrow{TP}$. Taking the plane rectangular coordinate system as an example, the EMAT or T_n is controllably launched for a certain direction. The coordinate is the known quantity $T_n(x_n, y_n)$. When the guided wave emission angle is θ_{nl}, the obtained vector length $\overrightarrow{TP}$ is TP_{nl}, and the corresponding scattering point P_{nl} coordinate $P_{nl}(x_{nl}, y_{nl})$ is

$$x_{nl} = x_n + TP_{nl}\cos\theta_{nl} \tag{5.22}$$

$$y_{nl} = y_n + TP_{nl}\sin\theta_{nl} \tag{5.23}$$

Similarly, the coordinates $P_{nl}(x_{nl}, y_{nl})$ of the scattering point P_{nl} are controlled according to the direction of the EMAT and its emission angle structure, forming a scattering point coordinate matrix P.

$$P = \begin{bmatrix} (x_{11}, y_{11}) & (x_{12}, y_{12}) & \cdots & (x_{1l}, y_{1l}) & \cdots & (x_{1L}, y_{1L}) \\ (x_{21}, y_{21}) & (x_{22}, y_{22}) & \cdots & (x_{2l}, y_{2l}) & \cdots & (x_{2L}, y_{2L}) \\ \vdots & \vdots & \ddots & \vdots & \ddots & \vdots \\ (x_{n1}, y_{n1}) & (x_{n2}, y_{n2}) & \cdots & (x_{nl}, y_{nl}) & \cdots & (x_{nL}, y_{nL}) \\ \vdots & \vdots & \ddots & \vdots & \ddots & \vdots \\ (x_{N1}, y_{N1}) & (x_{N2}, y_{N2}) & \cdots & (x_{Nl}, y_{Nl}) & \cdots & (x_{NL}, y_{NL}) \end{bmatrix}_{N\times L} \tag{5.24}$$

After obtaining the *TP* matrix, the above process of solving the scattering point position matrix P is given by taking the plane rectangular coordinate system as an example. However, the solution method is not limited to the plane rectangular coordinate system, and the polar coordinate system can be used according to actual needs. Since the position of the scattering point is determined after the *TP* matrix is solved, the solution of the position coordinates of the scattering point has no relationship with the selected coordinate system. Also, the selection of the origin of the coordinates has no effect on the solution of the coordinates of the position of the scattering point, and the relative relationship of the coordinates of the positions of the scattering points is constant, so it is irrelevant to the selection of the origin of the coordinates.

After the scattering point position matrix P is obtained, each scattering point can be connected into a curve by an interpolation method to form a defect contour curve.

Figure 5.3 shows a schematic diagram of the directional emission-omnidirectional accommodating magnetic acoustic array guided wave scattering imaging calculation model, which is clear. Only the *TP* solution results and the scatter point calculation results of the controllable emission EMAT in each direction are given, and the omni-directional receiving EMAT position and the corresponding scatter guide beam are not given.

This section is based on an analytical solution model and method for the position of the scattering point under directional emission-omnidirectional reception. A computational model for guided wave scattering imaging of defect contour directional emission-omnidirectional receiving magneto-acoustic array with controlled emission EMAT with multiple directions and its emission angle is established. The calculation method of *TP* matrix and scattering point coordinate matrix is given. In the following, based on the defect contour directional emission-omnidirectional receiving magnetic acoustic array guided wave scattering imaging model proposed

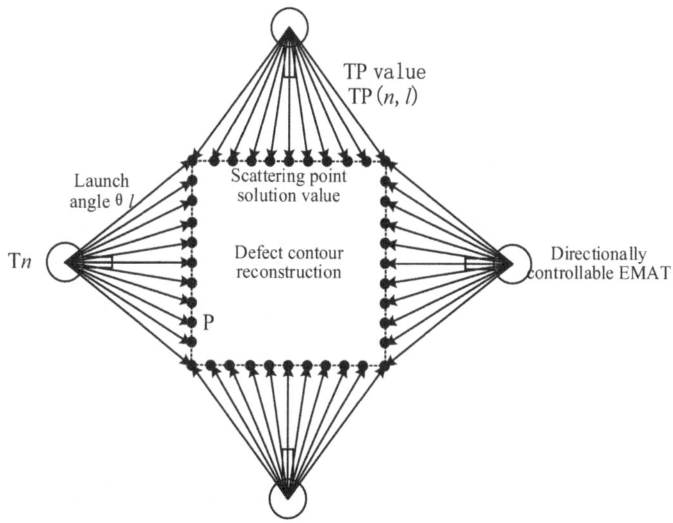

Fig. 5.3 Schematic diagram of the directional emission-orthogonal receiving magnetic acoustic array guided wave scattering imaging calculation model

in this section, the directional emission-omnidirectional receiving magnetic acoustic array guided wave scattering imaging detection method and its steps are specifically described.

5.1.2 Directional Emission-Omnidirectional Reception Magnetic Acoustic Array Guided Wave Scattering Imaging Method Steps

This section is aimed at the actual detection process of steel plate defects, combined with the directional emission-omnidirectional receiving magnetic acoustic array guided wave scattering imaging model proposed in the previous section. The directional emission-omnidirectional receiving magnetic acoustic array guided wave scattering imaging method and its steps are given in the actual detection process.

Figure 5.4 shows a block diagram of a specific implementation principle of the directional emission-omnidirectional receiving magnetic acoustic array guided wave scattering imaging method. This block diagram is only to explain the specific steps of the directional emission-omnidirectional receiving magnetic acoustic array guided wave scattering imaging method. However, it is not limited to the shape of the magneto-acoustic array, the number of direction-controllable emission EMATs, the number of omnidirectional receiving EMATs, the type of coordinate system, and the position of the coordinate origin.

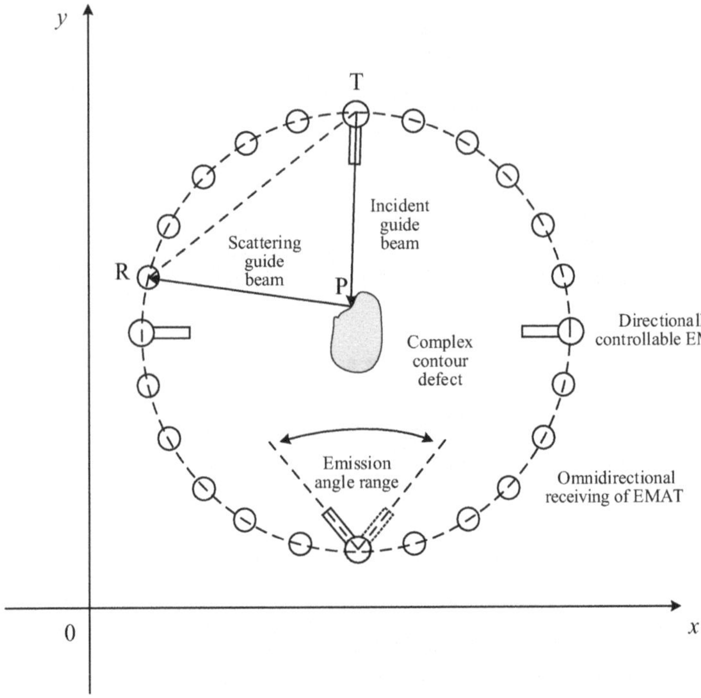

Fig. 5.4 Directional emission-omnidirectional receiving magnetic acoustic array guided wave scattering imaging method array schematic

The direction-controllable emission EMAT is used as the excitation transducer, and the omnidirectional receiving EMAT is used as the receiving transducer, which is evenly arranged in a circular array around the detection area of the steel plate to be tested. The number of omnidirectional receiving EMATs between the controllable emission EMATs in the two adjacent directions is the same; the EMAT in each direction controls the guided waves according to the preset range of the angle and the angle step. The omnidirectional receiving EMAT is responsible for receiving the guided wave signal each time there is a guided wave excitation. Combined with the directional emission-omnidirectional receiving magnetic acoustic array guided wave scattering imaging model, the steps of the directional emission-omnidirectional receiving magnetic acoustic array guided wave scattering imaging method are as follows:

(1) The N-direction controllable EMAT is used as the excitation transducer, and M omnidirectional EMATs are used as the receiving transducers, which are evenly arranged in a circular array around the detection area of the steel plate to be tested. Moreover, the number of omnidirectional receiving EMATs between the two adjacent controllable EMATs is the same and satisfies $M = K * N$, where N, M, and K are positive integers. Set the direction of the EMAT's emission

angle range $\theta_1 \sim \theta_2$ and the angle step θ_s, and the total number of emission angles is $L = (\theta_2 - \theta_1)/\theta_s + 1$.

(2) Selecting the nth direction-controllable emission EMAT as the excitation transducer T_n of the current detection, wherein $n = 1, 2, \ldots, N$.

(3) Selecting the emission angle θ_l of the controllable EMAT in the direction of the step (2), and using the direction of the controllable emission EMAT, the ultrasonic guided wave is excited in the steel plate to be tested along the emission angle, where $l = 1, 2, \ldots, L$.

(4) The omnidirectional receiving EMAT receives the ultrasonic guided wave signal in the steel plate, assuming that a total of $M1$ omnidirectional receiving EMATs receive the guided wave signal, denoted as R_{m1}, where $m1 = 1, 2, \ldots, M1$. Determining whether R_{m1} and the direction of the ultrasonic guided wave in a specific direction in step (3) controllable emission EMAT constitute a scattering group (T_n, R_{m1}), and if so, proceeding to step (5); if not, proceeding to step (6). The omnidirectional receiving EMAT (R_{m1}) of the guided wave signal is received by the TOF of the ultrasonic guided wave, and the direction-controllable emission EMAT (T_n) of the ultrasonic guided wave emitted in a specific direction at this time constitutes a scattering group. The measured TOF of the ultrasonic guided wave from T_n to R_{m1} is t_r, the propagation velocity of the ultrasonic guided wave in the steel plate is v, and the plane rectangular coordinate system is established. The position of the direction-controllable emission EMAT is T, and the position of the omnidirectional receiving EMAT is R. The theoretical time t_s used by the guided wave to travel directly from position T to position R along a straight line is

$$t_s = \frac{\left|\overrightarrow{TR}\right|}{v} \tag{5.25}$$

If $t_r > t_s$, then R_{m1} and T_n form a scattering group; otherwise, R_{m1} and T_n do not constitute a scattering group.
The above judgment of the scattering group is based on the fact that if the omnidirectional receiving EMAT receives the scattered guided wave detection signal, the TOF should be longer than the TOF of the direct wave signal or the transmitted wave signal received by the omnidirectional receiving EMAT.

(5) For the scattering group (T_n, R_{m1}), the spacing of the T_n and R_{m1}, the emission angle, and the TOF of the received guided wave signal to solve the position P of the scattering point, the specific solution method can be based on Formula (5.16).

(6) Judging whether guided wave excitation and reception have been performed along all the emission angles θ_l, and if so, proceeding to step (7); if not, the emission angle is changed to θ_{l+1}, and returning to step (3).

(7) Determine whether all directions of the controllable EMAT have been excited by the ultrasonic guided wave and if so, proceed to step (8); if not, the direction controlled emission EMAT becomes T_{n+1} and return to the step (2).

(8) Using the interpolation method, all the scattered points obtained are connected into a curve to obtain a clear contour image of the complex defect.

Based on the above method steps, Fig. 5.5 shows a flow chart of a directional emission-omnidirectional receiving magnetic acoustic array guided wave scattering imaging method.

This section elaborates the implementation steps of directional emission-omnidirectional receiving magnetic acoustic array guided wave scattering imaging method and gives a specific implementation form and array structure of directional emission-omnidirectional receiving magnetic acoustic array guided wave scattering imaging method, which illustrates a method of determining whether the omnidirectional receiving EMAT and the direction-controllable emission EMAT constitute a scattered group. In the following, specific steel plate defect contour imaging work will be carried out for the steel plate regular contour defect and complex contour defect, respectively, and the proposed directional emission-omnidirectional receiving magnetic acoustic array guided wave scattering imaging method will be tested and verified.

5.1.3 Experimental Verification of Steel Plate Regular Contour Defect Guided Wave Scattering Imaging

In order to verify the proposed directional emission-omnidirectional receiving magnetic acoustic array guided wave scattering imaging method, the accuracy of defect contour imaging is studied. On the 4-mm-thick steel plate, a directional emission-oriented omnidirectional receiving magnetic acoustic array guided wave scattering imaging test platform was established, and a series of related testing tests were carried out on this test platform. Steel plate regular contour defect directional emission-omnidirectional receiving magnetic acoustic array guided wave scattering imaging detection test platform block diagram is shown in Fig. 5.6.

The direction controlled emission EMAT proposed in Chap. 2 is used as the excitation transducer, and the magnetostrictive omnidirectional receiving EMAT based on the nickel ribbon and the densely wound coil is used as the receiving transducer. It is uniformly arranged in the form of a circular array around the regular contour defects of the steel sheet, and the circular magnetic acoustic array is centered on the defects. A total of four direction-controllable emission EMATs are used, which are evenly distributed in the circumferential 0°, 90°, 180°, and 270° positions. A total of five omnidirectional receiving EMATs are evenly arranged circumferentially between the two adjacent direction-controllable emission EMATs, so a total of 20 omnidirectional receiving EMATs are used. The absolute angle range of the direction-controllable emission EMAT transmitting ultrasonic guided wave is 0° ~ 90°; that is, the maximum absolute angle of the ultrasonic guided wave emitted to the left and right sides is 45° based on its centerline. The emission angle step is 5°. Therefore, the EMAT of each direction can be used to excite the ultrasonic guided

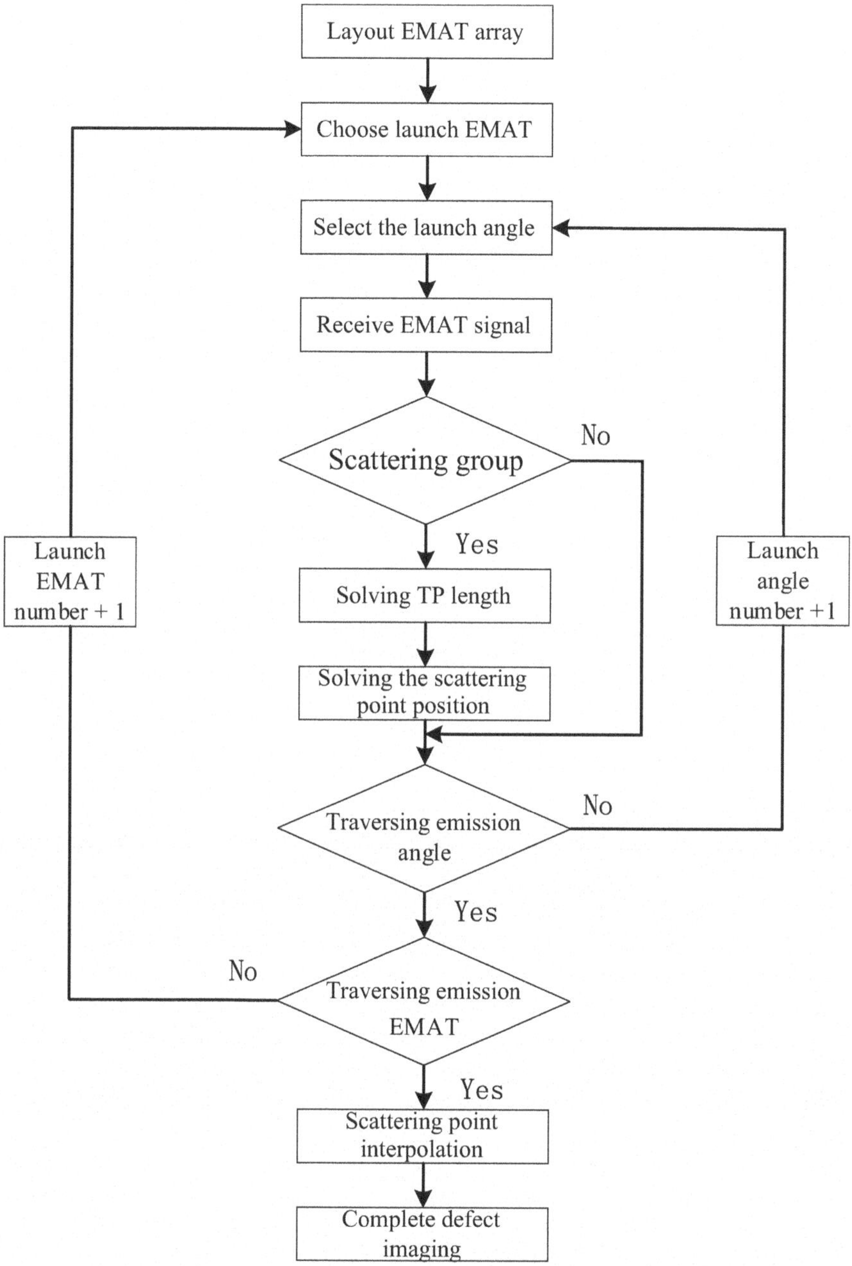

Fig. 5.5 Flow chart of directional emission-omnidirectional receiving magnetic acoustic array guided wave scattering imaging method

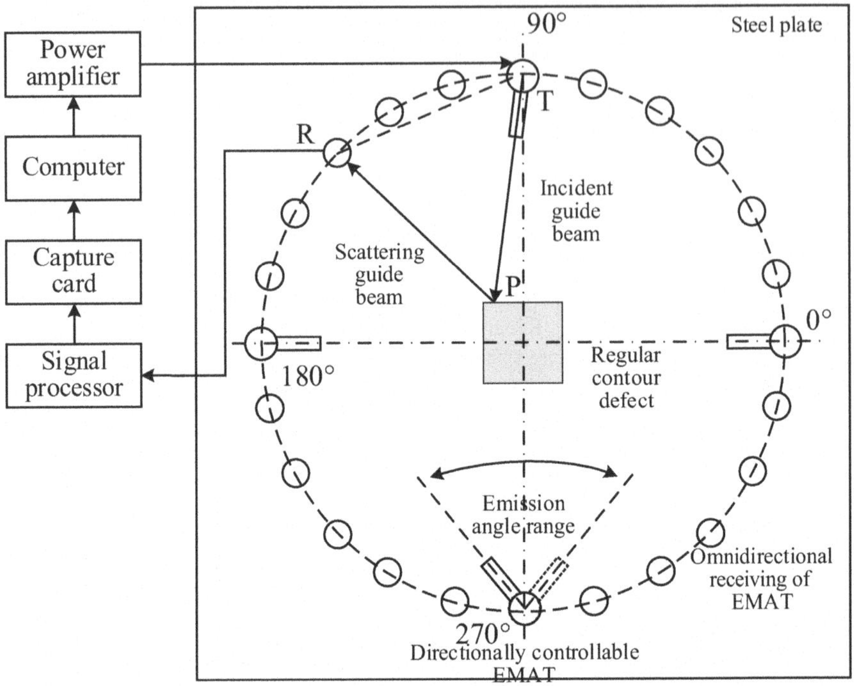

Fig. 5.6 Steel plate regular contour defect directional emission-omnidirectional receiving magnetic acoustic array guided wave scattering imaging detection test platform principle block diagram

waves according to 19 absolute angles. The whole test process includes 76 ultrasonic guided waves. When the EMAT-excited ultrasonic guided wave is controlled to be emitted in each direction, all omnidirectional receiving EMATs are ready to receive ultrasonic guided waves. That is, the direction controlled emission EMAT and the omnidirectional receiving EMAT are in a state of being overcharged each time.

The steel plate regular contour defects targeted in this test include square contour defects, regular triangle contour defects, and circular contour defects (Fig. 5.7).

All three types of rule defects were subjected to contour imaging test under the same test parameters. The test parameter settings during the test are shown in Table 5.1.

In the above test parameter setting, the design of the diameter of the circular magnetic acoustic array is based on the sensitivity analysis of the time–frequency energy density precipitation TOF extraction method and the relative error on the guided wave propagation distance. As far as possible, the scattering guided wave propagation distance is set to the time–frequency energy density precipitation walking time extraction method relative error stable region, to extract the traveling time of the scattered guided wave detecting signal with higher precision. The selection of operating frequency and guided wave mode, on the one hand, considers the operating frequency and guided mode of the controlled emission EMAT proposed in Chap.

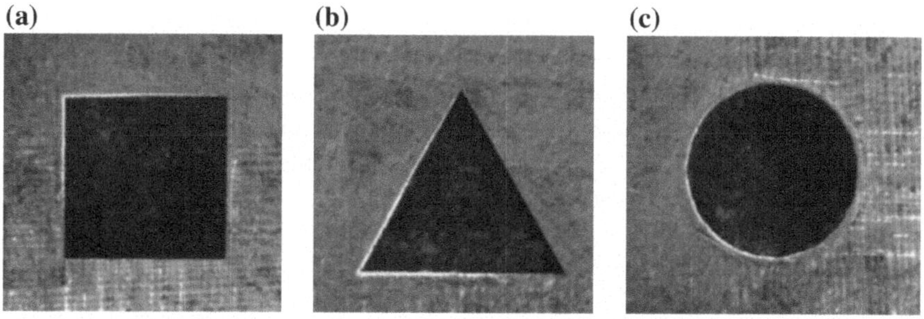

Fig. 5.7 Physical diagram of steel plate regular contour defect: **a** square defect; **b** positive triangle defect; **c** circular defect

Table 5.1 Steel plate regular contour defect directional emission-omnidirectional receiving magnetic acoustic array guided wave scattering imaging detection test parameter setting

Parameters	Value
Excitation voltage peak-to-peak value (V)	320
Excitation frequency f (kHz)	320
Number of incentive cycles (a)	12
Steel plate thickness (mm)	4
Square defect side length * depth (mm)	100 * 1.5
Triangle defect side length * depth (mm)	100 * 1.5
Round defect diameter * depth (mm)	100 * 1.5
Circular magnetic acoustic array diameter (mm)	900
Ultrasonic guided wave mode	SH0
Guide wave propagation velocity (m/s)	3200
Signal processor amplification	5000
Signal processor narrowband filter center frequency (kHz)	320
Signal processor narrow band filter bandwidth (kHz)	20

2, so that the operational performance of the direction-controllable EMAT is in a better state. On the other hand, combined with the sensitivity analysis of the working frequency and guided wave dispersion characteristics of the time–frequency energy density precipitation TOF extraction method in Chap. 3, the TOF information of the scattered guided wave detection signal is obtained with higher precision.

According to the above test parameters, the steel regular contour defect directional emission-omnidirectional receiving magnetic acoustic array guided wave scattering imaging detection test is carried out. The test results are analyzed and discussed below.

According to the directional emission-omnidirectional receiving magneto-acoustic array guided wave scattering imaging method steps proposed in the previous section, under the test mentioned above parameters, the contour detection tests of the square defects, the regular triangle defects, and the circular defects of the steel plates are respectively performed. The position of the scattering point is calculated separately, and a calculated defect profile is formed. For the sake of comparison, the calculated contour of the defect and the contour of the defect are respectively drawn onto an image, and the calculated scattering points of the three defects and the calculated contour are as shown in Fig. 5.8.

In the rule mentioned above defect contour imaging results, in general, the calculated contour curve basically reflects the corresponding true defect contour, and most of the calculated scattering points can also maintain a good fit with the real defect contour. Specifically, (1) for the square defect contour imaging result, the calculated scattering point near the four corners of the square defect is more deviated from the true defect contour concerning the scattering point near the square defect side. Scattering points near the square defect side have a higher degree of fit to the real defect; (2) the result of the square triangle defect contour imaging is similar to the square defect contour imaging result. The calculated scattering points near the three corners of the equilateral triangle defect deviate significantly from the real defect contour concerning the scattering points near the positive triangle defect side. Moreover, concerning the scattering point near the square defect angle, the scattering point near the positive triangle defect angle is more deviated from the true defect contour, and the scattering point near the positive triangle defect side has a higher degree of fit with the real defect.

In order to quantitatively describe the defect contour imaging accuracy of the magneto-acoustic array guided wave scattering imaging method and the deviation relationship between the imaging result and the true contour of the defect, try to define the relative error between the defect calculation contour and the real contour. Taking the plane rectangular coordinate system as an example, the coordinates of the scattering points on the contour are calculated as $C_j(x_{cj}, y_{cj}), j = 1, 2, 3, \ldots S$, where S is a positive integer, indicating the total number of scattering points on the contour. Find the closest point $A_j(x_{aj}, y_{aj})$ to the scattering point on the true contour of the defect

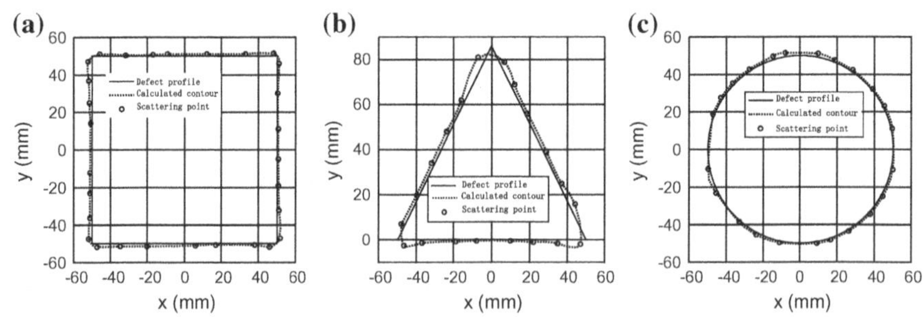

Fig. 5.8 Steel plate regular contour defect directional emission-omnidirectional receiving magnetic acoustic array guided wave scattering imaging results: **a** square defect contour imaging results; **b** positive triangle defect contour imaging results; **c** circular defect contour imaging results

$$A_j(x_{aj}, y_{aj}) = \arg(x_{aj}, y_{aj}) \text{ s.t. } \min d(A_j, C_j) \tag{5.26}$$

where $d(A_j, C_j)$ represents the Euclidean distance between the point A_j and the point C_j, which is defined herein as the absolute error between the scattering point C_j and the point A_j on the true contour of the corresponding defect,

$$d(A_j, C_j) = \sqrt{(x_{aj} - x_{cj})^2 + (y_{aj} - y_{cj})^2} \tag{5.27}$$

To represent the ratio of the absolute error $d(A_j, C_j)$ of the absolute error between the scattering point C_j and the point A_j on the true contour of the corresponding defect relative to the true contour of the defect, the average contour size d_a of the defect is defined to describe the average contour size of the real defect

$$d_\text{a} = \frac{\left|x_{a\max} - x_{a\min}\right| + \left|y_{a\max} - y_{a\min}\right|}{2} \tag{5.28}$$

where

$$\begin{cases} x_{a\max} = \arg\max x_{ai} & \text{for } i = 1, 2, \ldots, S \\ x_{a\min} = \arg\min x_{ai} & \text{for } i = 1, 2, \ldots, S \\ y_{a\max} = \arg\max y_{ai} & \text{for } i = 1, 2, \ldots, S \\ y_{a\min} = \arg\min y_{ai} & \text{for } i = 1, 2, \ldots, S \end{cases} \tag{5.29}$$

Then, the relative error e_j between the scattering point C_j and the point A_j on the true contour of the corresponding defect is defined as

$$e_j = \frac{d(A_j, C_j)}{d_\text{a}} \times 100\% \tag{5.30}$$

Calculate the relative error between the contour and the true contour according to the defects proposed above. Calculate the relative error between the square defect, the regular triangle defect, and the circular defect calculated contour and the true contour in Fig. 5.8, respectively. The normalized relative errors are respectively plotted on the 2D contour plane according to the defect real contour interpolation, as shown in Fig. 5.9.

The distribution of the relative error between the calculated contour and the true contour shown in Fig. 5.9 is consistent with the corresponding relationship between the calculated contour and the true contour of the defect shown in Fig. 4.8. That is, (1) for square defects and regular triangle defects, the relative error of the defect contour imaging is mainly concentrated at the corners of the contour, and the contour imaging error on the straight side of the defect is relatively small. (2) The contour imaging error of the circular defect is mainly concentrated in the vicinity of the circumferential 0°, 90°, 180°, and 270° positions, and the contour imaging error between the four positions is relatively small.

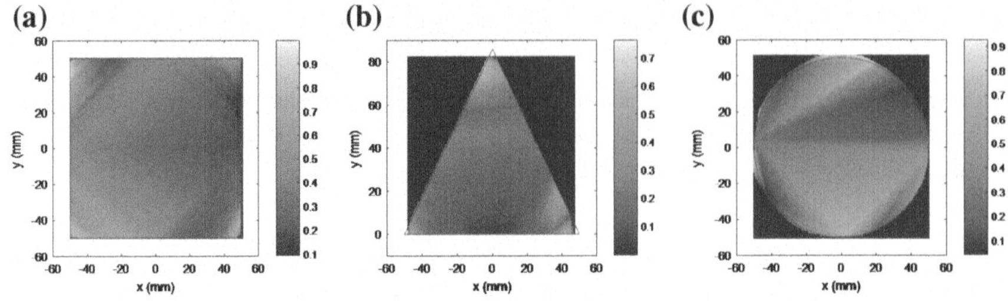

Fig. 5.9 Steel plate regular contour defect directional emission-omnidirectional receiving magnetic acoustic array guided wave scattering imaging relative error results: **a** square defect contour relative error. **b** Relative error of the contour of the triangle defect. **c** Round defect profile relative error

The main reason for the above error distribution is the arrangement of the magneto-acoustic array used in this test and the inherent scattering characteristics of the guided wave. For square defects and equilateral triangle defects, the sharp corner regions cannot allocate enough guided wave incident beams and scattered beams, and their vertices belong to typical scattering singular points, so the scattering points of the contour vertices cannot be obtained. The contour of the local region of the vertex can only be obtained by interpolating the adjacent scattering point, and the variation of the contour of the local region is larger than that of other flat or gently contoured portions of the defect. Therefore, the imaging error of the contour of the local region relative to other straight or flat portions is large. For a circular defect, the tangent of the circumferential 0°, 90°, 180°, and 270° positions is perpendicular to the incident guide beam. When the guided wave is incident on the above position, most of the guided wave energy will be in the opposite direction. Therefore, most of the scattered beam energy returns to the direction of the controllable emission EMAT, and the omnidirectional receiving EMAT cannot be reached, so the scattering point cannot be obtained in the above location area. The locally calculated contour can only be obtained by interpolating its adjacent scattering points, thus producing a relative error with the true defect contour.

In order to evaluate the relative error between the contour of the defect and the real contour, the average value of the relative error of each point is calculated as the average contour error E_{a} of the guided acoustic scattering imaging method of the magneto-acoustic array.

$$E_{\mathrm{a}} = \frac{1}{S}\sum_{j=1}^{S} e_j \tag{5.31}$$

Table 5.2 Steel plate regular contour defect directional emission-omnidirectional receiving magnetic acoustic array guided wave scattering imaging average contour error

Imaging error	Square defect (%)	Triangular defect (%)	Round defect (%)
E_a	1.13	1.89	1.91

Calculate the average contour error of the square defect, the equilateral triangle defect, and the circular defect calculation contour and the real contour according to the above formula. The calculated results are shown in Table 5.2.

According to Table 5.2, the directional emission-omnidirectional receiving magnetic acoustic array guided wave scattering imaging method has a smaller average imaging error for the contour defect of the steel plate. The average profile error for square defects, equilateral triangle defects, and circular defects is within 2%, showing a high degree of regular contour defect contour reconstruction accuracy. In the following, the proposed directional emission-omnidirectional receiving magnetic acoustic array guided wave scattering imaging method is tested and verified for the complex contour defects of steel plates.

5.1.4 Experimental Verification of Guided Wave Scattering Imaging of Complex Contour Defects of Steel Plates

In the previous section, the proposed directional emission-omnidirectional receiving magneto-acoustic guided wave scattering imaging method was tested for the proposed contour defects of the steel plate. This section will test the complex contour defects of the steel plate. The test platform used in this section has the same principle as the guideline directional emission-orthogonal receiving magneto-acoustic guided wave scattering imaging detection test platform shown in Fig. 5.6. Only the detection object is a complex contour defect of the steel plate, and the depth of the defect is also 1.5 mm, as shown in Fig. 5.10. The thickness of the steel plate is also 4 mm. The arrangement of the magnetic acoustic array and the working mode and the relevant test parameters are also the same as the previous one and will not be described here.

According to the proposed directional emission-omnidirectional receiving magnetic acoustic array guided wave scattering imaging method step, under the above test parameters, the contour detection of the complex contour defect of the steel plate is performed, the position of the scattering point is calculated, and the defect calculation contour curve is formed. For comparison purposes, the calculated contour of the complex contour defect and the contour of the defect itself are plotted onto an image, which calculates the scattering point and calculates the contour as shown in Fig. 5.11.

In the above complex defect contour imaging results, in general, the calculated contour curve basically reflects the corresponding true defect contour, and most of the calculated scattering points can also maintain a good fit with the real defect

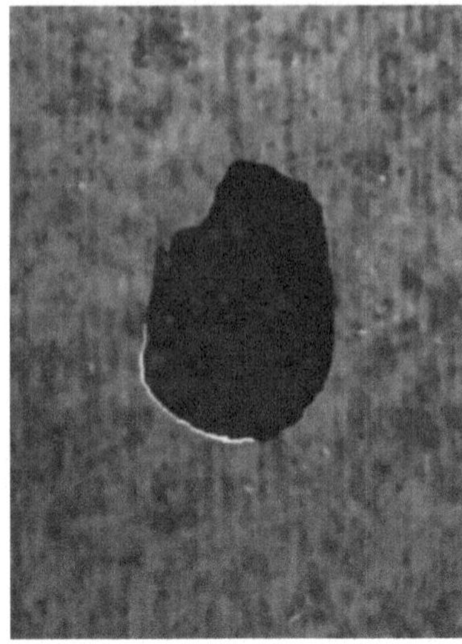

Fig. 5.10 Physical map of complex contour defects of steel plates

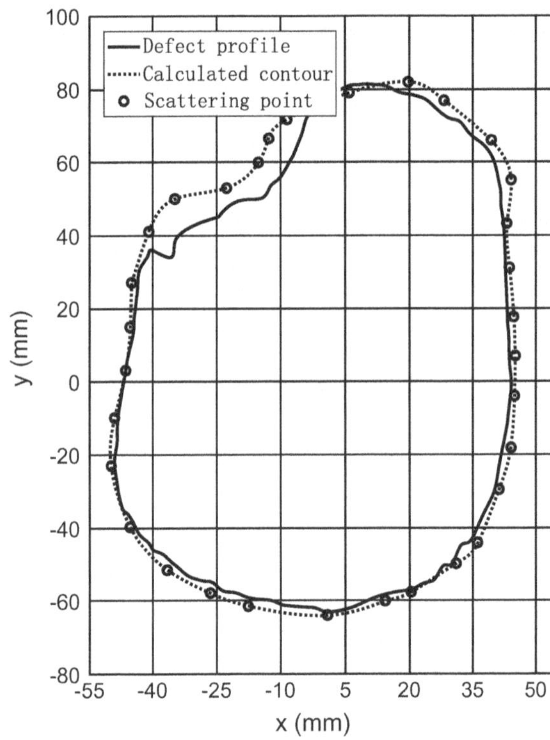

Fig. 5.11 Directional emission of complex contour defects of steel plates-guided scatter imaging results of omnidirectional receiving magneto-acoustic array

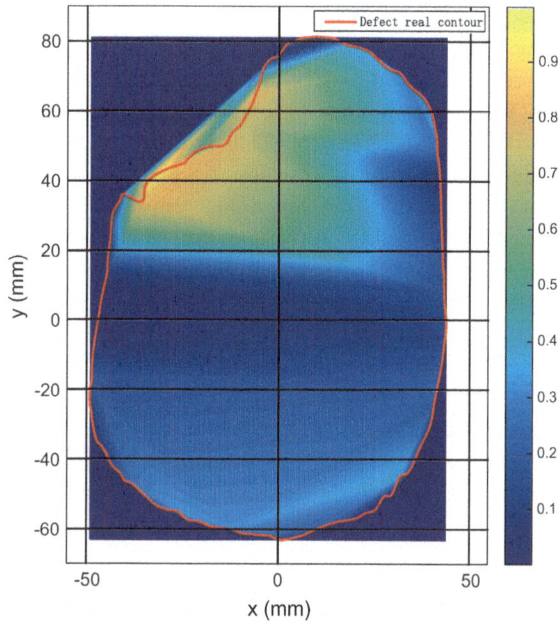

Fig. 5.12 Directional emission of complex contour defects of steel plate-relative error results of guided wave scattering imaging of omnidirectional receiving magneto-acoustic array

contour. In the upper left part of the complex contour defect, there is a certain error between the calculated contour and the true defect contour, and the scattering point of the local region deviates from the true defect contour. In order to quantitatively describe the defect contour imaging accuracy of the magnetic acoustic array guided wave scattering imaging method and the defect contour, the deviation relationship between the imaging result and the true contour of the defect is calculated. Calculate the relative error between the complex contour defect and the real contour, and plot the normalized relative error on the 2D contour plane according to the defect real contour interpolation, as shown in Fig. 5.12.

The distribution of the relative error between the calculated contour and the true contour of the complex contour defect shown in Fig. 5.12 is consistent with the corresponding relationship between the calculated contour and the true contour of the defect shown in Fig. 5.11. The relative error between the calculated contour and the true contour of the upper left part of the complex contour defect is relatively large relative to other parts, and the main reasons include: (1) The contour curve of the upper left part of the complex contour defect fluctuates drastically, which easily causes secondary reflection of the incident guide beam. The directional emission-omnidirectional receiving magnetic acoustic array guided wave scattering imaging model calculates the corresponding scattering point position based on the guided single reflection. Therefore, when the incident guide beam emits two or more reflections, it will bring a certain error to the solution of the position of the scattering point. (2) The upper left part of the complex contour defect contains a plurality of concave partial contours, the incident guide beam can hardly reach the inside of the concave defect contour, and the scattering point of the internal contour of the concave defect cannot be obtained. Therefore, it is impossible to describe the profile change inside the concave defect.

The average contour error of the complex contour defect and the real contour is calculated by using Eq. (5.30), and the average contour error is 2.21%, which shows high defect contour reconstruction accuracy.

In this section, the directional emission-omnidirectional receiving magnetic acoustic array guided wave scattering imaging method is used to carry out a contour reconstruction test on complex contour defects of steel plates. The experimental results show that the directional emission-omnidirectional receiving magnetic acoustic array guided wave scattering imaging method has high defect contour imaging accuracy for the complex contour defects of the steel plate, and the average contour error is only 2.21%. It is shown that the directional emission-omnidirectional receiving magnetic acoustic array guided wave scattering imaging method proposed in this chapter can reconstruct the contour of complex contour defects of steel plate with high precision.

5.2 Omnidirectional Emission-Omnidirectional Receiving Magnetic Acoustic Array Structure Optimization and Guided Wave Scattering Imaging Method

The directional emission-omnidirectional receiving magnetic acoustic array guided wave scattering imaging method proposed in the above section uses the direction controlled emission EMAT proposed in Chap. 2 for the excitation transducer. The size of the space occupied by the EMAT is larger than that of the omnidirectional receiving EMAT. The requirements for the plate-mounted transducer area during the inspection process are high, and at least, there is sufficient space to install the controllable EMAT. Therefore, some places reserved for the transducer installation space cannot meet the requirements of the direction-controllable emission EMAT on the installation space of the transducer. Also, the directional emission-omnidirectional receiving magneto-acoustic array guided wave scattering imaging method requires the direction-controllable emission EMAT to excite the ultrasonic guided wave at a certain angle during each calculation of the scattering point. The operation process in the actual detection process is relatively cumbersome, and the amount of data of the guided wave detection signal to be processed is large.

Given the above requirements and problems, this chapter proposes an omnidirectional transmit-omnidirectional receive magneto-acoustic array guided wave scattering imaging model and method. Both the EMAT and the receiving EMAT use a magnetostrictive omnidirectional EMAT based on a nickel strip and a tightly wound coil. That is, all EMATs in the magneto-acoustic array are of the same type. Also, this chapter will reconstruct features based on defect contours based on the results of defect contour reconstruction of omnidirectional transmit-omnidirectional receive magneto-acoustic array guided wave scattering imaging. The magnetic acoustic array structure optimization method is studied in order to reduce the error of defect contour reconstruction.

In this section, we first try to establish a guided wave scattering imaging model of omnidirectional transmit-omnidirectional receiving magneto-acoustic array. It is proposed to introduce the relationship between the guided signal detection signal strength and the guided wave propagation distance. Moreover, try to quantitatively describe the attenuation effect of the defect boundary contour scattering on the guided signal strength by the guided wave scattering attenuation coefficient. Secondly, an optimization model of magneto-acoustic array structure for omnidirectional emission-omnidirectional receiving magnetic acoustic array guided wave scattering imaging is proposed. The data fusion method of guided wave scattering defect calculation contour curve before and after magnetic acoustic array structure adjustment is established. Thirdly, it is proposed to construct an omnidirectional emission-orthogonal receiving magneto-acoustic array guided wave scattering imaging and a magnetic acoustic array structure optimization adjustment model and method test verification platform for regular contour defects and complex contour defects. The omnidirectional emission-omnidirectional receiving magnetic acoustic array guided wave scattering imaging and magneto-acoustic array structure optimization adjustment model and method are tested and verified.

5.2.1 Omnidirectional Emission-Omnidirectional Receiving Magnetic Acoustic Array Guided Wave Scattering Imaging Model

The omnidirectional emission-omnidirectional receiving magnetic acoustic array guided wave scattering imaging is still based on the basic idea of scattering points. The scattering point is used to link the defect contour with the guided wave scattering detection signal. The key to the imaging model is still the solution of the scattering point position.

In the directional emission-omnidirectional receiving magnetic acoustic array guided wave scattering imaging model, the excitation and reception of the guided waves are all omnidirectional; that is, the propagation direction of the guided waves is not determined when the guided waves are excited. Rather, the EMAT is centered on the guided wave excitation around it. Compared with the directional emission-omnidirectional receiving magnetic acoustic array guided wave scattering imaging model, the a priori condition of the guided wave propagation direction is lacking. The TOF of the scattered guided wave detection signal is still an important input to the model, but other conditions must be added to solve the position of the scattering point.

The amplitude of the SH guided wave displacement equation decays with the propagation distance, and its amplitude is inversely proportional to the one-half of the propagation distance. The amplitude of the SH guided wave displacement equation can be expressed by the guided wave detection signal strength in the specific guided wave detection signal. In this way, the attenuation relationship between the signal strength of the guided wave detection signal and the guided wave propagation distance is introduced into the model construction, which can compensate for the loss of the prior condition of the guided wave propagation direction. Furthermore, a scattering point solving the model for omnidirectional emission-omnidirectional reception is established.

According to the guided wave excitation, propagation, scattering, and receiving processes under directional emission conditions, the omnidirectional excitation, propagation, scattering, and receiving processes of guided waves under omnidirectional emission conditions are studied (Fig. 5.13).

In the omnidirectional emission mode, the propagation direction of the excitation guided wave is no longer limited to a certain direction, so the excited guided wave has both the incident and defect guiding beams propagating to the defect contour. There is also a direct-fed waveguide that is transmitted directly to the omnidirectional receiving EMAT and directly received by the omnidirectional receiving EMAT. Therefore, for the omnidirectional receiving EMAT, there may be three types of guided wave detection signals that may be received: a straight waveguide beam, a scattered guide

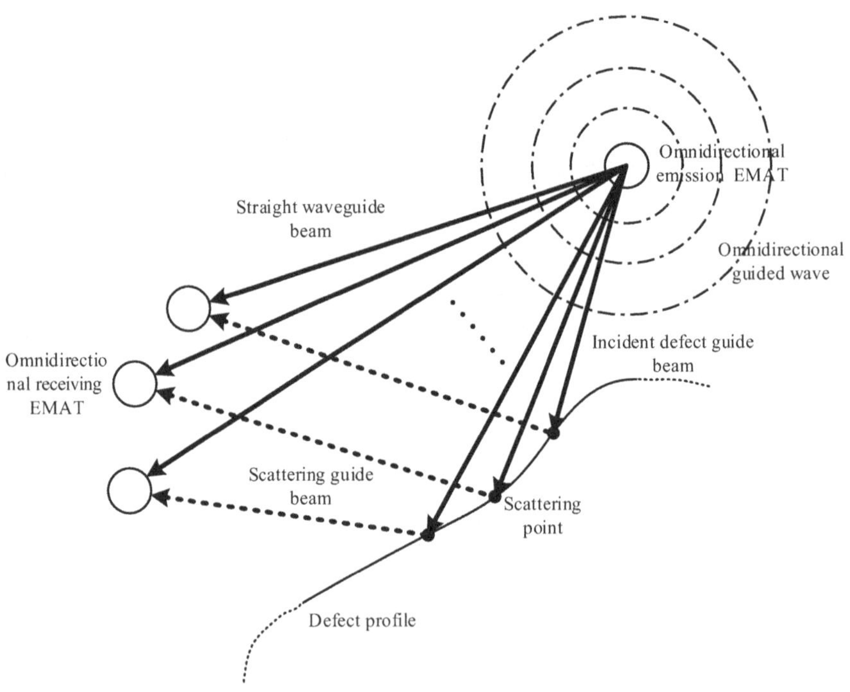

Fig. 5.13 Schematic diagram of guided wave excitation, propagation, scattering, and reception in omnidirectional emission

beam, and a transmission waveguide beam. Combined with the attenuation relationship of the guided signal strength with the guided wave propagation distance, the guided wave excitation, propagation, scattering, and receiving processes under omnidirectional emission conditions are abstracted into the plane rectangular coordinate system (Fig. 5.14).

In the plane rectangular coordinate system, the position of the omnidirectional emission EMAT is *T*, the guided wave incident on the scattering point *P* is $\overrightarrow{TP}$, the position of the scattering point is *P*, the guided wave $\overrightarrow{PR}$ position propagating from the scattering point *P* to the omnidirectional receiving EMAT is *R*, the position of the omnidirectional receiving EMAT is *R*, and the direct wave propagating directly from the omnidirectional EMAT excitation to the omnidirectional receiving EMAT is $\overrightarrow{TR}$. The gray shaded part of the figure indicates the change in the guided signal strength. The intensity of the guided wave excited at the omnidirectional EMAT is *A*. In the guided beam $\overrightarrow{TP}$, the guided signal strength decreases with distance. After reaching the scattering point *P*, the energy of the guided wave signal scattered by the scattering point *P* is lost due to the scattering effect of the defect profile, and

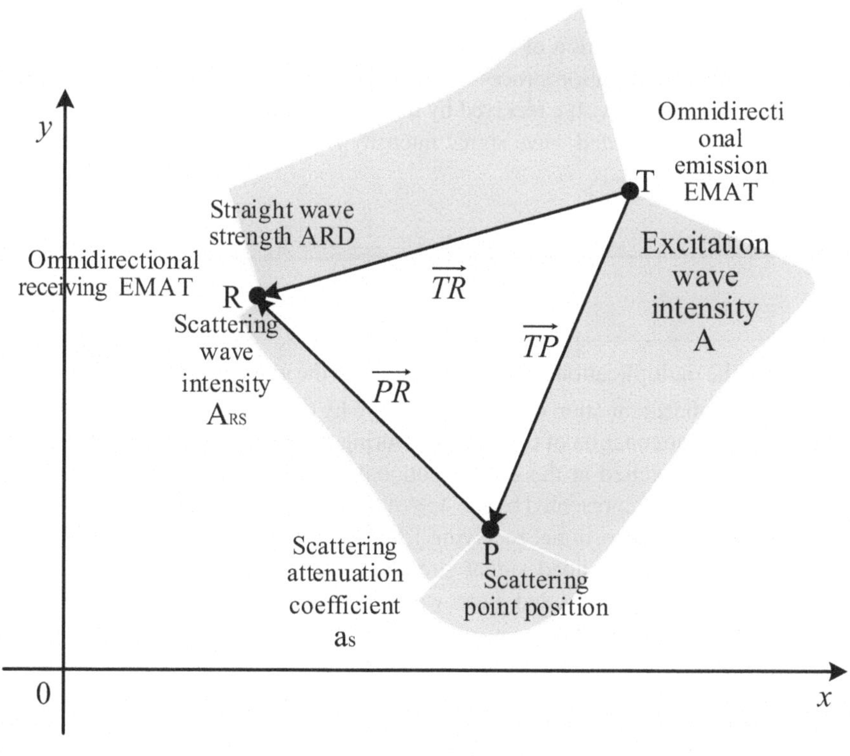

Fig. 5.14 Schematic diagram of the solution model for the scatter point position of omnidirectional emission-omnidirectional receiving magnetic acoustic array guided wave scattering imaging

the energy loss of the part of the guided wave signal is described by the scattering attenuation coefficient a_s. It is numerically equal to the ratio of the energy of the scattered guided signal to the energy of the guided signal reaching the scattering point. The scattered guided wave propagates along $\overrightarrow{PR}$ to the position R of the omnidirectional receiving EMAT, and the intensity of the scattered wave received by the omnidirectional receiving EMAT is A_{RS}; also, the intensity of the incoming signal from the omnidirectional transmitting EMAT received by the omnidirectional receiving EMAT is A_{RD}.

The ultrasonic guided wave is first excited by the direction-controllable emission EMAT of the T position and is scattered by the scattering point P and received by the omnidirectional receiving EMAT of the R position. The propagation distance S experienced by it is the sum of the length $\left|\overrightarrow{TP}\right|$ of the vector $\overrightarrow{TP}$ and the length $\left|\overrightarrow{PR}\right|$ of the vector $\overrightarrow{PR}$, $S = \left|\overrightarrow{TP}\right| + \left|\overrightarrow{PR}\right|$, and the relationship between the propagation distance and the propagation velocity v, and the scattered guided wave detection signal TOF t_r received by the omnidirectional receiving EMAT is as follows:

$$S = \left|\overrightarrow{TP}\right| + \left|\overrightarrow{PR}\right| = v * t_r \tag{5.32}$$

According to the variation of the guided wave detection signal intensity during the guided wave propagation process and the scattering process, the guided wave scattering signal intensity A_{RS} received by the omnidirectional receiving EMAT can be expressed by the guided wave signal intensity A excited by the omnidirectional emission EMAT.

$$A * \frac{1}{\sqrt{\left|\overrightarrow{TP}\right|}} * a_s * \frac{1}{\sqrt{\left|\overrightarrow{PR}\right|}} = A_{\mathrm{RS}} \tag{5.33}$$

Three of the multiplication coefficients represent the guided wave $\overrightarrow{TP}$ attenuation of the process, the attenuation of the guided wave by the scattering process $\overrightarrow{PR}$, and the guided wave attenuation of the process. During the detection process, the guided wave intensity A excited at the omnidirectional emission EMAT is unknown and therefore needs to be represented by the A_{RD} of the guided direct wave signal strength received at the omnidirectional receiving EMAT position R. The A_{RD} relationship between the excitation guided signal strength A at the omnidirectional emission EMAT position T and the guided direct wave signal strength at the omnidirectional receiving EMAT position R is

$$A * \frac{1}{\sqrt{\left|\overrightarrow{TR}\right|}} = A_{\mathrm{RD}} \tag{5.34}$$

Eliminate the unknown A, get

$$\left|\overrightarrow{TP}\right| * \left|\overrightarrow{PR}\right| = \left(\frac{A_{\mathrm{RD}} * a_{\mathrm{s}}}{A_{\mathrm{RS}}}\right)^2 * \left|\overrightarrow{TR}\right| \tag{5.35}$$

In the above formula, A_{RD}, A_{RS}, and $\left|\overrightarrow{TR}\right|$ are the known value in a single detection process; the guided wave scattering attenuation coefficient a_{s} can be solved by the intensity of the transmitted guided wave signal received by the omnidirectional receiving EMAT. The specific solution process is as follows:

Consider the omnidirectional emission-omnidirectional receiving magnetic acoustic array guided wave scattering imaging model, in the detection process of an omnidirectional EMAT excitation ultrasonic guided wave. The guided waves excited by them are divided into three categories: straight-laid waves (not shown in the figure), scattered guided waves that encounter the boundary of the defect contour (dashed guide beam in the figure) through the transmitted guided wave of the defect (solid beam in the figure), the solution of the guided wave scattering attenuation coefficient a_{s} will utilize the transmitted guiding beam. For the guided wave detection signal received by the omnidirectional receiving EMAT, it is first determined whether the guided wave detecting signal is a transmitted guiding beam. Let the intensity of the guided wave signal other than the scattered wave received by the omnidirectional receiving EMAT labeled R_i be A_{R_i}, where i is the number indicating the counting, and then, the intensity of the guided signal may be a straight wave or a transmitted wave. By comparing the intensity of the guided wave signal with the magnitude of the theoretical straight-wave signal strength to determine whether it is a transmitted wave (Fig. 5.15).

The method for solving the theoretical direct-wave signal strength $A_{\mathrm{Th}i}$ received by the omnidirectional receiving EMAT labeled R_i is

$$A_{\mathrm{Th}i} = A * \frac{1}{\sqrt{\left|\overrightarrow{TR_i}\right|}} \tag{5.36}$$

That is, the direct theoretical wave signal strength is the result of the attenuation of the excitation guided wave signal intensity A.

The guided wave simulated transmission coefficient β is defined, which indicates the ratio of the transmitted wave signal intensity to the incident wave signal intensity, which is generally 0.95 for the sake of conservatism. The guided wave is used to simulate the transmission coefficient β to act on the theoretical straight-wave signal intensity $A_{\mathrm{Th}i}$, the neighborhood where δ is the radius is taken near the applied signal strength, and the guided wave transmission intensity comparison threshold A_{Hi} is obtained in the neighborhood.

$$A_{\mathrm{Hi}} \in N_{\delta}(\beta * A_{\mathrm{Th}i}) = (\beta * A_{\mathrm{Th}i} - \delta,\ \beta * A_{\mathrm{Th}i} + \delta) \tag{5.37}$$

Among them, $N_\delta(\beta * A_{\mathrm{Th}i})$ is a neighborhood centered on $\beta * A_{\mathrm{Th}i}$ and having a radius of δ.

Thus, if the omnidirectional receiving EMAT receives the guided wave signal strength, A_{Ri} is smaller than the defined guided wave transmission intensity comparison threshold A_{Hi}. The guided wave detection signal is determined as a transmitted wave signal, and all of the omnidirectional receiving EMATs with the number R_i satisfying this condition are counted, and the total number K of omnidirectional receiving EMATs receiving the transmitted guided wave signals is obtained.

$$K = \mathrm{Count}(i)\ \text{s.t.}\ (A_{\mathrm{Ri}} < A_{\mathrm{Hi}}) \tag{5.38}$$

where Count(i) is a function that counts i.

Then, the scattering attenuation coefficient a_{s} of the signal intensity when the guided wave is scattered is solved by

$$a_{\mathrm{s}} = 1 - \frac{1}{K}\sum_{i=1}^{K}\frac{A_{\mathrm{Ri}}}{A_{\mathrm{Th}i}} \tag{5.39}$$

That is, by obtaining the average value of the transmission coefficient of the transmitted guided wave detection signal received by the omnidirectional receiving

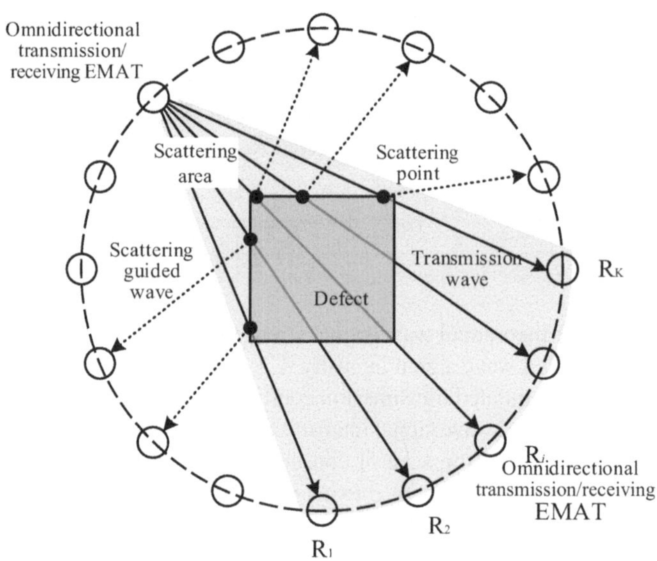

Fig. 5.15 Asymmetric emission-attenuation coefficient and scattering region solving model for omnidirectional emission-oriented receiving magneto-acoustic array guided wave scattering imaging

EMAT, the scattering attenuation coefficient a_s as the signal intensity at which the guided wave is scattered can be obtained.

Further, the fan shape R_1TR_K is obtained from the position R_i of the K omnidirectional receiving EMAT and the position T of the omnidirectional emission EMAT. The sector region R_1TR_K is defined herein as a scattering region, and it is known from the definition that in the detection of the omnidirectional emission EMAT transmitting ultrasonic guided wave labeled T today, the possible scattering point position must be located in the scattering region. Thus, the scattering region defined herein is used to determine whether the obtained scattering point is an effective scattering point; that is, if the obtained scattering point is located in the scattering region, the scattering point is an effective scattering point. Otherwise, this scattering point is an invalid scattering point.

Then

$$\begin{cases} \left|\overrightarrow{TP}\right| + \left|\overrightarrow{PR}\right| = v * t_r \\ \left|\overrightarrow{TP}\right| * \left|\overrightarrow{PR}\right| = (\frac{A_{RD} * a_s}{A_{RS}})^2 * \left|\overrightarrow{TR}\right| \end{cases} \quad (5.40)$$

In a process in which an omnidirectional emission EMAT excitation ultrasonic guided wave and an omnidirectional receiving EMAT receive a scattered guided wave detecting signal, the position T of the omnidirectional emission EMAT and the position R of the omnidirectional receiving EMAT are both known amounts. Taking the plane rectangular coordinate system as an example, the position T and the position R appear as two fixed points in the plane rectangular coordinate system, and the position of the scattering point P is the amount to be determined. The sum of the scattering point P to the omnidirectional emission EMAT position T and the omnidirectional receiving EMAT position R is a fixed value $v * t_r$, so the position of the scattering point P is located on the ellipse with the position T and the position R as the focus. Similarly, the product of the scattering point P to the omnidirectional emission EMAT position T and the omnidirectional receiving EMAT position R is also a fixed value. Therefore, the position of the scattering point P is located on the Cassini oval line with the position T and the position R as the focus. The position of the scattering point P is such that both equations are satisfied, so the position of the scattering point P is the intersection of the elliptic curve and the Cassini oval line. The elliptic curve has at most four intersections with the Cassini oval line. According to the arrangement of the magneto-acoustic array, two points outside the magneto-acoustic array can be eliminated. The two points inside the magneto-acoustic array can be screened by judging whether they are located in the scattering region. The position of the scattering point P can be obtained, and all the scattered point positions P obtained are subjected to three smooth spline interpolations to obtain a contour curve of the defect to be detected.

In this section, the omnidirectional transmit-omnidirectional receiving magnetic acoustic array guided wave scattering imaging model is established. It is pointed out that the calculation of the scattering point position is still the key to the model solution, and the relationship between the guided wave detection signal strength and the guided wave propagation distance is introduced. The attenuation effect of the boundary contour scattering on the guided signal strength is quantitatively described by the guided wave scattering attenuation coefficient, and the calculation method of the guided wave scattering attenuation coefficient is given. The calculation process of the guided wave scattering attenuation coefficient is completely dependent on the various types of guided wave detection signals received by the omnidirectional receiving EMAT and does not depend on any other function, so it has good adaptability and versatility. A simultaneous equation for solving the position of the scattering point is established by the TOF constraint and signal strength constraint of the scattered guided wave detection signal. The concept of the guided wave scattering region is defined to determine whether the obtained scattering point is an effective scattering point, and then, the scattering point solution result is screened.

In the following, for the specific detection process, based on the defect contour omnidirectional emission-omnidirectional receiving magnetic acoustic array guided wave scattering imaging model proposed in this section, the omnidirectional emission-omnidirectional receiving magnetic acoustic array guided wave scattering imaging detection method and steps are described.

5.2.2 Omnidirectional Emission-Omnidirectional Receiving Magnetic Acoustic Array Guided Wave Scattering Imaging Method and Steps

This section is aimed at the actual detection process of steel plate defects, combined with the omnidirectional emission-omnidirectional receiving magnetic acoustic array guided wave scattering imaging model proposed in the previous section. The omnidirectional emission-omnidirectional receiving magnetic acoustic array guided wave scattering imaging method and its steps are given in the actual detection process.

Figure 5.16 shows a block diagram of a specific implementation principle of the omnidirectional emission-omnidirectional receiving magnetic acoustic array guided wave scattering imaging method. This block diagram is only to explain the specific steps of the omnidirectional emission-omnidirectional receiving magnetic acoustic array guided wave scattering imaging method. However, it is not limited to the shape of the magneto-acoustic array, the number of omnidirectional EMATs, the type of coordinate system, and the position of the coordinate origin.

The omnidirectional EMAT is used as the excitation transducer and the receiving transducer. When the omnidirectional EMAT is used as an excitation transducer, it is responsible for exciting the ultrasonic guided wave, and all other omnidirectional EMATs are used as receiving transducers. Combined with the omnidirectional emission-omnidirectional receiving magnetic acoustic array guided wave scattering imaging model, the steps of the omnidirectional emission-omnidirectional receiving magnetic acoustic array guided wave scattering imaging method are as follows:

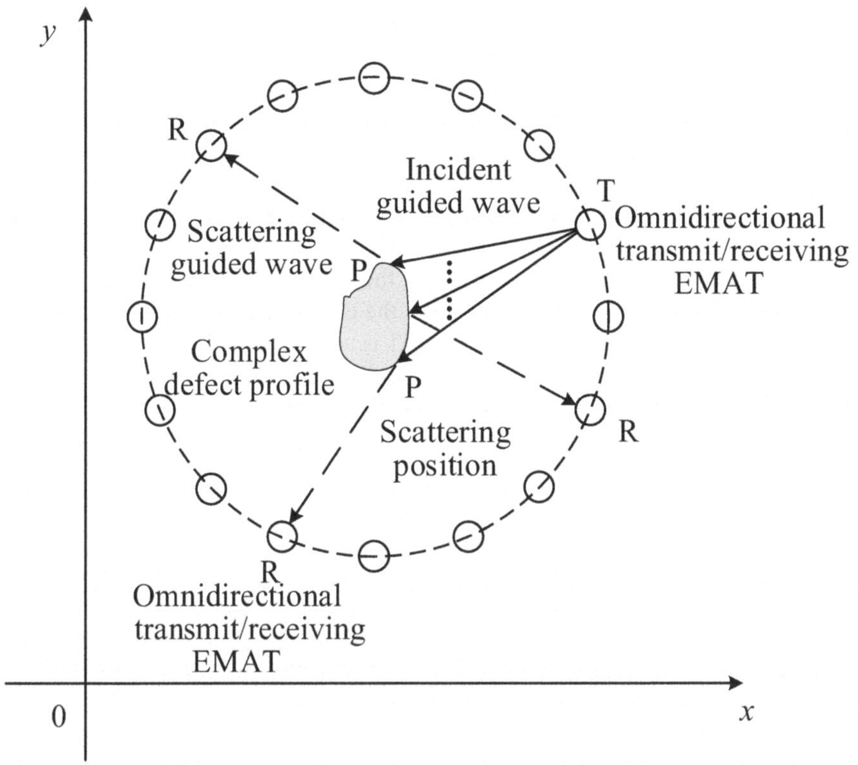

Fig. 5.16 Schematic diagram of omnidirectional emission-omnidirectional receiving magnetic acoustic array guided wave scattering imaging method

(1) N omnidirectional EMATs are evenly arranged in a circular array around the detection area of the steel sheet to be tested, wherein N is a positive integer; each omnidirectional EMAT is responsible for exciting the omnidirectional ultrasonic guided wave in a certain order in the steel plate and is also responsible for receiving the ultrasonic guided wave detecting signal in the omnidirectional direction when there is ultrasonic guided wave in the steel plate.
(2) Selecting the nth omnidirectional emission EMAT as the excitation magnetic acoustic transducer T_n of the present detection, wherein $n = 1, 2, \ldots, N$, and exciting the omnidirectional ultrasonic guided wave in the steel plate.
(3) The omnidirectional receiving EMAT receives the ultrasonic guided wave detection signal in the steel plate, assuming that a total of M omnidirectional receiving EMATs receive the guided wave signal. Expressed as R_m, where $m = 1, 2, \ldots, M$, M are positive integers; calculate the TOF and signal strength of each guided wave detection signal received by the omnidirectional receiving EMAT.

(4) It is judged one by one whether R_m in step (3) and T_n in step (2) constitute a scattering group (T_n, R_m). If yes, proceed to step (5); if not, the omnidirectional receiving EMAT number becomes R_{m+1} and returns to step (3).
The TOF judgment of the ultrasonic guided wave detection signal is used to determine whether the omnidirectional receiving EMAT (R_m) of the guided wave detection signal and the omnidirectional emission EMAT (T_n) for performing the omnidirectional excitation at this time constitute a scattering group. For an ultrasonic guided wave detection signal received by the omnidirectional receiving EMAT, the traveling time is t_r, the ultrasonic guided wave propagates in the steel plate at a velocity v, and in the planar rectangular coordinate system, the omnidirectional emission EMAT is at a position T. The position of the omnidirectional receiving EMAT is R, and the theoretical time t_s used for the ultrasonic guided wave signal to propagate directly from the position T to the position R along a straight line is

$$t_s = \frac{\left|\overrightarrow{TR}\right|}{v} \tag{5.41}$$

If $t_r > t_s$, R_m and T_n form a scattering group for the ultrasonic guided wave signal; otherwise, R_m and T_n do not constitute a scattering group for the ultrasonic guided wave signal. The above judgment of the scattering group is based on the fact that if the omnidirectional receiving EMAT receives the scattered guided wave detection signal, the TOF should be longer than the TOF of the direct wave signal or the transmitted wave signal received by the omnidirectional receiving EMAT.

(5) For the scattering group (T_n, R_m), using each guided wave detection signal TOF, each guided wave detection signal strength, and the corresponding scattering group position, calculate the scattering point position P.

(6) It is judged whether or not the scattering point position P obtained in step (5) is located in the scattering region, and if so, the solution result is the effective scattering point, and the step (7) is performed. If not, the solution result is an invalid scattering point, and the result is eliminated from the scattering point solution result set.

(7) It is judged whether all the omnidirectional EMATs have been excited by the omnidirectional ultrasonic guided waves on the steel sheet, and if so, the step (8) is performed. If not, the omnidirectional emission EMAT becomes T_{n+1} and returns to step (2).

(8) Using the interpolation method, all the scattered points obtained are connected into a curve to obtain a clear contour image of the complex defect.

Based on the above method steps, Fig. 5.17 shows a flow chart of the omnidirectional emission-omnidirectional receiving magnetic acoustic array guided wave scattering imaging method.

This section elaborates on the specific implementation steps of the omnidirectional emission-omnidirectional receiving magnetic acoustic array guided wave scattering

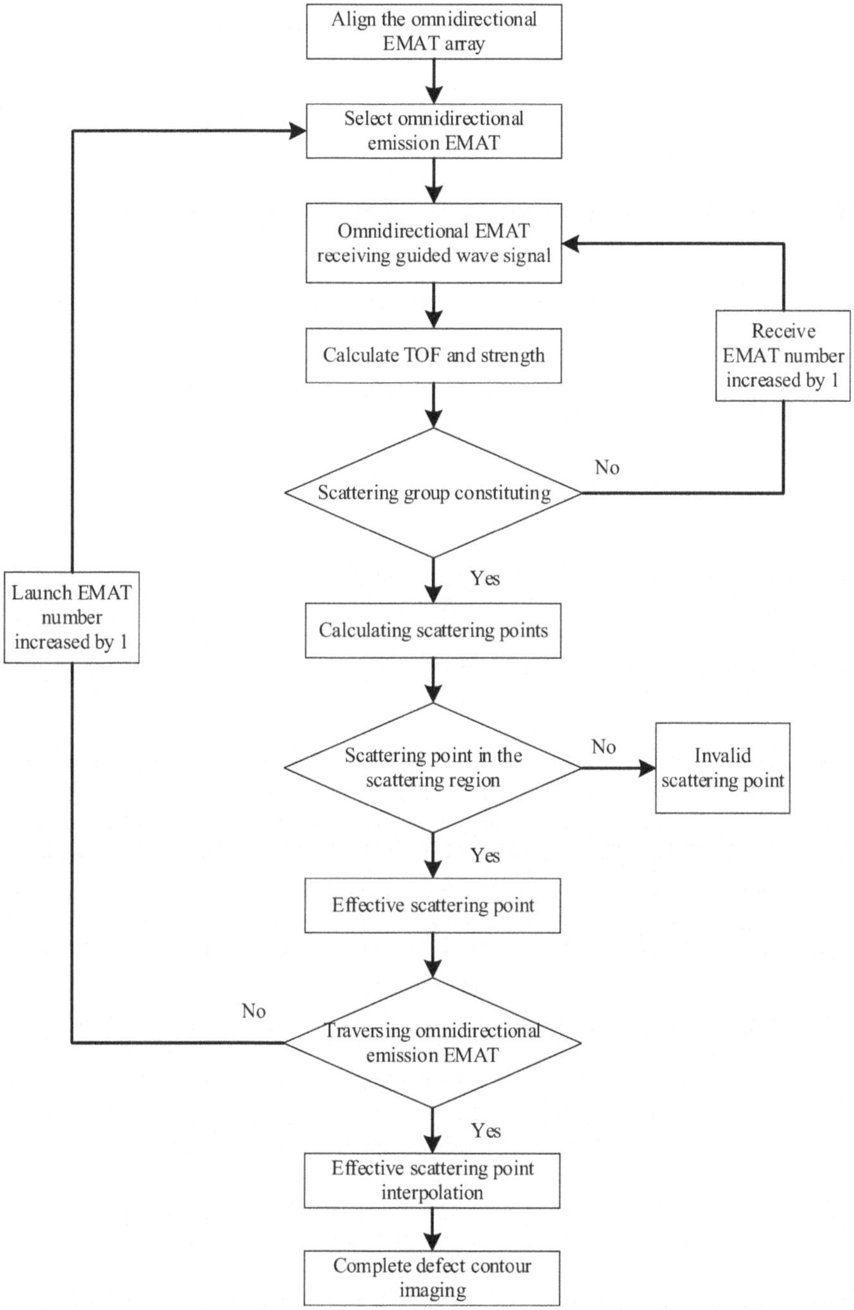

Fig. 5.17 Flow chart of omnidirectional emission-omnidirectional receiving magnetic acoustic array guided wave scattering imaging method

imaging method. A specific implementation form and array structure of an omnidirectional transmit-omnidirectional receiving magnetic acoustic array guided wave scattering imaging method are presented.

The actual defect contours are complex and different in shape, and the sensor array geometry of a specific regular shape cannot maintain high sensitivity to defects of various contour shapes. To this end, based on the defect contour results of omnidirectional emission-omnidirectional receiving magnetic acoustic array guided wave scattering imaging, the magnetic acoustic array structure optimization method is studied in order to reduce the defect contour reconstruction error.

5.2.3 Omnidirectional Transmission-Omnidirectional Receiving Magnetic Acoustic Array Structure Optimization Method

The purpose of adjusting the structure of the magnetic acoustic array is to improve the imaging accuracy of the defect contour. Therefore, the basic starting point or adjustment basis of the array structure adjustment is the relative error between the defect calculation contour and the real contour. For the actual inspection project, when the defect profile is imaged, the true contour of the defect is unknown in many cases. The basis for evaluating the defect and the safety evaluation of the structural member is generally only the defect profile detection result. For this book, it is the calculated contour of the defect. The actual engineering inspection experience shows that there is a certain relationship between the relative error between the defect calculation contour and the real contour and the calculated contour curve of the defect. That is, the local contour region with a relatively large relative error corresponds to a local region where the calculated contour curve of the defect changes sharply, and the region where the defect calculation contour curve is relatively gentle has a relatively small relative error between the corresponding defect calculation contour and the real contour. Therefore, the degree of change in the local area of the defect calculation profile can be used to some extent to characterize the relative error that may exist between the defect calculation profile and the real profile. Therefore, the degree of variation of the local area of the defect calculation profile curve can be considered as a reference for the adjustment of the structure of the magneto-acoustic array.

Defect calculation of the curvature of the local region of the contour curve can more accurately reflect the degree of change of the calculated contour curve in the local region; that is, the local region with large curvature of the contour curve is calculated, and the degree of curve change is relatively large. The local area with a small curvature of the contour curve is calculated, and the degree of curve change is relatively small. Based on the above analysis, this section uses the defect to calculate the curvature of the local region of the contour curve as an important basis for the structure adjustment and optimization of the magneto-acoustic array.

Using the omnidirectional emission-omnidirectional receiving magnetic acoustic array guided wave scattering imaging model and method proposed in the previous section, the position of the scattering point and its coordinates are obtained. A total of S scattering points (S is a positive integer) are obtained, and the coordinates are expressed as a polar coordinate form $P_i(\theta_i, r_i)$, where $i = 1, 2, \ldots, S$. Three smooth spline interpolations are performed on the scattering point polar coordinate data to form a defect contour curve in polar coordinates and its function $\mathrm{PC}(\theta_j, r_j)$:

$$\mathrm{PC}(\theta_j, r_j) = \mathrm{CSplineI}[P_i(\theta_i, r_i)] \tag{5.42}$$

where $j = 1, 2, \ldots, S1$, $S1$ are positive integers, indicating the total number of points on the defect contour curve after interpolation. CSplineI is a function of three smooth spline interpolations on the scattering point polar coordinates $P_i(\theta_i, r_i)$.

The curvature of the point on the defect calculation profile is calculated by solving the defect calculation of the contour curve interpolation function $\mathrm{PC}(\theta_j, r_j)$ at the first and second derivatives of the point. For a point $P_{Cj}(\theta_j, r_j)$ on the defect calculation curve, the calculation method of the curvature $C_j(\theta_j, r_j)$ is

$$C_j(\theta_j, r_j) = \frac{\left| r_j^2 + 2 * (\frac{\mathrm{d}r}{\mathrm{d}\theta}|\theta_j)^2 - r_j * (\frac{\mathrm{d}^2 r}{\mathrm{d}\theta^2}|\theta_j) \right|}{[r_j^2 + (\frac{\mathrm{d}r}{\mathrm{d}\theta}|\theta_j)^2]^{\frac{3}{2}}} \tag{5.43}$$

According to the above formula, the curvature $C_j(\theta_j, r_j)$ of each point on the contour curve of the defect calculation can be calculated. The magnetic acoustic array structure is defined to adjust the curvature threshold C_{TH}. When the calculated curvature value of a point on the defect calculation profile exceeds the structural adjustment curvature threshold C_{TH}, the point is included in the magneto-acoustic array structure adjustment area. The magnetic acoustic array adjustment region is represented by a set of angular coordinates in a polar coordinate system, represented as one or more angular collection ranges. Therefore, in the polar coordinate system, the magnetic acoustic array structure adjustment region $R(\theta)$ is defined as

$$R(\theta) = \arg\{\theta_j\} \text{ s.t. } C_j(\theta_j) > C_{\mathrm{TH}} \tag{5.44}$$

Based on the solved magnetic acoustic array structure adjustment region $R(\theta)$, by the basic structure of the original magneto-acoustic array, it is located outside the basic structure adjustment region of the original magneto-acoustic array. The array structure is not changed; for the structural adjustment region $R(\theta)$, the omnidirectional EMAT number density in the magneto-acoustic array adjustment region $R(\theta)$ is increased, and the adjacent omnidirectional EMAT spacing is reduced. The omnidirectional EMAT is mainly concentrated inside the array structure adjustment area. Figure 5.18 shows a schematic diagram of the structure optimization of the magneto-acoustic array.

Based on the adjusted structure of the magneto-acoustic array, the omnidirectional emission-omnidirectional receiving magneto-acoustic array guided wave scattering

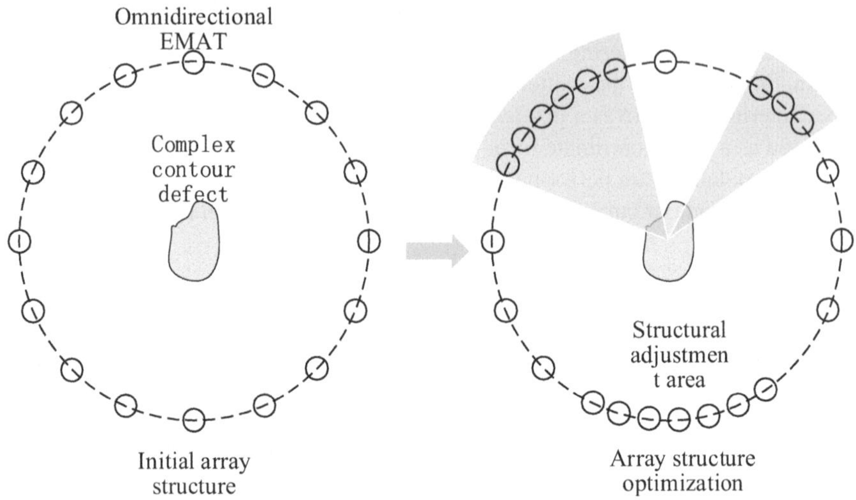

Fig. 5.18 Schematic diagram of magnetic acoustic array structure optimization adjustment

imaging method is used to detect the defect contour again, and the scatter point data after the array structure adjustment are obtained. The data of the guided wave scattering defect calculation contour curve before and after the adjustment of the magnetic acoustic array structure are performed: For the two defect calculation contour curves, the scattering points in the structural adjustment region are calculated by using the optimized array structure, and the contour curve points are calculated. The scattering points outside the structural adjustment area are calculated using the initial array structure. Specifically, in the polar coordinate system, the defect calculation contour curve and function calculated after the array adjustment are PC(θ_k, r_k), and the points on the defect calculation contour curve are represented as $P_{Ck}(\theta_k, r_k)$, where $k = 1$, 2, …, K, K are positive integers, indicating the total number of points on the defect calculation profile calculated after the array is adjusted. The data fusion method for calculating the contour curve of the guided wave scattering defect before and after the adjustment of the array structure is

$$\mathrm{PC}_F(\theta_m, r_m) = \{P_{Ck}(\theta_k, r_k)\ \text{s.t.}\ \theta_m \in R(\theta)\} \bigcup \{P_{Cj}(\theta_j, r_j)\ \text{s.t.}\ \theta_m \notin R(\theta)\} \tag{5.45}$$

Among them, $\mathrm{PC}_F(\theta_m, r_m)$ is a set of points for the defect after data fusion and a set of points on it, $m = 1, 2, \ldots, M$, M are positive integers. It indicates the total number of points on the contour curve of the defect calculation after data fusion.

This section presents a magneto-acoustic array structure optimization model for omnidirectional emission-omnidirectional receiving magnetic acoustic array guided wave scattering imaging. It demonstrates the necessity and rationality of determining the structural adjustment area and structural optimization based on the curvature of the contour curve. In the polar coordinate system, the calculation method of the curvature of any point of the defect calculation curve is given. The adjustment area of the magneto-acoustic array structure is studied and defined, and the optimization method of the structure of the magneto-acoustic array based on the structure adjustment area is proposed. The data fusion method of the contour curve of the guided wave scattering defect before and after the adjustment of the structure of the magneto-acoustic array is given.

In the following, for the specific detection process of the steel plate defect, based on the magnetic acoustic array structure optimization model proposed in this section, the steps of the omnidirectional emission-omnidirectional receiving magnetic acoustic array structure optimization adjustment method are specifically described.

This section is based on the actual detection process of steel plate defects, based on the optimization model of the magnetic acoustic array structure proposed in the previous section, and gives the steps of the optimization adjustment method of the omnidirectional emission-omnidirectional receiving magnetic acoustic array structure in the actual detection process, which is described as follows:

(1) N omnidirectional EMATs are evenly arranged in a regular array around the steel plate detection area to be tested, wherein N is a positive integer; each omnidirectional EMAT is responsible for exciting the omnidirectional ultrasonic guided wave in a certain order in the steel plate and is also responsible for receiving the ultrasonic guided wave signal omnidirectionally when there is the ultrasonic guided wave in the steel plate.
(2) Solving the scattering point position $P_i(\theta_i, r_i)$ by using the scatter scattering waveguide amplitude and TOF received by the omnidirectional receiving EMAT. Performing three smooth spline interpolations on the scattering point coordinate data to form a defect calculation contour curve PC(θ_j, r_j).
(3) In the plane polar coordinate system, a point $P_{Cj}(\theta_j, r_j)$ on the contour curve is calculated for the defect. Calculating the curvature $C_j(\theta_j, r_j)$ of the point on the defect contour curve by calculating the defect contour calculation curve interpolation function at the first and second derivatives of the point.
(4) Determining whether the curvature of the point on the defect calculation curve in step (3) exceeds the structural adjustment curvature threshold, and if so, the point enters the adjustment area; if not, proceeding to step (5).
(5) Judging whether the curvature of all points on the defect calculation curve has been calculated and compared with the structural adjustment curvature threshold, and if so, proceed to step (6). If not, the point on the defect calculation contour curve becomes $P_{Cj+1}(\theta_{j+1}, r_{j+1})$ and returns to step (3).
(6) In the plane polar coordinate system, determine the magnetic acoustic array adjustment region $R(\theta)$, optimize the magnetic acoustic array structure, and calculate the guided wave scattering defect calculation contour curve again.
(7) The contour curve of the guided wave scattering defect before and after the adjustment of the structure of the magneto-acoustic array is data-fused to form a high-precision image of the defect contour.

Based on the above method steps, Fig. 5.19 shows a flow chart of the omnidirectional transmission-omnidirectional receiving magnetic acoustic array structure optimization adjustment method.

This section elaborates on the specific implementation steps of the omnidirectional emission-omnidirectional receiving magnetic acoustic array structure optimization adjustment method. In the following, specific steel plate defect contour imaging inspection work will be carried out for steel plate regular contour defects and complex contour defects. The proposed omnidirectional emission-omnidirectional receiving magnetic acoustic array guided wave scattering imaging model method and magnetic acoustic array structure optimization adjustment method are tested and verified.

5.2.4 Steel Plate Regular Contour Defect Guided Wave Scattering Imaging and Array Adjustment Test Verification

In order to verify the proposed omnidirectional emission-omnidirectional receiving magnetic acoustic array guided wave scattering imaging method and magnetic acoustic array structure optimization adjustment method, on the 4-mm-thick steel plate, a test platform for the omnidirectional emission-to-omnidirectional receiving magneto-acoustic guided wave scattering imaging inspection of steel plates with regular contour defects was built, and a series of related testing tests were carried out on this test platform. Steel plate regular contour defect omnidirectional emission-omnidirectional receiving magnetic acoustic array guided wave scattering imaging detection test platform principle block diagram is shown in Fig. 5.20.

All omnidirectional EMATs are magnetostrictive omnidirectional EMATs based on nickel strips and densely wound coils in Chap. 2. A total of 16 omnidirectional EMATs are arranged in a circular array uniformly around the regular contour defects of the steel sheet. Moreover, the circular magnetic acoustic array is centered on the defect. Each omnidirectional EMAT excites the ultrasonic guided waves in a certain order. The other 15 omnidirectional EMATs are responsible for receiving the guided wave detection signals when there is an ultrasonic guided wave signal. Therefore, 240 sets of guided wave detection signals are obtained. The working principle of the power amplifier, computer, acquisition card, and signal processor in the above test platform is the same as that of the corresponding components in the test platform in Fig. 2.13 and will not be described here.

The regular contour defects of the steel plate in this test include square contour defects, regular triangle contour defects, and circular contour defects. All three types of regular defects are subjected to contour imaging test under the same test parameters. The test parameter setting during the test is the same as that of the steel plate regular contour defect directional emission-omnidirectional receiving magnetic acoustic array guided wave scattering imaging test parameter and will not be described here. According to the above test parameters, the omnidirectional emission

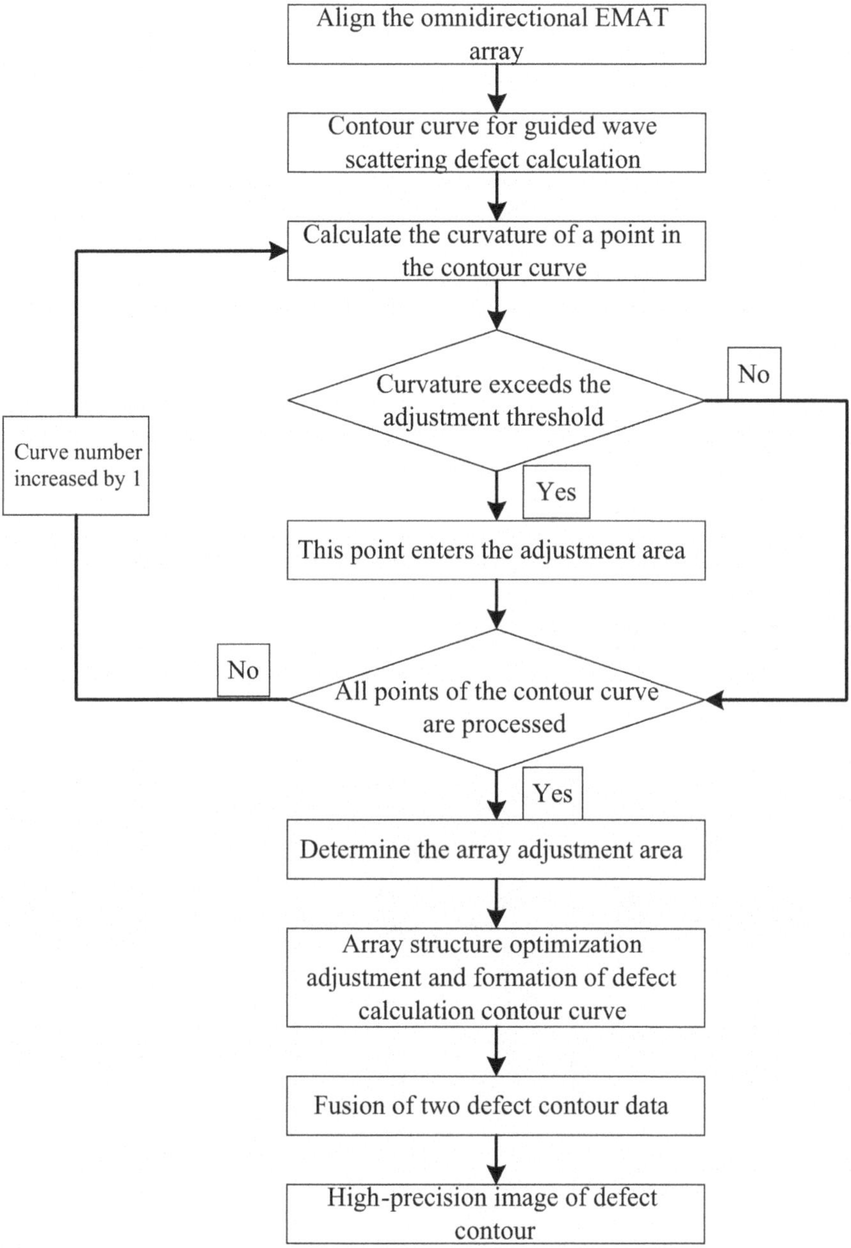

Fig. 5.19 Flow chart of the omnidirectional transmission-omnidirectional receiving magnetic acoustic array structure optimization adjustment method

of the steel plate regular contour defect omnidirectional receiving magnetic acoustic array guided wave scattering imaging detection and array adjustment test is carried out. The test results are analyzed and discussed below.

According to the above section, the omnidirectional emission-omnidirectional receiving magnetic acoustic array guided wave scattering imaging and magnetic acoustic array structure adjustment method steps. Under the above test parameters, the image detection test was performed on the square defects of the steel plate, the triangle defects and the contours of the circular defects. The position of the scattering point is calculated separately and a calculated defect profile is formed. For the sake of comparison, the calculated contour of the defect and the contour of the defect are respectively drawn onto an image, and the calculated scattering points of the three defects and the calculated contour are as shown in Fig. 5.21.

In the rule mentioned above defect contour imaging results, in general, the calculated contour curve reflects the corresponding true defect contour, and most of the calculated scattering points can also maintain a good fit with the real defect contour. Compared with Fig. 4.8, the omnidirectional emission-omnidirectional receiving magneto-acoustic array guided wave scattering imaging method obtains more scattering points than the directional emission-omnidirectional receiving magnetic acoustic array guided wave scattering imaging method. Moreover, there is a better fit for the contour of the circular defect.

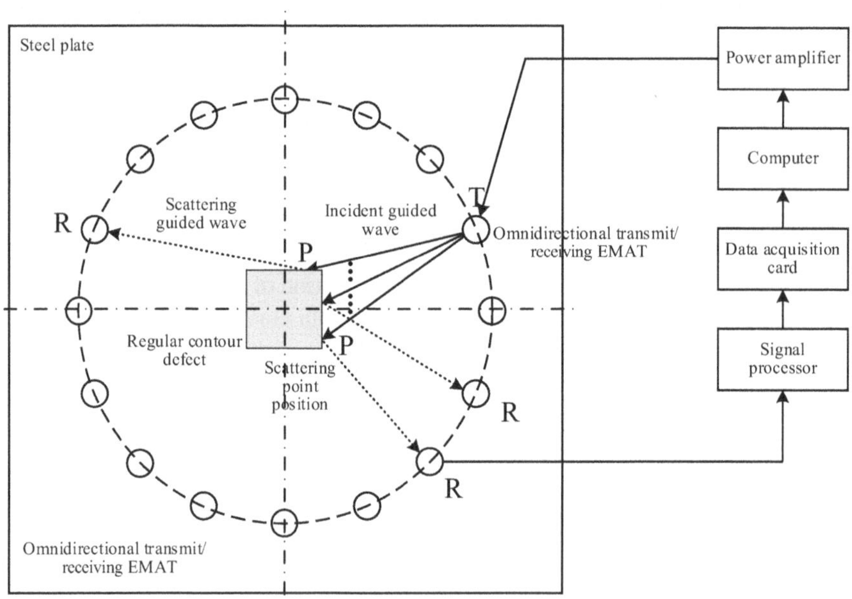

Fig. 5.20 Steel plate regular contour defect omnidirectional emission-omnidirectional receiving magnetic acoustic array guided wave scattering imaging detection test platform block diagram

In the following, the omnidirectional emission-omnidirectional receiving magnetic acoustic array structure optimization adjustment model and method proposed in this chapter are adopted, and the magnetic acoustic array structure is optimized and re-imaged for the steel mentioned above plate regular contour defect imaging result. According to the proposed omnidirectional emission-omnidirectional receiving magnetic acoustic array structure optimization adjustment method steps, the curvature of each point on the calculated contour curve of the square defect, the equilateral triangle defect, and the circular defect is respectively obtained. Moreover, in the Cartesian coordinate system, draw the positional relationship between the curvature value and the coordinates, as shown in Fig. 5.22.

The curvature values of the square defect and the equilateral triangle defect at the corner of the calculated contour are relatively large; the curvature value of each point on the contour curve of the circular defect calculation is significantly smaller than that of the square defect and the equilateral triangle defect, and the curvature coordinate dimension is only 0.15. The corresponding square defects and regular triangle defects are 0.8 and 20, respectively; the reason why the curvature distribution of the circular defect calculation curve is sharply changed is also that the curvature

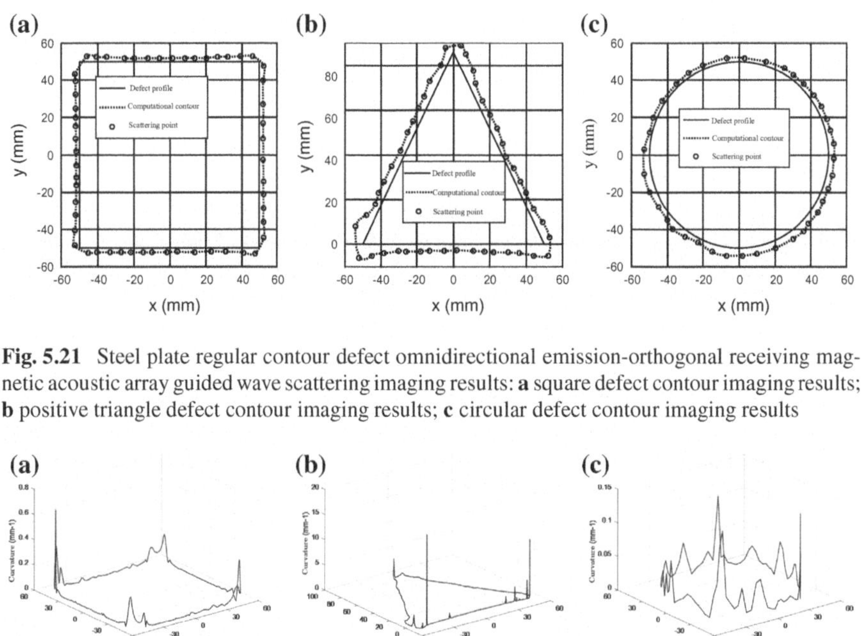

Fig. 5.21 Steel plate regular contour defect omnidirectional emission-orthogonal receiving magnetic acoustic array guided wave scattering imaging results: **a** square defect contour imaging results; **b** positive triangle defect contour imaging results; **c** circular defect contour imaging results

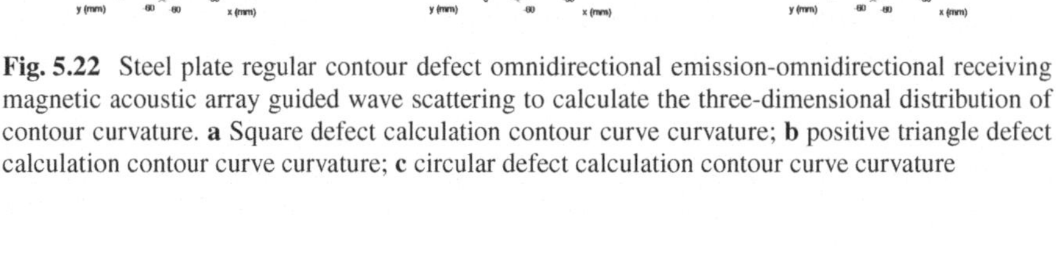

Fig. 5.22 Steel plate regular contour defect omnidirectional emission-omnidirectional receiving magnetic acoustic array guided wave scattering to calculate the three-dimensional distribution of contour curvature. **a** Square defect calculation contour curve curvature; **b** positive triangle defect calculation contour curve curvature; **c** circular defect calculation contour curve curvature

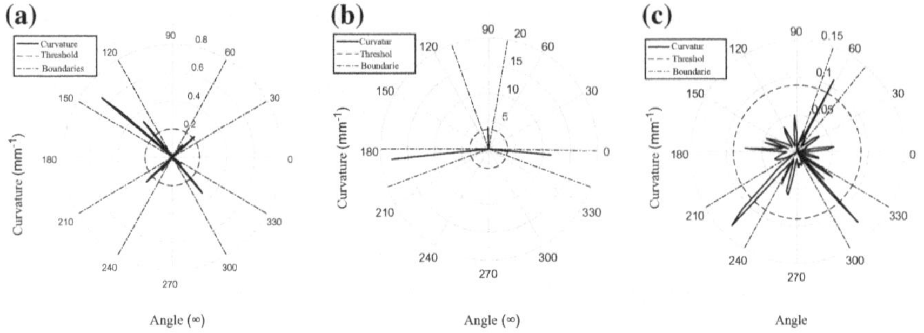

Fig. 5.23 Steel plate regular contour defect omnidirectional emission-omnidirectional receiving magnetic acoustic array structure adjustment area. **a** Square defect array structure adjustment area; **b** equilateral triangle defect array structure adjustment area; **c** circular defect array structure adjustment area

Table 5.3 Steel plate regular contour defect omnidirectional emission-omnidirectional receiving magnetic acoustic array structure adjustment area

Defect type	Array structure adjustment area (circumferential angle a ° ~ circumferential angle b °)
Square defects	30 ~ 60, 120 ~ 150, 210 ~ 240, 300 ~ 330
Normal triangle defect	80 ~ 110, 180 ~ 200, 340 ~ 360
Round defects	50 ~ 70, 210 ~ 240, 300 ~ 330

of each point is small. Therefore, the change with little curvature will still show more severe fluctuations in its curvature profile.

According to the array structure adjustment region solving method proposed in this chapter, the array structure adjustment regions of the above three kinds of defects are calculated and plotted in the polar coordinate system. The solution results are shown in Fig. 5.23.

The circumferential angle range of the array structure adjustment area of the three defect calculation profile curves is shown in Table 5.3.

According to the three kinds of defects obtained above, the circumferential angle range of the array structure adjustment region of the contour curve is calculated, and the magneto-acoustic array structure is adjusted according to the magnetic acoustic array structure optimization adjustment method proposed herein; the adjusted magnetic acoustic array structure is adopted. The omnidirectional emission-omnidirectional receiving magneto-acoustic array guided wave scattering imaging method is used to perform defect contour reconstruction on the three kinds of defects, respectively, and the scattered point data after the array structure adjustment are obtained. The contours of the guided wave scattering defects before and after the adjustment of the structure of the magneto-acoustic array are respectively data fusion, and the three kinds of defects after data fusion are calculated to calculate the scattering points and the calculated contour curves. Similarly, for the sake of comparison, the calculated contour of the defect and the contour of the defect are respectively

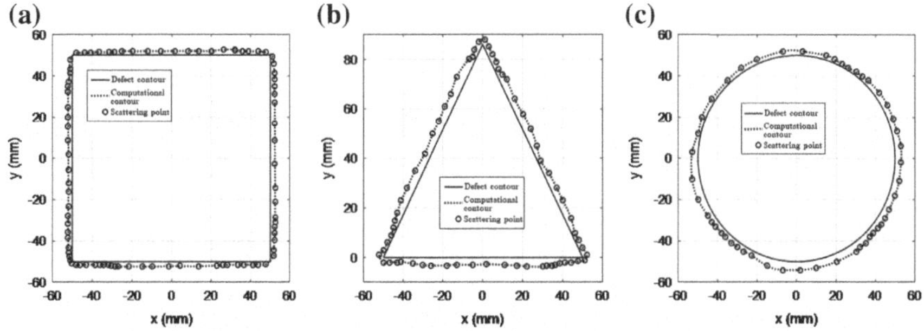

Fig. 5.24 Steel plate regular contour defect omnidirectional emission-orthogonal receiving magnetic acoustic array adjusted guided wave scattering imaging results. **a** Square defect contour imaging results; **b** positive triangle defect contour imaging results; **c** circular defect contour imaging results

drawn onto an image, and the calculated scattering points of the three defects and the calculated contour are shown in Fig. 5.24.

According to the results of the regular defect guided wave scattering imaging after the array structure adjustment, the calculated scattering point position in the array structure adjustment region is closer to the true defect contour than the initial array guided wave scattering imaging result. The number of calculated scatter points in the array structure adjustment area increases. The imaging accuracy at the corners of square and equilateral triangle defects is improved, especially for the three vertices of the equilateral triangle defect, and the calculated contour curve and the scattering point position of the array structure are significantly improved relative to the initial array structure imaging results.

In order to quantitatively describe the error of the contour curve and the real contour, the relative error between the contour and the real contour is calculated by using the defined defect, and the relative error between the three defect calculation contours and the real contour before and after the array structure adjustment optimization is calculated, respectively. The error data before and after the adjustment of each defect array structure are uniformly normalized, and the normalized relative error is plotted on the two-dimensional contour plane according to the defect real contour interpolation, as shown in Fig. 5.25.

Similarly, in order to evaluate the relative error between the defect calculation contour and the real contour point, the average contour error of the contour curve and the real contour before and after optimization of the three defect magneto-acoustic array structures are calculated according to the average contour error. The calculation results are shown in Table 5.4.

According to Table 5.4, the average contour error of the three rule defects is reduced after the optimization of the magneto-acoustic array structure, as low as 2.5%.

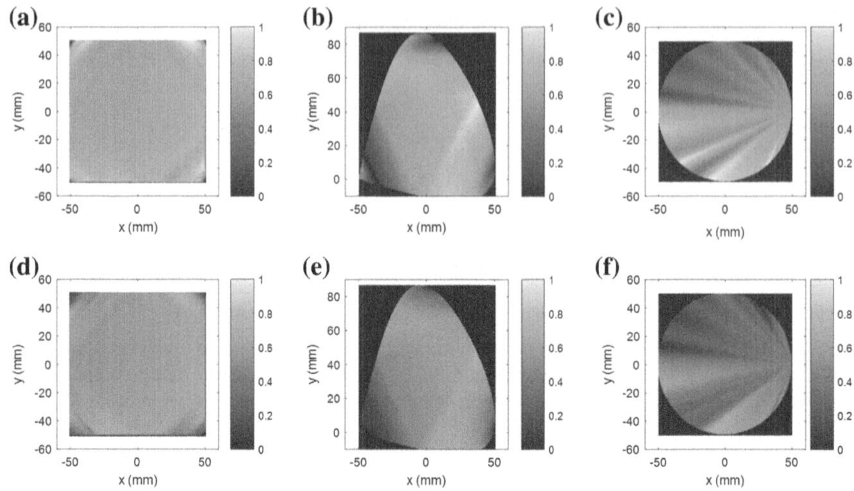

Fig. 5.25 Steel plate regular contour defect omnidirectional emission-comparison of relative error before and after adjustment of omnidirectional receiving magnetic acoustic array. **a** The relative error of the square defect profile before the array structure adjustment; **b** the relative error of the regular triangle defect profile before the array structure adjustment; **c** relative error of circular defect profile before array structure adjustment; **d** relative error of square defect profile after array structure adjustment; **e** relative error of normal triangle defect profile after array structure adjustment; **f** the relative error of circular defect profile after array structure adjustment

Table 5.4 Steel plate regular contour defect directional emission-average contour error before and after omnidirectional receiving magnetic acoustic array optimization

Imaging error	Square defect (%)	Triangular defect (%)	Round defect (%)
Initial array structure	2.11	3.06	3.15
Optimize array structure	1.84	2.19	2.24

In summary, the omnidirectional emission-omnidirectional receiving magneto-acoustic array guided wave scattering imaging method has a smaller average imaging error for the contour defect of the steel plate and exhibits a higher regular contour defect contour reconstruction accuracy. The magnetic acoustic array structure optimization adjustment method proposed in this chapter is used to further reduce the contour reconstruction error of the region with large reconstruction error under the initial array structure and effectively improve the contour reconstruction accuracy of the regular contour defect.

In the following, for the complex contour defect of steel plate, the proposed omnidirectional emission-omnidirectional receiving magnetic acoustic array guided wave scattering imaging and its magnetic acoustic array structure optimization adjustment method are tested and verified.

5.2.5 *Steel Plate Complex Contour Derivative Guided Wave Scattering Imaging and Array Adjustment Test Verification*

In the previous section, the proposed omnidirectional emission-omnidirectional receiving magneto-acoustic array guided wave scattering imaging and its magnetic acoustic array structure optimization adjustment method are tested and verified. This section will test the complex contour defects of steel plates. The test platform used in this section has the same principle as the omnidirectional emission-to-one-direction accommodating magneto-acoustic guided wave scatter imaging detection test platform, except that the test object is a complex contour defect of the steel plate. The defect depth is also 1.5 mm, and the thickness of the steel plate is also 4 mm. The initial magnetic acoustic array layout and working mode, and the relevant test parameter settings are also the same as in the previous section and will not be described here.

According to the proposed omnidirectional emission-omnidirectional receiving magneto-acoustic array guided wave scattering imaging method step, under the test mentioned above parameters, the contour detection of the complex contour defect of the steel plate is carried out. Calculate the position of the scattering point and form a defect calculation profile. For the sake of comparison, the calculated contour of the complex contour defect and the contour of the defect itself are drawn onto an image, and the calculation of the scattering point and the calculation of the contour are shown in Fig. 5.26.

In the above complex defect contour imaging results, in general, the calculated contour curve reflects the corresponding true defect contour, and most of the calculated scattering points can also maintain a good fit with the real defect contour. The omnidirectional emission-omnidirectional receiving magnetic acoustic array guided wave scattering imaging method obtains more scattered points than the directional emission-omnidirectional receiving magnetic acoustic array guided wave scattering imaging method. However, there are many serious bending trends in the distribution of scattering points.

In the following, the omnidirectional emission-omnidirectional receiving magneto-acoustic array structure optimization adjustment model and method proposed in this chapter are used to optimize the magneto-acoustic array structure and re-image the steel mentioned above plate complex contour defect imaging results. According to the proposed omnidirectional emission-omnidirectional receiving magnetic acoustic array structure optimization adjustment method step, the complex contour defect is calculated to calculate the curvature of each point on the contour curve, and the positional relationship between the curvature value and the coordinate is drawn in the Cartesian coordinate system. At the same time, according to the array structure adjustment region solving method proposed in this chapter, the array structure adjustment region of the complex contour defect calculation contour curve is obtained and plotted in the polar coordinate system; the solution result is shown in Fig. 5.27.

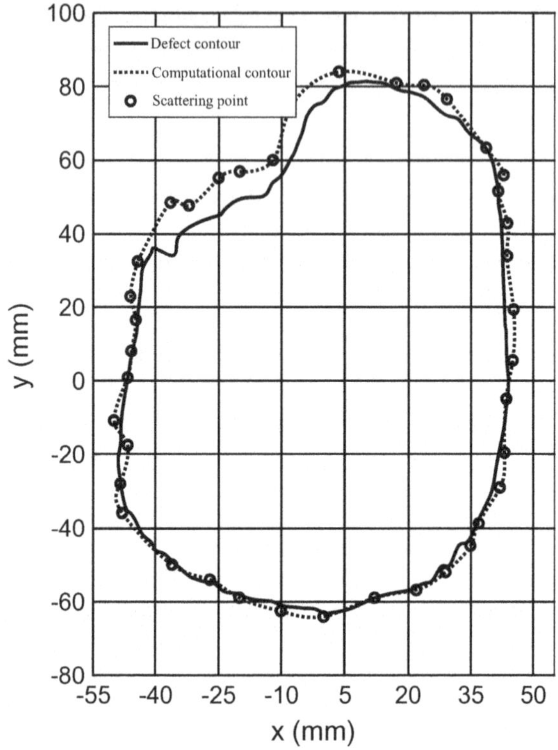

Fig. 5.26 Omnidirectional emission of complex contour defects of steel plates-relative error results of guided wave scattering imaging of omnidirectional receiving magneto-acoustic array

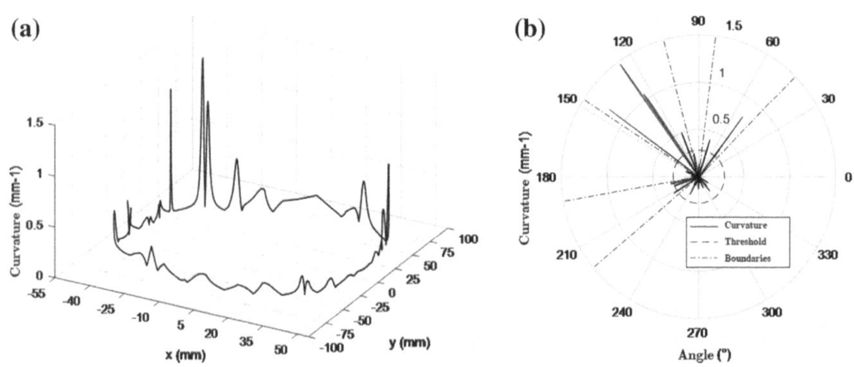

Fig. 5.27 Steel plate complex contour defect omnidirectional emission-omnidirectional receiving magnetic acoustic array guided wave scattering calculation contour curve curvature distribution: **a** calculate the contour curve curvature three-dimensional distribution; **b** array structure adjustment area

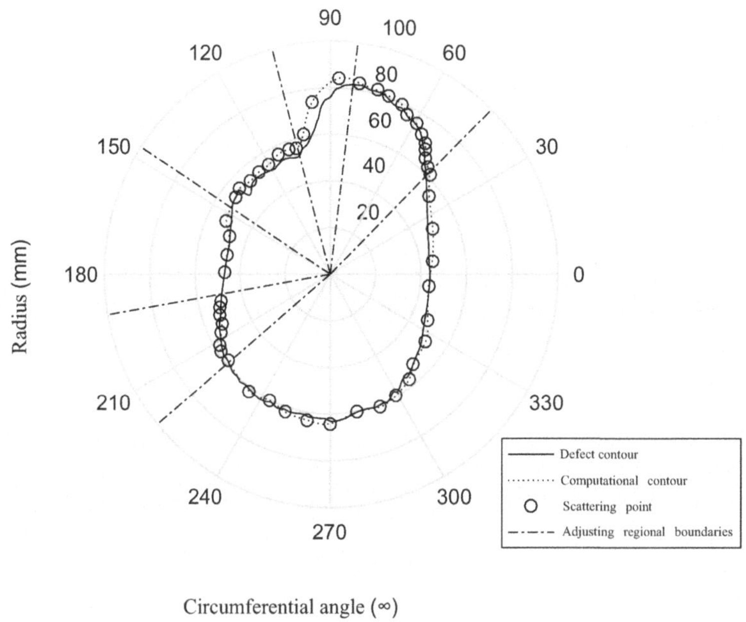

Fig. 5.28 Steel plate complex contour defect omnidirectional emission-omnidirectional receiving magnetic acoustic array structure optimized imaging results

The complex contour defect has a relatively large curvature value in the upper left part; the structural adjustment area of the complex contour defect calculation contour curve has a circumferential angle range of 45°–83°, 105°–147°, and 190°–220°.

According to the complex contour defect obtained above, the circumferential angle range of the array structure adjustment region of the contour curve is calculated, and the structure of the magneto-acoustic array is adjusted according to the magnetic acoustic array structure optimization adjustment method proposed in the present invention, based on the adjusted magneto-acoustic array structure. The omnidirectional emission-omnidirectional receiving magnetic acoustic array guided wave scattering imaging method is used to perform a defect contour reconstruction test on the complex contour defect, and the scatter point data after the array structure adjustment are obtained. The contour curve of the guided wave scattering defect before and after the adjustment of the structure of the magneto-acoustic array is data-fused, and the scattered point of the complex contour defect calculated by the data fusion and the calculated contour curve are obtained. Similarly, for comparison, the calculated contour of the complex contour defect and the contour of the defect itself are plotted onto an image, and the calculated scattering point and the calculated contour are shown in Fig. 5.28.

According to the results of the array structure adjusted complex contour defect guided wave scattering imaging, the calculated scattering point position in the array structure adjustment region is closer to the true defect contour than the initial array guided wave scattering imaging result. The number of calculated scatter points in the array structure adjustment area increases. The calculated contour curve and the position of the scattering point after the array structure adjustment are significantly improved compared with the initial array structure imaging results.

To quantitatively describe the error of the contour curve and the true contour of the complex contour defect, the relative error between the contour and the real contour is calculated using the defects defined by Eqs. (4.26) to (4.30). Calculate the relative error between the contour and the real contour of the defect before and after the optimization of the array structure adjustment, and uniformly normalize the error data before and after the adjustment of the defect array structure. The normalized relative error is plotted on the 2D contour plane according to the defect real contour interpolation, as shown in Fig. 5.29.

After the optimization and adjustment of the structure of the magnetic acoustic array, the relative error of the calculated contour of the complex contour defect is reduced compared with the initial array structure, especially the magnetic acoustic array adjustment area, and the relative error is more obvious. The optimal adjustment of the magnetic acoustic array plays an important role in reducing the relative error of calculating the contour curve and improving the imaging accuracy of the defect contour.

Similarly, in order to evaluate the relative error between the defect calculation contour and the real contour point, the average contour error of the contour curve and the real contour before and after optimization of the defect magneto-acoustic array structure are calculated according to the average contour error. The average

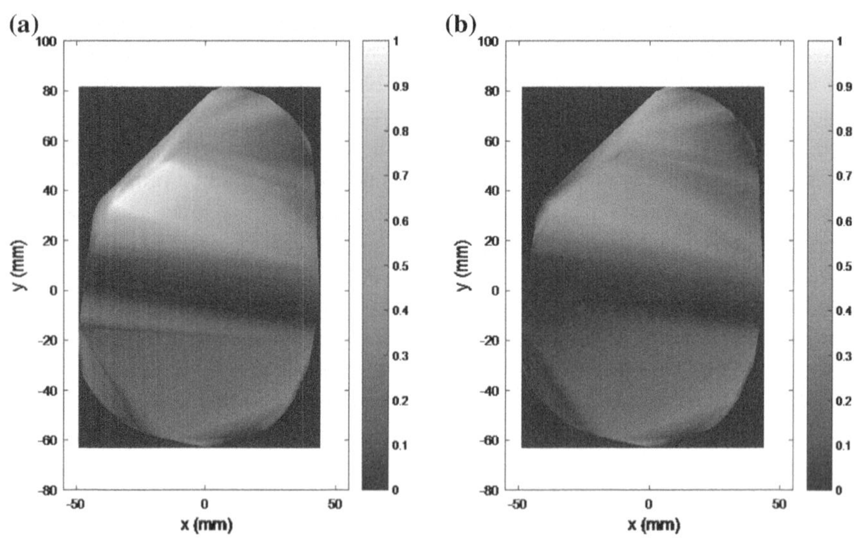

Fig. 5.29 Comparison of relative error before and after adjustment of omnidirectional emission and omnidirectional receiving magneto-acoustic array of steel plate: **a** relative error of initial array defect profile; **b** relative error of defect profile after array structure adjustment

contour error under the initial magneto-acoustic array structure is 3.24%, the average contour error of the array structure is 2.67%, and the average contour error is reduced by 0.57%.

After the above test, the omnidirectional emission-omnidirectional receiving magneto-acoustic array guided wave scattering imaging method has a small average error of contour imaging of complex contour defects of steel plates and shows high precision of contour defect reconstruction. The magnetic acoustic array structure optimization adjustment method proposed in this chapter is used to further reduce the contour reconstruction error of the region with large reconstruction error under the initial array structure and effectively improve the contour reconstruction accuracy of the regular contour defect.

References

1. Y. Zhang, S. Huang, S. Wang, W. Zhao, Time-frequency energy density precipitation method for time-of-flight extraction of narrowband lamb wave detection signals. Rev. Sci. Instrum. **87**(5), 054702, 1–8 (2016)
2. S. Huang, Y. Zhang, S. Wang, W. Zhao, Multi-mode electromagnetic ultrasonic lamb wave tomography imaging for variable-depth defects in metal plates. Sensors **16**(5), 628, 1–10 (2016)
3. Y. Zhang, S. Huang, S. Wang, Z. Wei, W. Zhao, Recognition of overlapped lamb wave detecting signals in aluminum plate by EMD-based STFT flight time extraction method. Int. J. Appl. Electromagnet. Mech. **52**(3–4), 991–998 (2016)
4. Y. Zhang, S. Wang, Q. Wang, S. Huang, W. Zhao, Profile imaging of actual defect in steel plate based on electromagnetic ultrasonic SH guided wave scattering. Insight
5. Y. Zhang, S. Huang, W. Zhao, S. Wang, Q. Wang, Electromagnetic ultrasonic guided wave long-term monitoring and data difference adaptive extraction method for buried oil-gas pipelines. Int. J. Appl. Electromagnet. Mech.
6. S. Wang, S. Huang, Y. Zhang, W. Zhao, Modeling of an omni-directional electromagnetic acoustic transducer driven by the Lorentz force mechanism. Smart Mater. Struct. **25**(12), 125029 (2016)
7. S. Wang, S. Huang, Y. Zhang, W. Zhao, Multiphysics modeling of a Lorentz force-based meander coil electromagnetic acoustic transducer via steady-state and transient analyses. IEEE Sens. J. **16**(17), 6641–6651 (2016)

The manufacturer's authorised representative in the EU is Springer Nature Customer Service Centre GmbH, Europaplatz 3, 69115 Heidelberg, Germany. If you have any concerns regarding our products, please contact ProductSafety@springernature.com

Printed and bound by CPI Group (UK) Ltd, Croydon, CR0 4YY

24/07/2026

02175970-0003